HOSTILE INTENT AND COUNTER-TERRORISM

Human Factors in Defence

Series Editors:
Dr Don Harris, Managing Director of HFI Solutions Ltd, UK
Professor Neville Stanton, Chair in Human Factors at the University of Southampton, UK
Dr Eduardo Salas, University of Central Florida, USA

Human factors is key to enabling today's armed forces to implement their vision to "produce battle-winning people and equipment that are fit for the challenge of today, ready for the tasks of tomorrow and capable of building for the future" (source: UK MoD). Modern armed forces fulfil a wider variety of roles than ever before. In addition to defending sovereign territory and prosecuting armed conflicts, military personnel are engaged in homeland defence and in undertaking peacekeeping operations and delivering humanitarian aid right across the world.

This requires top-class personnel, trained to the highest standards in the use of first class equipment. The military has long recognised that good human factors is essential if these aims are to be achieved.

The defence sector is far and away the largest employer of human factors personnel across the globe and is the largest funder of basic and applied research. Much of this research is applicable to a wide audience, not just the military; this series aims to give readers access to some of this high-quality work.

Ashgate's *Human Factors in Defence* series comprises of specially commissioned books from internationally recognised experts in the field. They provide in-depth, authoritative accounts of key human factors issues being addressed by the defence industry across the world.

Hostile Intent and Counter-Terrorism
Human Factors Theory and Application

Edited by

ALEX STEDMON
Coventry University, UK

GLYN LAWSON
The University of Nottingham, UK

CRC Press is an imprint of the
Taylor & Francis Group, an **informa** business

CRC Press
Taylor & Francis Group
6000 Broken Sound Parkway NW, Suite 300
Boca Raton, FL 33487-2742

First issued in paperback 2019

ISBN-13: 978-1-4094-4521-0 (hbk)
ISBN-13: 978-0-367-37785-4 (pbk)

Visit the Taylor & Francis Web site at
http://www.taylorandfrancis.com

and the CRC Press Web site at
http://www.crcpress.com

Contents

List of Figures

List of Tables

List of Contributors

The Editors

Dr Alex Stedmon FIEHF CPsychol CSci FRSA is a Chartered Psychologist, Chartered Scientist, Fellow of the Institute of Ergonomics and Human Factors and Fellow of the Royal Society of Arts. He is a Reader in Human Factors at Coventry University. He worked for the Ministry of Defence before moving into academia and explores human factors issues of technology use in security applications as well as contextual methods for investigating suspicious behaviours in various security-related domains. Alex was one of the technical leads for a strategic security consortium (the Engineering and Physical Sciences Research Council project 'Shades of Grey') and received Centre for Defence Enterprise funding for projects on human factors of automated CCTV; identifying human pheromones associated with deception; and collaborative intelligence information-gathering. Alex has co-edited special issues of *Applied Ergonomics* (44/2013): 'Detecting terrorist activities: hostile intent and suspicious behaviours', and *The Journal of Police & Criminal Psychology* (28(2)/2013) *Terrorism Psychology: Theory and Application.*

Dr Glyn Lawson FIEHF is a Lecturer within the Faculty of Engineering and a member of the Human Factors Research Group at The University of Nottingham. His research expertise includes the human-centred development of new technologies in design and engineering applications. Glyn has particular expertise in the evaluation of methods for predicting behaviour in emergency situations. He has also conducted research on deception detection and worked on requirements capture within the security domain. Glyn is a Fellow of the Institute of Ergonomics and Human Factors and sits on the Education and Training panel. Glyn co-edited the 2012 special issue of the *Applied Ergonomics* journal focused on detecting terrorist activities: hostile intent and suspicious behaviours; he also co-edited the 2013 special issue of *The Journal of Police & Criminal Psychology* which focused on terrorism psychology: theory and pplication.

Contributing Authors

Professor Babak Akhgar is Professor of Informatics and Director of CENTRIC at Sheffield Hallam University and a Fellow of the British Computer Society. He graduated from Sheffield Hallam University in Software Engineering, and later gained considerable commercial experience as a strategy analyst and methodology director for several companies. He consolidated this experience by obtaining a Masters degree (with distinction) in Information Systems in Management and a Ph.D. in Information Systems. Babak has more than 100 referred publications in international journals and conferences on information systems with specific focus on knowledge management (KM). He is member of the editorial boards of three international journals, and Chair and programme committee member of several international conferences. He has extensive and hands-on experience in development, management and execution of large international security initiatives (e.g. combating terrorism and organised crime, cyber security, public order and cross-cultural ideology polarisation). He has co-edited two recent books on intelligence management – *Intelligence Management: Knowledge Driven Frameworks for Combating Terrorism and Organised Crime* (Springer: London, 2013), and *Strategic Intelligence Management: National Security Imperatives and Information and Communications Technologies* (Butterworth-Heinemann: Oxford, 2013).

Dr Simon Andrews is a Reader in Computer Science and is the Technical Lead of CENTRIC at Sheffield Hallam University, where his main roles are to provide technical visions for new project ideas and manage the technical work in current projects. Simon is an international expert on Formal Concept Analysis (FCA) and is co-Editor in Chief of the *International Journal of Conceptual Structures and Smart Applications* (IJCSSA). He was Programme Chair of the Nineteenth International Conference on Conceptual Structures and has presented his work on fast concept-mining algorithms worldwide, including events in Moscow and Japan. As well as being a lead researcher for the prestigious European CUBIST project, Simon has recently been involved in writing three successful proposals in CENTRIC for further major European projects that will incorporate his FCA work. The new projects are in the areas of crisis management through the use of social media (ATHENA), environmental scanning for organised crime detection (ePOOLICE) and setting a European agenda for research in Cyber Crime and Cyber Terrorism (COuRAGE).

Professor Les Baillie is Professor of Microbiology at Cardiff University. He has varied research interests but his main focus is understanding the biology of *Bacillus anthracis*, the causative agent of anthrax. Professor Baillie leads a team including the Queen's University of Belfast, Ercisyes University Turkey, National Center for Disease Control in Georgia, the Defence Science Technology Laboratory in Porton Down and the US Army Medical Research Institute at Fort Detrick, MD. This NATO Science for Peace and Security Programme includes a research project entitled 'A multi-task investigation on the human immune response to anthrax aimed at developing more efficient vaccines' aims to tackle the menace posed by anthrax.

Dr Valentina Bartolucci is a Lecturer at the University of Pisa, Italy, and an Associate Research Fellow of the University of Bradford, UK. Apart from her doctoral thesis on the governmental discourse on terrorism, Valentina has written numerous articles in the fields of politics, security, critical discourse analysis and American and Middle East and North African (MENA) politics. She has dealt with various aspects of critical discourse analysis as well as with States' representations of various contemporary phenomena, among which are terrorism and migration, as existential threats, and on the consequences of those representations. At the moment, she is particularly interested in the supposed trade-off between security and liberty, and in particular on the issues democracies face when dealing with matters of security especially since the September 11, 2001 attacks against the US and subsequent terrorist events.

Dr Gordon Baxter is a Senior Research Fellow in the School of Computer Science at St Andrews University, Scotland with an interdisciplinary background in human factors and computer science. His main research interest lies in how to improve the development of sociotechnical systems to make them more useful, more usable, and more acceptable to the end-users, as well as more resilient. He has published over 50 articles covering research in a variety of domains including aviation, financial trading, neonatal intensive care, and telecare, and is currently co-authoring a book on the foundations for designing user-centred systems.

Professor Mia Bloom is Professor of Security Studies at the University of Massachusetts, Lowell and the author of *Dying to Kill: The Allure of Suicide Terror* (Columbia University Press: New York, NY, 2005) and *Living Together After Ethnic Killing with Roy Licklider* (London: Routledge, 2007) and *Bombshell: Women and Terror* (University of Pennsylvania Press: Philadelphia, 2011). Bloom has held research/teaching appointments at Cornell, Harvard, Princeton, and McGill University and her research focuses on suicide terrorism, women and terrorism, and the growing phenomenon of children's recruitment by terrorist organisations. Bloom has a Ph.D. in political science from Columbia University, a Masters degree from the School of Foreign Service (in Arab Studies) at Georgetown University and graduated Magna Cum Laude from McGill University in Russian History, Islamic and Middle East Studies.

Dr Kurt Braddock is a post-doctoral Teaching Fellow in the Department of Communication Arts and Sciences at Penn State University. He has also worked as a researcher at Penn State's former terrorism research centre, The International Center for the Study of Terrorism, as well as Penn State's Applied Research Laboratory. His research explores the nexus of communication, psychology, and terrorism, particularly as it occurs on the Internet. Specifically, he has performed research related to terrorist use of narratives and other strategic communication to radicalise potential recruits. He has published refereed articles in *Terrorism and Political Violence*, the *Journal of Personality*, and *Communication Quarterly*, and he has co-edited a volume with Dr John Horgan entitled *Terrorism Studies: A Reader* (Routledge, London, 2011).

Dr Emma Bradford completed her Ph.D. at the University of Liverpool, subsequently working on research projects funded by the EPSRC as well as the US Department of Homeland Security, via START. Her Ph.D. research focused on revenge crimes in a series of studies ranging from non-criminal acts to mass homicide incidents. Comparable mass casualty terrorist and non-terrorist incidents were examined, allowing for the exploration of both non-political and politically motivated revenge behaviours. Dr Bradford has presented her research at numerous international conferences worldwide and has published on the group dynamics of terrorism, terrorist shooting sprees and terrorist attacks on schools.

Dr Alex Braithwaite is Associate Professor in the School of Government and Public Policy at the University of Arizona and Senior Research Associate at University College London. His research interests are concerned with many aspects of the causes and geography of violent political conflict – including terrorism, protests and riots, civil war, and international wars. He has published two dozen articles and book chapters, including at *Journal of Politics*, *British Journal of Political Science*, *International Studies Quarterly*, and *Journal of Peace Research*. He has also published a book, *Conflict Hotspots: Emergence, Causes, and Consequences* (Ashgate, 2010).

Professor Kathleen M. Carley (Harvard Ph.D.) is a Professor of Computation, Organizations and Society in the School of Computer Science at Carnegie Mellon University and the director of the centre for Computational Analysis of Social and Organizational Systems (CASOS). Research areas include social network analysis, dynamic network analysis, big-data network analytics, agent-based modelling, adaptation and evolution, social-network text-mining, cyber-security, information diffusion, social media and telecommunication, disease contagion, and disaster response. She and members of her centre have developed novel tools and technologies for rapidly extracting networks from texts (AutoMap) and analysing large scale geo-temporal multidimensional networks (ORA) and agent-based simulations (Construct).

Ben Dalton is a computational design and experimental media researcher, currently on research leave from Leeds Metropolitan University, to work with the AHRC Creative Exchange hub and as a Ph.D. at the Royal College of Art. He has recently been guest Professor at the Bergen National Academy of Art and Design, co-investigator on several EPSRC projects on interactive urban spaces and co-director of the VERDIKT-funded Data is Political symposium. He has a background in ubiquitous computing and mobile sensor design from MIT Media Lab, Århus Electron-Molecular Interaction group, Leeds Spintronics and Magnetic Nanostructures lab, and Jim Henson's Creature Shop.

Dr Coral J. Dando is a Forensic Cognitive Psychologist and Reader in Applied Cognition at the University of Wolverhampton, UK. Her primary research interests are eyewitness memory and verbal deception in interview settings with particular interest in how contemporary theories of episodic memory, cognitive demand and split attention can inform the practices and procedures of the criminal justice system. Coral teaches intelligence interviewing methods to police officers and security staff worldwide (e.g. in the UK, USA, Sweden, Norway, Malaysia). She has published widely, and her research has attracted in excess of £1m of UK and US government funding. Coral is a Health and Care Professionals Council registered (HCPC) Consultant Forensic Psychologist, and Chartered Scientist.

Professor Tanya Dronzina holds a Ph.D. in political science and is a Professor at the department of political science of Kliment Ohridski University of Sofia. Her research field is gender and terrorism. She has lectured as a guest Professor in the Carlos Tercero de Madrid University and Granada University (Spain). Currently she is teaching in the Kokshe Academy in Kokshetau, Kazakhstan, and is director of the Institute for Peace, Conflicts and Mediation at the same university. She has been an external consultant of the Program to counter terrorism and religious extremism of Kazakhstan government (2013) and a leading scholar in the project 'Labor conflicts in Western Kazakhstan'.

Dr Peter Eachus is Director of Psychology and Public Health within the School of Health Sciences at the University of Salford. He holds a BSc (Hons) Psychology, MSc, Ph.D., PGCE. Peter is a Chartered Psychologist and an Associate Fellow of the British Psychological Society, a Chartered Scientist and Internet Editor of the journal *Health Education*. Research interests include detecting hostile intent, the psychology of intelligence failure and decision-making within the context of intelligence analysis. He is a member of the Society for Terrorism Research and has presented papers on the likelihood of a CBRNE terrorist attack in the UK.

Dr Dawn L. Eubanks is an Associate Professor at Warwick Business School at University of Warwick. Dawn's research interests are primarily in the areas of leadership and innovation, with particular interest in destructive leadership. Her current focus is on leader errors and follower reactions to errors. Innovation research includes a focus upon how to foster this characteristic across a range of contexts. Dawn is a member is the editorial board for *The Leadership Quarterly* and *Journal of Occupational and Organizational Psychology*. Dawn has been successful earning grants in excess of £100k from government funding bodies such as the Engineering and Physical Sciences Research Council (EPSRC).

Detective Chief Inspector Dave Fortune. After completing 30 years policing service in the UK, he has taken up a role as principal research officer with CENTRIC and is a founding member of the CENTRIC Advisory Board. Dave spent his last few years of police service leading the regional police engagement with Europe, he was involved in the development and delivery of numerous projects in the security domain. He currently sits on the advisory board of three EU FP7 projects and has worked with CENTRIC to develop the ATHENA and COuRAGE proposals which will start shortly. He is currently lead evaluator of the EU DG Home Affairs funded project SMART CV which has a counter-terrorism focus, and is working on the development of future security-related proposals utilising national and EU funding. He is an experienced proposal evaluator having worked for the European Commission undertaking the evaluation of several security funding proposals and is currently contracted with the EC to act as a project reviewer in the security domain.

Dr Lara A. Frumkin is a Senior Lecturer in the School of Psychology at the University of East London. Lara has previously worked outside of academia at the American Psychological Association and the US Justice Department. Her current research is on marginalised groups and eyewitness testimony. Eyewitness research involves assessing how individuals are viewed within legal contexts. Her research on marginalised groups looks at how individuals within the groups are perceived and judged. Lara has won grants from local authorities and government bodies such as the Engineering and Physical Sciences Research Council (EPSRC).

Professor Pete Fussey is a Professor of Sociology in the Department of Sociology at the University of Essex, UK. Professor Fussey's main research interests focus on security, social control and the city. He has published widely in these areas and is currently working on two large-scale ESRC and EPSRC projects researching counter-terrorism in the UK's crowded spaces and the future urban resilience until 2050. His other work focuses on organised crime in the EU with particular reference to the trafficking of children for criminal exploitation (monograph due to be published by Routledge in 2014). Recent books include *Securing and Sustaining the Olympic City* (Ashgate, 2011) and *Terrorism and the Olympics* (Routledge, 2011).

Dr Mils Hills is currently Associate Professor in Risk, Resilience and Corporate Security at the University of Northampton Business School (UK) alongside his role as Head of the online MBAplus Programme and a management role in the Centre for Citizenship, Enterprise and Governance (CCEG). Mils graduated with a Ph.D. (1998) and MA (Hons) (1995) in Social Anthropology from the University of St Andrews. He joined the research agency of the Ministry of Defence in 1998. Within a couple of years, he rose from being a contributing researcher in the area of information warfare to leading the national research capability in targeting/defending decision-making and business processes (Information Operations). He was later seconded to the Cabinet Office, helping establish the Civil Contingencies Secretariat, building the resilience and security of UK plc as well as supporting the development of foreign policy, legislation and operational interventions in civilian, defence, intelligence and other activities. After running his own risk and security consultancy (2005–10), Mils has worked in academia.

Professor John Horgan is Professor of Security Studies at the School of Criminology and Justice Studies of the University of Massachusetts Lowell where he is also Director of the Center for Terrorism and Security Studies. An applied psychologist by training, his research focuses on terrorist behavior. His books include *The Psychology of Terrorism*, *Walking Away from Terrorism* (Routledge, 2009), and *Leaving Terrorism Behind* (Routledge, 2008). His latest book *Divided We Stand: The Strategy and Psychology of Ireland's Dissident Terrorists* (Oxford University Press, 2013). He is Special Editions Editor of *Terrorism and Political Violence*, Associate Editor of *Dynamics of Asymmetric Conflict*, and serves on the Editorial Boards of several further journals, including *Studies in Conflict and Terrorism* and *Journal of Strategic Security*.

Dr Cale Horne is Assistant Professor of political studies at Covenant College, where his research involves social movements, authoritarian politics, and public opinion. His recent work appears in the *Journal of Elections, Public Opinion & Parties*, *Studies in Conflict and Terrorism*, and *International Studies Perspectives*. Dr Horne previously held a post-doctoral research position at Penn State University's International Center for the Study of Terrorism, where he managed terrorism-related research projects sponsored by the Office of Naval Research.

Professor David M. Howard BSc(Eng) Ph.D. CEng FIET FIOA MAES researches in speech, singing and music, including their analysis, synthesis, production and perception and teaches aspects of music technology. He is currently Head of the Department of Electronics at the University of York, UK where he also leads the Audio Laboratory and is a member of the York Centre for Singing Science. Key research areas include the perception of speech and singing, intonation in unaccompanied singing, the acoustics of singing voice development, and computers and iPads in voice training and natural voice synthesis. David presented on the BBC4-TV programs *Castrato* and *Voice*. He plays the organ and conducts the Vale of York Voices who sing evensong in York Minster once a month.

Dr Ron Iphofen AcSS is an independent consultant advising public and independent agencies at UK and international levels. Until 2008 he was Director of Postgraduate Studies in Healthcare Sciences, Bangor University. He was scientific consultant on the European Commission's (EC) RESPECT project – establishing pan-European standards in the social sciences. He advises the UK Research integrity Office and is a science and ethics evaluator on the EC's Seventh Framework Programme (FP7). He chairs the Ethics and Societal Impact Advisory Group for a major EC Project on mass passenger transport security (SECUR-ED).

He is a Visiting Senior Research Fellow with the Open University, an Academician of the Academy of Social Sciences, and a Fellow of the Royal Society of Medicine.

Dr Joan H. Johnston received her Ph.D. in Industrial and Organizational Psychology from the University of South Florida. She is presently the Chief of the US Army Research Institute Training Technologies Research Unit in Orlando. Prior to this, she was a senior research psychologist at the Naval Air Warfare Center Training Systems Division. Over 22 years, Dr Johnston conducted applied research on tactical decision-making under stress and team training. Dr Johnston was a member of the research team that evaluated the effectiveness of profiling and tracking training in the Border Hunter Research Project; the team received the 2010 Modeling and Simulation Team Training Award from the National Training System Association.

Dr Michael Kenney is Associate Professor of International Affairs in the Graduate School of Public and International Affairs at the University of Pittsburgh. He is the author of numerous publications on network analysis, terrorism, counter-terrorism, and drug trafficking. Dr Kenney has conducted fieldwork on terrorism and drug trafficking in Colombia, Spain, the United Kingdom, Morocco, and Israel. He has held research fellowships at Stanford University and the University of Southern California. His research has been supported by the Office of Naval Research, the National Institute of Justice, and the National Science Foundation.

Dr Christin Kirchhübel is a forensic caseworker and consultant at J.P. French Associates and works mainly within the areas of speaker comparison, transcription and disputed content analysis. She has carried out research into the acoustic and temporal characteristics of deceptive speech and lectures in forensic speech science at postgraduate level. She is a member of the International Association for Forensic Phonetics and Acoustics (IAFPA) and the British Association of Academic Phoneticians (BAAP) and is involved in editorial duties for the *International Journal of Speech Language and the Law*. Her research interests include forensic phonetics, sociophonetics and sociolinguistics. One of her recent papers 'Analysing deceptive speech' co-authored with Dr Alex Stedmon and Professor David Howard, received the Best Paper Award at the Tenth International Conference on Engineering Psychology and Cognitive Ergonomics (Las Vegas, July 21–26, 2013).

Dr Sharon Leal is a Senior Research Fellow for the International Centre for Research in Forensic Psychology, University of Portsmouth (UK). Her doctoral research focused on the central and peripheral physiology of attention and cognitive demand. Her current research focuses on the behavioural and physiological effects of cognitive load during deception. This research is highly relevant for detecting how people engaging in 'high stake' deception respond verbally, non-verbally and physiologically. Her work involves cooperation with national and international governments, police and insurance fraud detection companies.

Dr Claire McAndrew is a Research Associate and Chartered Psychologist at The Bartlett School of Graduate Studies, University College London. Her research combines psychology and design in the study of the built environment, which includes security contexts: Safer Spaces: Communication design for counter-terrorism (EPSRC, ESRC, AHRC) and Designing With Intent: Influencing behaviour in transitional spaces (Centre for Defence Enterprise). Claire has published a number of book chapters and journal articles on decision-making 'in the wild' and the role of interactive communication design in public spaces, recently co-authoring a set of design guidelines for Centre for the Protection of National Infrastructure that are relevant to communication designers, architects, academics and policy-makers.

Dr Samantha Mann is a Senior Research Fellow at the International Centre for Research in Forensic Psychology, University of Portsmouth (UK). Her doctoral research investigated the behaviour of real-life high-stakes liars in their police interviews, and police officers' ability to detect those lies. Her current research focuses on developing interview techniques that enhance ability to detect deceit by increasing cognitive load, and how best to conduct an interview with more than one interviewer or interviewee. Her work is very applied and involves working with governments as well as with police and other law enforcers.

Karen Martin is Research Assistant in the Centre for Architecture and Sustainable Environment, University of Kent. Karen undertakes research into interaction in public space. She develops methods for exploring how public space is created, negotiated and understood. Karen is an EngD candidate at University College London on a joint programme with the departments of Architecture and Computer Science. The subject of her doctoral research is design strategies for urban computing in in-between spaces. Karen is design unit tutor at Plymouth University for Unit In-between which explores strategies of community engagement for architects.

Tarek Menacere is a Senior Software Developer and Systems Analyst at the University of Liverpool. He is currently developing a national animal disease surveillance network (SAVSNET) which will not only monitor disease in small animals, but will also impact on both research and clinical practice. Tarek graduated from the University of Liverpool with a first-class degree in Computer Science. Previously, he worked at Lancaster University creating software and electronic infrastructure both for data collection and real-world analysis. These project areas included deception, insider threat and linguistic analysis. To date he has been in charge of delivering software for projects totalling over £1.3m of investment from government and commercial institutions.

Ashwin Mehta started his career with a Master's Degree in Chemistry from the University of Liverpool. After several years as a synthetic chemist in the private sector, Ashwin moved into consultancy and emergency response at the UK National Chemical Emergency Centre (NCEC). More recently, Ashwin is the Corporate Resilience Leader at the UK Medical Research Council. Ashwin has considerable experience with safety and security in controversial business; crisis management; and organisational resilience in scientific and high-hazard environments. Ashwin's other interests include blended learning in environments with unstable infrastructure, and the effects of technology application on front-line resilience.

Professor Marialena Nikolopoulou BEng MPhil Ph.D. is Director of the Research Centre for Architecture and Sustainable Environment at the University of Kent. She has participated in numerous EU and UK/US-funded research projects on sustainability, use of open spaces and environmental quality, including perception of environmental stimuli. Her work on outdoor environments has received various awards from diverse bodies (such as the Royal Institute of British Architects and the International Society of Biometeorology), as well as best papers prizes. Marialena has regularly worked at the interface between different disciplines, and she has been invited to give talks internationally in workshops, research seminars and conferences.

Professor Thomas C. Ormerod is Professor of Cognitive Psychology and Associate Dean for Research at The University of Surrey. He has published over 100 refereed articles and three books, and been principal investigator on over £3m of UK and US Government-funded projects, exploring expertise in creative design, education, human–computer interaction, and criminal investigation. He has held large research grants from RCUK, MoD and the US Department for Homeland Security to develop and test new ways of detecting deception indicative of security threats. Recent research focuses on developing effective methods for evaluating human behaviour for security screening.

John Parkinson OBE commenced his career with West Yorkshire Police in 1979. He completed the Counter Revolutionary Warfare and Counter Insurgency Module at the British Military Command Course in 1996. He has been a Senior Investigating Officer leading investigations into complex homicide, major crime, kidnap and extortion, undercover and specialist operations. In 2004 he became accredited as an ACPO Counter Terrorism SIO. Following the events of the London bombings on 7 July 2005, he was appointed as the SIO in West Yorkshire alongside the Metropolitan Police SIO, investigating those responsible for the bombings. John set up, established and became the Head the North East Counter Terrorism Unit. He led on many areas of development, operational practices and threat mitigation. John also directs and oversees national counter-terrorism exercises facilitated between the Home Office and police forces across the UK with responsibility for identifying issues of capability or capacity. John retired from policing during 2013 after 34 years service and is now the Steering Group Chair of the Centre of Excellence for Terrorism, Resilience, Intelligence and Organised Crime Research (CENTRIC).

Dr Simon Polovina is the Enterprise Architect for CENTRIC, and is a Principal Investigator for the European Commission Seventh Framework Programme project CUBIST (Combining and Unifying Business Intelligence with Semantic Technologies). He engages in roles that draw upon his expertise in enterprise architecture and conceptual structures, which harmonises the human conceptual approach to problem-solving with the formal structures that computer applications need to bring their productivity to bear. Simon is also a Senior Lecturer in Business Computing at Sheffield Hallam University, and leads the CCRC's Conceptual Structures Research Group. He has many years of industrial experience in accounting and information and communication technologies, and his interests include the use of smart applications and how they can detect novel or unusual transactions that would otherwise remain as lost business opportunities or represent illicit business or criminal activity. He has published widely, with over 90 publications to date, and is editor-in-chief of the *International Journal of Conceptual Structures and Smart Applications* (IJCSSA) published by IGI-Global.

Dr Brendan Ryan BSc.Hons. Ph.D. MIEHF CMIOSH is a Lecturer in Human Factors at the University of Nottingham, with a broad range of experience in rail human factors research and specific expertise in access to railways (e.g. in relation to suicide/trespass or access for rail engineering work). He has carried out and supervised research on use of scenarios to evaluate human and organisational risks in rail activities, research to understand difficulties of predicting error in railway activities, resilience in rail planning, investigation of culture and factors affecting behaviour of staff, strategy for sustainability in the rail industry, and analyses of rail incidents. Brendan was formerly a National Accident Investigator at Network Rail and has expertise in interview and survey development and administration.

Dr Alexandra L. Sandham is an applied psychologist at Defence Science and Technology Laboratory (Dstl). Formerly a Research Associate at Lancaster University where, while completing her Ph.D. part-time, she worked on a number of projects looking at detecting deception in various security environments. These projects included working at national and international airports carrying out ethnographic study and training security personnel. Alex's Ph.D. thesis addressed hypothesis generation in investigative contexts, in particular the effect of heuristics and biases on the hypotheses generated by individuals, dyads and triads. Investigative reasoning including investigative interviewing and detecting deception remain her primary research interests.

Dr Rose Saikayasit has been a Research Fellow of the Human Factors Research Group in the Faculty of Engineering at The University of Nottingham. Rose's research has focused on understanding and eliciting user requirements of front-line security personnel and stakeholders to develop requirement specifications for further applications in security and counter-terrorism, as part of the EPSRC 'Shades of Grey' project. Her research interests include understanding and supporting collaboration in co-located and virtual engineering and design teams through the use of user-centred design for novel collaborative technologies. She is experienced in user requirements elicitation, scenario development, and human factors evaluation of technologies in a variety of security and non-security sectors.

Professor Ian Sommerville is Professor of Software Engineering at St Andrews University, Scotland. His principal research interest is in large-scale complex IT systems and how these are used in enterprises. For many years, he has worked in the area of sociotechnical systems and is interested in how methods of social analysis may be applied to improve the requirements and the design of enterprise systems. Ian has published almost 200 papers and is the author of a widely used textbook on software engineering.

Dr V. Alan Spiker received his Ph.D. in experimental psychology from the University of New Mexico and is presently an independent consultant. For over 30 years, Dr Spiker was a principal scientist at Anacapa Sciences, a human factors firm in Santa Barbara, California where he conducted field studies in such areas as aircrew simulation, mission planning, crew resource management, and team training. Dr Spiker was a member of the research team that evaluated the effectiveness of profiling and tracking training in the Border Hunter Research Project; the team received the 2010 Modeling and Simulation Team Training Award from the National Training System Association.

Detective Inspector Dr Andrew Staniforth is a serving police Detective Inspector and former Special Branch detective. He has extensive operational experience across multiple counter-terrorism disciplines, now specialising in security-themed research as Programme Manager for the International Projects Programme Team at West Yorkshire Police. In this role Andrew is the Project Coordinator of SMART CV (Social Media Anti Radicalisation Training for Credible Voices) and project ATHENA, both funded by Internal Security and Framework Programme Seven grants of the European Commission Research Executive Agency. Prior to completing his doctorate in 2013, *Counter-Terrorism: The Role of Law Enforcement Agencies from National Security Perspectives*, he authored Blackstone's *Counter-Terrorism Handbook* (Oxford University Press 2009, 2010, 2013), the *Routledge Companion to UK Counter-Terrorism* (Routledge, 2012), and Blackstone's *Handbook of Ports and Border Security* (Oxford University Press, 2013). Andrew continues his applied research as a Senior Research Fellow at the Centre of Excellence for Terrorism, Resilience, Intelligence and Organised Crime Research (CENTRIC).

Professor Teal Triggs is Professor of Graphic Design and an Associate Dean, School of Communication, Royal College of Art, London. She is also Adjunct Professor in the School of Media and Communication at RMIT, Australia. As a graphic design historian, critic and educator she has lectured and broadcast widely and her writings have appeared in numerous edited books and international design publications. She was co-Principle Investigator with Professor Mike Press on Safer Spaces: Communication Design for Counter Terror (EP/F008503/1) funded by EPSRC/ESRC/AHRC. She is a Fellow of the International Society of Typographic Designers and the Royal Society of Arts.

Peter Vining is a Ph.D. student in New York University's Department of Politics and is a National Science Foundation Graduate Research Fellow. His research focuses on civilian targeting during war, authoritarian regime behaviour, and rebel mobilisation. He holds an MA degree from Duke University, and undergraduate degrees from Pennsylvania State University.

Professor Aldert Vrij is a Professor of Applied Social Psychology in the Department of Psychology at the University of Portsmouth (UK). He has published more than 400 articles and book chapters, mainly on non-verbal and verbal cues to deception (i.e., how do liars behave and what do they say), and lie detection. He gives invited talks and workshops on lie detection to practitioners and scholars across the world, including police, homeland security, defence, judges, solicitors, social workers, fraud investigators, insurers and bankers. He has held research grants from research councils and national governments totalling more than £3 million.

Dr Margaret A. Wilson has been conducting research on the psychology of terrorism for over 20 years, developing theoretical models of terrorist behaviour which have implications for the management, investigation and resolution of terrorist incidents. Her research has been funded by START, a centre of excellence of the US Department of Homeland Security. Dr Wilson is frequently invited to present her research findings to academic conferences, as well as professional meetings in law enforcement, security services and hostage negotiation worldwide. Dr Wilson can be contacted via the Institute of Security Science and Technology at Imperial College London.

Ke Zhang is a Ph.D. Candidate at Warwick Business School at University of Warwick. Ke is mainly interested in interpersonal deception in various contexts. Her recent research concerns intentions and non-verbal indicators of deception, as well as situational factors influencing deception behaviour. Ke is also interested in the role of depletion of self-regulation in the leakage of cues to deception.

Acknowledgements

This book would not have been possible without the support of many people. In particular, we would like to thank the following:

- Professor Don Harris and Guy Loft for their encouragement, patience and invaluable guidance with this book.
- Lianne Sherlock and Kevin Selmes for their help and support once we delivered the manuscript to Ashgate.
- Professor John Wilson, Dr Craig Bennell and Dr Brent Snook for their support and in providing us with the opportunities as Guest Editors of special issues for *Applied Ergonomics* and the *Journal of Police & Criminal Psychology*.
- Davide Salanitri and Richard Tew with the preparation and design of the cover for this edition.
- Donna Stedmon who worked extremely hard to improve all the figures in the book.

We would also like to thank all the contributors of each chapter for their hard work and dedication; the reviewers for their valuable feedback; and various funding bodies that have supported many aspects of the research reported in this book.

Alex would like to acknowledge his colleagues at Coventry University in the Human Systems Integration Group for their support with this book. Finally he would like to thank his family and in particular his wife, Donna, who has been a tremendous support through the development and writing of this book. Thank you for helping to make the time fly!

Glyn would like to acknowledge the support from his colleagues in the Human Factors Research Group at The University of Nottingham. He would also like to express his gratitude for the never-ending encouragement from his wife, Dr Rose Saikayasit.

This book is for to anyone with an interest in human factors and security but is also particularly dedicated to Oscar and James.

Foreword

The nature of defence has changed considerably in the last two decades. Preparation for conventional combat operations is now only a part of the remit of the armed services and associated government departments. Defence, intelligence and police services are also tasked with providing the facilities to ensure the peacetime protection and security of a nation and its citizens. Counter-terrorism activities lie at the very heart of these efforts. Furthermore, the terrorist threat is becoming more diverse, innovative and is continually evolving. It is also less well understood than conventional military endeavours.

This book is a welcome and extremely timely addition to the Ashgate series 'Human Factors in Defence'. The work presented in these chapters describes contemporary thinking from a number of internationally recognised experts in this emerging domain. Human factors in counter-terrorism is still an area that is largely under-researched yet it is one that has immense potential for making significant inroads into improving safety and security; after all, terrorism is a fundamentally human activity.

Terrorism as a form of asymmetric warfare uses methods that lie outside *jus in bello*. This means that countering such activities requires not only an understanding of the tactics used by the belligerents but also the motivations and beliefs underpinning them. The UK government strategy adopts a four-pronged counter-terrorism approach:

- *Pursue* – stop terrorist attacks
- *Prevent* – stop people becoming terrorists or supporting violent extremism
- *Protect* – strengthen the protection against terrorist attack
- *Prepare* – where an attack cannot be stopped, to mitigate its impact.

The chapters in this volume, either implicitly or explicitly, address the human dimension in all these areas. It is also evident that human factors-related activities are becoming embedded into more systemic approaches to counter-terrorism and not only encompass contributions from scientific disciplines such as psychology, neuroscience, engineering and computing, but also other fields including criminology, sociology, social and scientific philosophy, art and design, political science, and ethics. This is to be encouraged. To be effective, human factors must neither be regarded as a stand-alone activity nor as one with rigidly defined subject boundaries.

I hope that you find the contents of this book interesting, stimulating, engaging and ultimately useful.

Professor Don Harris
Coventry University (UK)
January 2014

Preface

I've worked in the broad area of 'human factors' for the last 19 years. When I began, my focus was on making people's lives better, one button press at a time. Mobile phones were just beginning to become a mainstream device and, unlike today's smartphones, were incredibly engineering- and technology-focused rather than being developed from a person-centred perspective.

During those early years of research, colleagues and I tried to bring home the importance of good human factors by equating the time wasted, globally, when doing mundane tasks on these new devices with human lifespans. In lectures and talks, I'd often say that every day, lifetimes were being lost through bad design; I'd say that bad design was killing people.

Since that time, it has become rather apparent that push-button or touch screen frustrations and inefficiencies are not the greatest threat to a person's sense of well-being or longevity. Over those 19 years, all of us working in a range of research domains have had our eyes opened to the grand challenges of climate change, financial stress, increasingly ageing populations and, the topic of this book, terrorism. We've had to grow up, and so have our methods and approaches.

Four years ago, I along with 40 or so others – drawn from a broad number of research disciplines – spent the week in a secluded and charming country hotel. It was the sort of venue routinely ready to help families celebrate weddings and other important life events. For that week, though, the talk was not about hopeful plans for the future, the honeymoon or new home; instead, we sat down together with experts from the police and security services and confronted a different – but just as real – aspect of human life: the drive in some to bring fear, terror and death to individuals and communities. At this Engineering and Physical Sciences Research Council (EPSRC)-sponsored research sandpit we listened, and we learned that people intent on such 'work' approached their tasks with precision planning at individual and systems levels, with multiple visits to potential sites, rehearsals and contingency planning.

During the week, we were charged with developing research ideas that might lead to approaches to spot suspicious activities in public spaces, helping to thwart attacks some time before any incident occurred. Of course, there was much discussion of advanced technologies that seem to offer a fast, easy solution, appealing both to the public and to those who have the responsibility of keeping us safe. Better surveillance cameras, perhaps with highly sophisticated machine-learning algorithms for face recognition or gait identification; new forms of scanners that might detect concealed weapons at a distance; data-mining approaches that can trawl through human-to-human transactions over the Internet to determine the difference between Facebook status updates and hidden interactions or subtle pokes from serious plots.

Decades of human factors research, though, show that technological advances that aren't crafted to fit and respond to human capabilities, limitations and contexts are less effective than they might be or, as is often the case, fail to have their intended impact at all. The advanced features of mobile phones I worked with almost two decades ago were very often unused by the people who bought the devices because it was just too difficult to operate them.

While advanced technologies are a vital part of the toolkit against terrorism, the week at that country house highlighted the value of using this bigger human-centred frame to look at the problems. Those involved in terrorist activities are people, with lives, families and jobs. They target places and spaces that enable the rest of us to do things that are important or necessary: to shop, to relax, to exercise, to commute. Any technologies or methods we develop, then, need to be permeable to influences and shaped by understandings from a spectrum of disciplines including psychology, sociology, criminology, computer science, engineering and architecture.

This book explores the importance, as well as the pitfalls, of such a broad, human-factors orientation to terrorism threats. It draws not only on those projects that came out of the EPSRC week and subsequent funded projects but from a wider set of researchers and activities across the world; reminding us that as with more positive aspects of life that we can design for, like productivity and love, loss and pain caused by such activities are a global concern.

The human factors approach has three important underpinning concerns: to understand people and their wider contexts; to iteratively design to fit these needs; and to evaluate the impact of potential choices and ways forward. In this book, then, we encounter important perspectives and techniques to better see the problems with diverse lenses such as policy and modelling. There are also fascinating descriptions of intervention designs and evaluations carried out both in the controlled settings of laboratories or 'in the wild' in challenging, public and real-world contexts.

As you read through this book, you will see that there are no simple answers to the evolving problems of keeping communities safe. No one approach, no one technology will undermine the efforts of the very small number of people who can bring massive destruction and fear to our lives. This is a hopeful book though – the way it explores the complexities and difficulties do not lead to despair, leading us to think we are facing an impossible task – but rather help to set out a clear research agenda. By facing up to the messiness, uncertainty, ambiguity of the task – to embrace, that is, the 'human factor' – is the only way we will be able to get a purchase and make progress towards a safer society for us all.

Professor Matt Jones
Swansea University (UK)
January 2014

In Memoriam

In memory of Professor John Wilson, who mentored us both in the early stages of our careers and without whom this book would not have been possible. John supported this book, motivated by his belief that this is an important area where human factors can really make a difference.

Reviews for
Hostile Intent and Counter-Terrorism

This book has an important contribution to make to those seeking to develop counter-terrorism policy and practices informed by evidence-based scholarship. It contains a diverse set of reflections from around the world, inspired by a group of researchers who initially came together to consider ways of developing robust, reliable and ethical ways of detecting the covert activities of terrorists in crowded places. This book illustrates, in its scale and scope, the size and complexity of the challenge.

Tristram Riley-Smith, University of Cambridge, UK

The essays in this book provide an original set of insights into the genesis of terrorism and its actors. The exploration of terrorist psychology goes well beyond any legal textbook. Historical analysis is combined with demographic examination. Those who detect, disrupt and research terrorism on behalf of the public will find new and revealing material that provides important background for their work.

Lord Carlile of Berriew CBE QC,
former independent reviewer of terrorism legislation

In this insightful and incisive text, Stedmon, Lawson and their many colleagues and co-contributors grapple with one of the most pressing issues for our species and our survival on this planet. They undertake to show how the integration of people and technology is at once the genesis of and potential solution to the vexed problems of contemporary asymmetric conflict, expressed through terrorism. But more than this, their crucial collective deliberations mandate that we consider what our future society can and should look like. These are issues at the very heart of the human enterprise. Thus, while both a timely and important text for the declared central concern for counter-terrorism and the place of human factors and ergonomics in that struggle, their work forces us to examine the inherent sub-text which asks and addresses persistent and perennial questions about the individual and their place in a communal and technologically-driven society. Accessible to the general reader, yet of great value to the involved professional, this text is one that must be widely read in order that we understand what threats surround us and what avenues we all possess to resolve them.

Peter A. Hancock, University of Central Florida, USA

Chapter 1
Hostile Intent and Counter-terrorism: Strategic Issues and the Research Landscape

Alex Stedmon
Human Systems Integration Group, Faculty of Engineering & Computing, Coventry University, UK

Glyn Lawson
Human Factors Research Group, Faculty of Engineering, The University of Nottingham, UK

Introduction

September 11 2001 is etched into the memory of many people. On that day, the 9/11 attacks in which four commercial aircraft were hijacked by 19 terrorists with devastating consequences were carried out. American Airlines Flight 11 was deliberately flown into the North Tower of the World Trade Center in New York City, followed shortly after by United Airlines Flight 175 crashing into the South Tower. In addition, American Airlines Flight 77 was aimed and crashed into the Pentagon in Virginia. A fourth aircraft, United Airlines Flight 93, crashed in Pennsylvania although its intended target was assumed to be Washington DC.

The date was chosen to parody the 911 phone number for the emergency services whose personnel would later respond to the single most deadly terrorist attack on US soil, taking 2996 lives. Of these, over 400 emergency response personnel became victims when the Twin Towers collapsed a few hours later. This was one of the highest profile terrorist attacks in recent times, and perhaps the first to unfold in real-time across the media who were able to capture it as it happened.

It was only much later in the ensuing investigations that the degree of planning and training that the terrorists undertook became apparent and Al-Qa'ida eventually admitted responsibility. In planning their attack on that specific day, the terrorists had conducted 'hostile reconnaissance' of airports and specific flights; they had trained together and developed a carefully coordinated plan to hijack four aircraft from three different airports on the morning of 11 September. They had even taken flying lessons in order to take control of the aircraft once they had been hijacked after take-off. This level of organization and planning had not been seen before, nor had such an act been anticipated: that terrorists would use commercial aircraft as missiles laden with fuel shortly after take-off. These were not the typical random acts of a few militant individuals, but rather a strategic act of terrorism achieved through precision planning and a carefully planned deception.

On 7 July 2005 a new form of terrorism came to the UK mainland. Four suicide bombers attacked the London Underground network with devastating effects. Three bombs were detonated on the London Underground (between Liverpool Street and Aldgate, Kings Cross/St Pancras and Russell Square; and also at Edgware Road). A further bomb was detonated on a double-decker bus in Tavistock Square. The bombs had been constructed from homemade organic peroxide-based explosive devices hidden inside rucksacks and the terrorists had simply walked into stations and boarded the trains without arousing any suspicion. Only the bomber on the bus had aroused suspicion in the moments before he detonated his rucksack.

Post-incident investigations would later reveal that in the run up to the attack, the terrorists appeared to have been guided by the wider organization of Al-Qa'ida; spending time abroad; meeting and training together; and then conducting a rehearsal of the attack by travelling to London from Leeds days before the planned final attack. In addition, another key feature of this attack was that the terrorists were largely unknown to the authorities and held normal jobs, had families and even contributed to community activities. Three of them were British-born sons of Pakistani immigrants while the fourth was born in Jamaica. In many ways they were invisible to the security services and even members of their local community.

The degree of preparation that was required for both of these attacks raised the profile of these new forms of terrorism within security agencies on both sides of the Atlantic. With the emergence of Al-Qa'ida as a global terrorist threat, hostile intent and counter-terrorism has since become a major interdisciplinary research area. A key focus is in understanding the hostile intentions of terrorists and the means by which they might attempt to collect information about a target, while concealing their identity and deceiving others in order not to raise suspicion. This, in turn, underpins the need for counter-terrorism interventions that might be used to monitor and safeguard crowded public spaces, mass transit hubs and social venues and sporting arenas that are being targeted more often by terrorists. With recent attacks such as the Boston Marathon, the Oslo bombing and subsequent attack on the youth camp at Uotya Island, the Westgate Shopping Centre attack in Nairobi, and the intensification of home-grown terrorism at a time of resurgent levels of anti-Western rhetoric and global violence, security policy is focused upon maximizing the efficiency in counter-terrorism practices at local, national and international levels.

With the focus of many terrorist activities now moving towards indiscriminate, mass impact and high publicity activities their modus operandi often combines initial attacks with pre-planned secondary attacks to maximize the impact of an attack and expose the vulnerability of civilians. As with the events of 9/11, the London bombing attacks and the more recent killing of Drummer Lee Rigby in what appeared to be an execution by Islamic fundamentalists, the power of the media and in particular social media are being exploited by terrorists to undermine public safety and generate individual, social, economic and political fear.

There are many challenges to detecting terrorist activities and a major concern is how to monitor crowded public spaces while also protecting people going about their daily lives, free from fear. Even with advances in technology informed by the science of human behaviour, various forms of hostile intent present major challenges for those tasked with securing public safety. An underlying principle is that terrorism shares attributes with many lower and more common forms of crime, e.g., trespass, assault, petty theft, burglary. Just as other criminals will survey a public space for security measures and vulnerable targets, terrorists conduct hostile reconnaissance in the planning of an attack in a public space. This poses a range of questions about how science might inform counter-terrorism as well as underlying ethical issues of surveillance and public security. More fundamentally, from a research perspective, this translates into a need to detect suspicious behaviours and identify hostile intent.

It is comparatively recently that the approaches and applied nature of human factors has been incorporated into counter-terrorism by combining aspects of psychology, systems design and approaches, and identifying specific user requirements and limitations within the security domain. In order to tackle hostile intent (in all its guises from low-level criminal behaviours through to large-scale terrorist attacks) and develop a sociotechnical approach to counter-terrorism, fundamental research must be integrated with applied methods and approaches grounded in the practical issues faced by security personnel in the field.

Background and Aim of this Book

This book has been developed from a range of research activities and wider dissemination events over the last few years. One of the prime activities was the 'Shades of Grey' strategic security consortium funded through by the Engineering and Physical Sciences Research Council (EPSRC) from 2010 to 2013 (Grant no. EP/H02302X/1). Shades of Grey emerged in response to the requirement for novel surveillance interventions that might elicit robust, reliable and usable indicators of notable behaviours in crowded public spaces. A further aspect was that counter-terrorism interventions should not frustrate, degrade, or restrict the general experience of legitimate users these public spaces.

In the wider development of this book and in order to invite the contribution of leading researchers around the world, a range of special sessions and events were organized at the following: the Ninth International Conference on Engineering Psychology and Cognitive Ergonomics within HCI International 2011 Orlando, Florida, USA; the Institute of Ergonomics and Human Factors Annual Conference, Blackpool, UK (plenary lecture); the Fourth International Conference on Applied Human Factors and Ergonomics Conference 2012 San Francisco, USA; the Australia and New Zealand Society of Criminology Conference 2012, Auckland, New Zealand; and the Tenth International Conference on Engineering Psychology and Cognitive Ergonomics within HCI International 2013 Las Vegas, Nevada, USA.

These opportunities provided the basis for two special issues of leading international journals published in 2013 that focused on different areas of hostile intent and counter-terrorism:

- *Detecting terrorist activities: hostile intent and suspicious behaviours* (*Applied Ergonomics*). This special issue explored a range of topics within the spectrum of detecting terrorist activities, hostile intent, crowded public spaces and suspicious behaviour including: methodological issues in counter-terrorism, novel research methods and innovative technologies in counter-terrorism.
- *Terrorism psychology: theory and application* (*Journal of Police & Criminal Psychology*). This special issue focused on the application of psychology and related disciplines to understanding terrorism and for effective counter-terrorism policies and practices.

During the course of these activities it became apparent that the application of human factors methods and perspectives to the specific issues embedded within hostile intent and counter-terrorism have the power to inform new ways of thinking about the problems as well as supporting current perspectives. Individuals engaged in terrorist activities are difficult to detect and actively conceal their hostile intent. Approaches taken from sociotechnical systems perspectives, informed by the integration of science and fundamental understanding principles of human behaviour, can provide valuable insights into human behaviour in complex systems. From research and practitioner perspectives, the discipline of human factors has much to offer this endeavour, including:

- an applied knowledge of psychology which can be used to analyse suspicious behaviours in different contexts
- a systems-based approach to consider not only hostile intent but the groups and wider social networks in which terrorists operate
- methods and techniques to support the development of new initiatives and approaches for detecting terrorist activities
- the wider integration of counter-terrorism into security training, policy and underlying ethical debates surrounding public surveillance.

This book presents an extension of these ideas and provides a timely publication that draws together wide-reaching and high-impact contributions across the spectrum of hostile intent and counter-terrorism.

Scope of the Book

The book has been designed as a resource that is accessible to a wide readership, integrating leading research and knowledge from around the world across a diverse range of areas within the scope of human factors and psychology. However, a further aspect of this book that became apparent through special sessions at conferences (including those mentioned above) was the necessity to widen the scope beyond more traditional perspectives on human factors.

In order to achieve this, the editors have chosen a range of contributions that expand typical definitions of human factors, stimulating discussion by representing diverse disciplines that contribute knowledge to this area, such as psychology, criminology, sociology, social and scientific philosophy, political science, ethics, art and design, neuroscience, engineering and computer science. In doing so the aim has been to illustrate the multidisciplinary requirements of investigating the issues surrounding hostile intent that might then facilitate counter-terrorism strategies, policies and interventions.

The chapters interpret and translate the problem of hostile intent in different ways ranging from social and political theory, applied and social psychology, interaction design, and user-centred approaches. They explore the spectrum of these domains across a number of themes: contextualizing terrorism; deception and decision-making; modelling hostile intent, sociocultural factors; strategies and approaches for counter-terrorism. In providing a multidisciplinary resource for researchers and practitioners alike, the underlying theme across all chapters is the inclusion of end-users (e.g. security personnel, police, general public) within the various debates on hostile intent and counter-terrorism.

In the development of this book, the editors received chapter submissions from authors across Europe, the USA, the Far East and Australasia. The task was then to select chapters that explored a spectrum of hostile intent, identifying suspicious behaviours, detecting terrorist activities, and maintaining safety in crowded public spaces to develop topics around key research areas to present a global perspective for the widest audience appeal. The result is a truly international edition including monographs, reviews, and findings from empirical research. Across 20 contributed chapters, 27 senior academics and 13 professors represent those at the forefront of their subject areas. These are supplemented by valuable contributions from government advisors and leading figures in counter-terrorism policing who have led UK initiatives, training and the development of police handbooks as well as key members of a steering group of the newly formed Centre of Excellence for Terrorism, Resilience, Intelligence and Organised Crime Research (CENTRIC).

Conceptual Framework

A major challenge with this book has been the integration of a diverse range of topics into a coherent narrative on hostile intent and counter-terrorism. In order to assist the reader in navigating between the many different areas, a conceptual framework has been developed based on the three key areas that the book addresses (Figure 1.1).

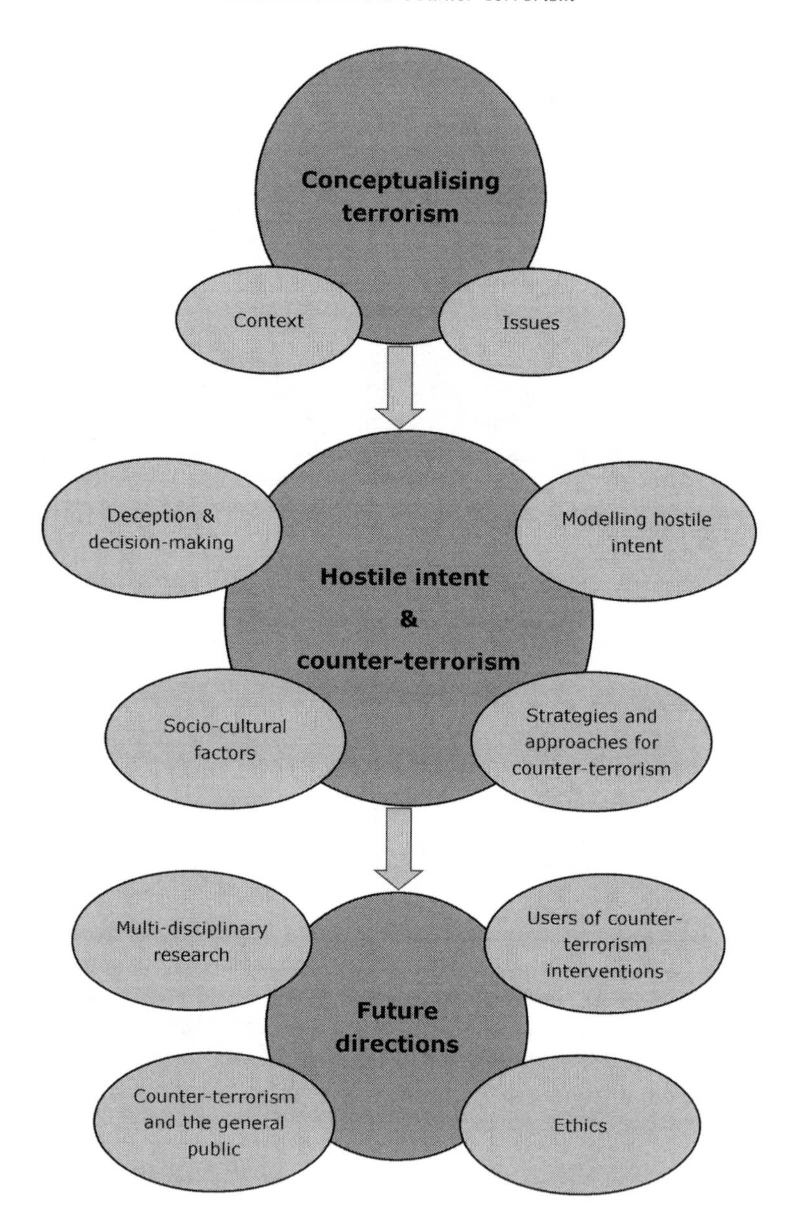

Figure 1.1 Conceptual framework of hostile intent and counter-terrorism

In this way the three areas provide a set of waypoints with associated sub-topics to assist the reader in focusing on specific areas. The book starts with a section that sets out the underlying issues of contextualizing terrorism. It deals with notions of terrorism, the ideas behind fear and issues of governance within society. Leading on from this, important questions are raised about the role of science and the nature of research into hostile intent, along with important ethical considerations embodied in surveillance and public safety.

From this foundation the next main section focuses on a diverse range of hostile intent and counter-terrorism factors. These are presented across four sub-sections:

- deception and decision-making
- modelling hostile intent
- sociocultural factors
- strategies and approaches for counter-terrorism.

Finally the book concludes with a section that looks to the potential future directions of hostile intent and counter-terrorism research. This section draws together the common themes from the preceding chapters and identifies specific research issues that will help inform future policy and interventions.

The conceptual framework is presented and expanded at the start of each section to provide an overview of the chapters that follow. Author biographies are included at the end of the book along with extensive reference lists and an index to assist readers with focused searches of the material presented.

Overview of Contents

This section provides a brief introduction to each of the chapters.

Part 1: Conceptualizing Terrorism

- *The role of fear in terrorism* – this chapter provides an overview of the ways in which the academic literature has tended to characterize the role of fear and psychological violence in the process of political terrorism. The role of fear in both common definitions and the logic that underpins terrorist violence are considered. Empirical findings of public opinion surveys and polls in the UK and US between 2001 and 2010 are drawn upon to illustrate 'fear in practice'. This chapter advocates the use of counter-terrorism measures that recognize the importance of fear and, accordingly, attempt to reduce the psychological responses of the public or, at least, to not exacerbate it.
- *Understanding terrorism through criminology? Merging crime control and counter-terrorism in the UK* – in this chapter enduring practices and underpinning principles of counter-terrorism are identified and questions are raised over their applicability to new, mutable and dynamic forms of terrorism. In doing so, the chapter first reports on the convergence of crime control and counter-terrorism and considers the implications of drawing diverse agencies and organizations together into wider ensembles

of security practice. The discussion then moves towards an identification and analysis of how, despite this plurality and diversity of practice, some homogenizing tendencies are visible. In particular specific forms of understanding transgressive behaviour based on utilitarian notions of rational choice theory are identified. The remainder of the chapter examines the applicability of such explanations of transgression across crime and terrorism.

- *Analysing the terrorist brain: neurobiological advances, ethical concerns and social implications* – terrorism is often perceived as the most existential threat humanity is currently facing, despite the fact that the actual danger is negligible. Terrorism is not an easy subject of enquiry and the serious researcher faces several issues when addressing it. In particular, terrorism is an emotive subject and a morally detached analysis is difficult, if not impossible, to achieve. This chapter focuses on neurobiological research into terrorism and the underlying ethical and societal issues that unravel as research is conducted and the findings are communicated back to a wider audience.
- *Ethical issues in surveillance and privacy* – this chapter explores issues involved in balancing public security against individual privacy. To illustrate this, specific reference is made to an ongoing project aimed at supporting mass public transport security, with surveillance exercises in complex travel interchanges across major European cities. Members of the public in mass transit hubs are notoriously vulnerable to terrorist attack as evidenced by bombings in subways in London, Paris and Madrid and the toxin attacks in Tokyo. While there is a need to find ways to protect the travelling public, many of the underlying issues highlight problems associated with surveillance and security in a range of other complex settings.

Part 2: Deception and Decision-making

- *Non-verbal cues to deception and their relationship to terrorism* – while the desire to detect deception has existed as long as the act of deception itself, vast amounts of research resources have been used in order to gain a better understanding of verbal and non-verbal cues (including speech tone, speed, pitch, body and facial movements). This chapter outlines the progress of research related to the use of non-verbal cues in deception detection within the specific context of terrorist activities. Conducting research in this area is a challenge because of the fortunate rarity of terrorist events. However, relying on our understanding of cognitive processes and extrapolating from other situations, we can begin to understand how non-verbal cues work in a terrorist domain.
- *Deception detection in counter-terrorism* – building on the previous chapter, for decades, deception researchers have attempted to improve

people's ability to detect deceit by teaching them the cues that liars typically display. In recent years, the emphasis has changed and attention is paid to interviewing techniques that elicit and enhance cues to deception. In today's political climate, researchers should pay particular attention to settings that are relevant for terrorism such as: lying about intentions; examining people when they are secretly observed; and interviewing suspects together. This chapter also considers physiological and neurological lie detection methods that are often discussed in the media as well as the theoretical underpinnings of non-verbal and verbal cues to deceit.

- *A field trial to investigate human pheromones associated with hostile intent* – while the psychology and physiology of stress is relatively well understood, our understanding of the biology of stress, in terms of the production of stress pheromones, is less well developed. With a specific focus on an ecologically valid environment simulating hostile intent in crowded public spaces, this chapter presents evidence that stressed individuals secrete a volatile steroid-based marker that could form the basis for remote detection of deception. In this way, a human pheromone linked to deception may exist and overall the findings provide a validated model of hostile intent that can be used by other researchers to test interventions aimed at detecting or deterring hostile intent.
- *On the trail of the terrorist: a research environment to simulate criminal investigations* – this chapter reports on the multidisciplinary Detecting Scent trails (DScent) project, that brought together technologists and behavioural scientists to design and evaluate novel proof-of-concept methods for countering terrorism in public places. Through a mixture of technology prototyping and empirical evaluations, an immersive simulation environment for detecting and investigating deceptive behaviours indicative of terrorist activities was developed and assessed. An underlying thesis of the DSCENT project was that the collective movements and communications of persons working together and co-operating in the planning and execution of a major terrorist event may provide either (i) an indication of their possible intent prior to its successful conclusion and therefore assist early intervention (ii) a reduction in the number of suspect groups or persons enabling more detailed direct surveillance of suspect groups or (iii) an ability to schedule direct surveillance or carry-out disruptive intervention actions based on suspect patterns of a single person or collaborative movement behaviours and communications.

Part 3: Modelling Hostile Intent

- *Safety and security in rail systems: drawing knowledge from the prevention of railway suicide and trespass to inform security interventions* – this chapter provides an overview of how the problems associated with rail trespass and suicide are being considered across Europe as well as examples from national prevention programmes that are implementing various interventions. The transfer of knowledge and experiences from the investigation of railway suicide and trespass to the wider arena of railway security is considered.

- *Tackling financial and economic crime through strategic intelligence management* – Serious Organised Economic Crime (SOEC) and the associated activity of fraud is a growing multinational business without respect to national borders. This chapter describes an approach for strategic intelligence management that would be capable of capturing even the low-level and low intensity crimes, thus providing Member States with a tool to counter-terrorism on a spectrum of criminal activities. By taking a systems perspective and appropriating and applying existing business tools and analysis techniques to the illegal businesses of SOEC and fraud, better intelligence is possible to help Member States target these crimes and criminals. However, to be effective at the multinational level these systems must be comprehensively integrated into one pan-European system.
- *Competitive adaptation in militant networks: preliminary findings from an Islamist case study* – in recent years a consensus has emerged in counter-terrorism studies that violent non-state militant groups and networks adapt in response to information and feedback. However, few studies move beyond facile metaphors to unpack how militants and terrorists actually adapt their tactics and strategies in response to counter-terrorism pressure. Even fewer studies examine how counter-terrorism agencies alter their own activities in response to terrorist behaviour. This chapter provides an interdisciplinary framework from which to study the behaviour of militant groups that either carry out acts of political violence themselves or support the use of violence by others. Specifically, data from news reports and interviews are analysed concerning the militant activist group Al-Muhajiroun. Using competitive adaptation as a comparative organizational framework, this chapter focuses on the process by which adversaries learn from each other in complex adaptive systems and tailor their activities to achieve their organizational goals in light of their opponents' action.
- *Evaluating emergency preparedness: using responsibility models to identify vulnerabilities* – preventing acts of terrorism requires being able to accurately and reliably predict when, where and how the terrorists will strike. When prevention is not possible the aim is to mitigate the severity of the consequences of the terrorist acts. Terrorists normally attempt to cause maximum disruption and achieve significant publicity by attacking systems that they perceive as being of great importance (physically and psychologically). This chapter investigates the notion of responsibility in sociotechnical systems that encompass both organizations and infrastructure. A graphical technique for modelling and analysing responsibilities in these sociotechnical systems is introduced and classes of vulnerabilities that are associated with responsibilities are discussed. These vulnerabilities can be

mapped to resources to support contingency, resilience planning, and help improve emergency preparedness.

Part 4: Sociocultural Factors

- *Unintended consequences of the 'War on Terror': home-grown terrorism and conflict-engaged citizens returning to civil society* – this chapter presents a series of case studies in terrorism investigations focusing on the emerging threat of home-grown terrorism as well as individuals returning from conflict abroad. During 2006 the International Institute for Strategic Studies estimated that as many as 10 per cent of the 20,000 insurgents fighting the conflict in Iraq were foreign-born. It was also revealed that up to 150 radicals from Britain had travelled to Iraq to join up with a 'British Brigade' that had been established by Al-Qa'ida leaders to fight coalition forces. This raised two key issues for the UK government: first, how were these men travelling to Iraq and receiving their training? The second was the realization that, while many of the volunteers may die in the theatre of conflict in Iraq, some may well survive their experiences and return to the UK with military training and hardened combat experience. Such conflict-engaged citizens could put their skills and experiences to unlawful use when they returned to their local communities in the UK continuing their violent jihadist crusade.
- *Parasites, energy and complex systems: generating novel intervention options to counter recruitment to suicide terrorism* – this chapter has been developed in response to a constant irritation by superficial media coverage that describes terrorists and other criminals as 'brainwashed'. This chapter challenges conventional explanations of terrorism by means of Analogical Research (AR) and elaborates around the analogy of parasitic infection shaping or driving undesirable behaviour. It is intended to once again catalyse reactions as well as suggest the potential for practical interventions in policy. AR seeks to shed new light on intractable problems by exploiting analogous contexts. In this case, viewing suicide terrorism as being the product of rational (to the perpetrator) actions driven by powerful bundles of memes that influence cognitive processes. Parasites change behaviour in animal models. Subtle mechanisms (changes at the endocrinal level) produce profound effects (from timid to aggressive traits). At the same time, the human brain can achieve extraordinary effects when driven by extreme belief: in other words, when infected by memes. This chapter argues that there is the potential for memes to drive transformational and problematic behaviour: such as that of the migration of a classroom assistant in a school to self-recruitment and sacrifice as a willing suicide bomber. In further advancing a challenging and unusual approach to understanding deviant actions, the authors seek to stimulate engagement with readers in order to

advance further novel ideas for development in policy-making and other applied communities.

- *Terrorist targeting of schools and educational establishments* – terrorist attacks on educational institutions have taken many forms: armed assaults, bombings, hostage takings, chemical attacks, and arson. This chapter focuses on armed assaults. The frequency of this type of attack on educational institutions has increased sharply since 2003; with the incident rate in 2007 being double that of the previous year. A comprehensive chronology of all armed assaults on educational institutions since 1980 was created with reference to the Global Terrorism Database, and other credible sources. Descriptive reports of each incident were subjected to content analysis according to a series of 143 variables. These variables describe various aspects of the incident including information about who the victim(s) were, where and when the attack took place, and who the offender(s) were. The data set was analysed using traditional inferential statistics which revealed that, over time, armed terrorist attacks on educational institutions have increased in lethality, but changed in terms of the main targets. Lethality is also associated with whether or not terrorist groups claim responsibility. The data were also analysed using Multidimensional Scalogram Analysis (MSA) to explore interrelationships between behaviours and identify the underlying dimensions in terrorist attacks on educational institutions. The analysis demonstrates that attacks can be classified on an expressive-instrumental continuum.
- *Female suicide terrorism as a function of patriarchal societies* – this chapter argues that there is a relationship between patriarchy and female suicide terrorism. On the basis of a comparative study of Turkey (PKK), Chechnya (Chechen Rebels) and Dagestan cases conclusions are drawn that, despite the different contexts, female suicide terrorism is likely to appear in societies which share at least five characteristics: a perception of acute ethnic conflict or foreign invasion (which poses ethnic or national identity in danger) that exists in part of the population; where there are radical organizations which have chosen female suicide terrorism as a means of political action and mobilization; where religious or cultural norms restrict women's access to the public sphere; structures of traditional society are in a process of disintegration (sometimes as a consequence of conflict); the process of disintegration causes changes of authorities, values and priorities. It is argued that motivation of individual perpetrators goes beyond religion and ideology and has to do with a reaffirmation of female identity as designed and tolerated by traditional society (referred to as patriarchy).

Part 5: Strategies and Approaches for Counter-terrorism

- *Designing visible counter-terrorism interventions in public spaces* – this chapter explores publicly visible counter-terrorism measures, uncovering the strategic role of design in creating controlled disruption in public spaces to reduce threat while at the same time reducing anxiety. A range of techniques, suitable for disrupting routine uses of 'public' crowded spaces, are reviewed. With a basis in psychology, where unexpected questioning can elicit observable errors in consistency or disrupt scripted responses, this chapter presents physical alternatives to unexpected questioning drawn from the fields of advertising, art, architecture and entertainment. These techniques may offer means to physically interrupt pre-planned and scripted actions, with the potential of eliciting detectable covert behaviour, and provide approaches to fostering calm, enjoyable public spaces. Such interventions are therefore a good response to addressing heightened security concerns in heavily used 'public' spaces.
- *A macro-ergonomics perspective on security: a rail case study* – this chapter outlines user requirements methods/approaches and key themes emerging from the 'Shades of Grey' project. It is written from a sociotechnical perspective identifying systemic issues and differences between security agencies and front-line security personnel.
- *Deception and speech: a theoretical overview to inform future research* – this chapter presents an overview of the theoretical foundations pertinent to the investigation of deception and speech. Using a parameter-centred description the reader is presented with a summary of the acoustic and temporal correlates of various affective and cognitive states. This overview provides an informative point of departure for those wishing to embark on the study of deception-related speech characteristics.
- *Evaluating counter-terrorism training using behavioural measures and theory* – since 2008, the US Department of Defense has placed Irregular Warfare (IW) on an equal footing with conventional warfare in future military planning and operations. Among IW mission objectives are developing counter-terrorism competencies to identify people with hostile intent before events become lethal. As part of IW mission readiness, small units are called upon to execute a full range of kinetic and non-kinetic operations, often within a single day. Whether practiced by military ground units or law enforcement personnel (e.g., Customs and Border Patrol), the knowledge, skills, and attitudes (KSAs) to read the human and physical terrain are an essential element of training. Constructing the behaviour profiles necessary to read terrain and infer hostile intent is now considered every bit as important as body armour and weaponry, so that profiling KSAs have become a valuable addition to small unit tactics, techniques, and procedures.

Part 6: Future Directions

- *Hostile intent and counter-terrorism: future research themes and questions* – this chapter reviews the themes of the book and identifies a range of issues to be taken forward in future research and development of counter-terrorism measures across the human factors domain.

Thoughts from the Editors

We have developed this book to provide a diverse resource of current thinking surrounding key human factors issues of hostile intent and counter-terrorism. We hope you will enjoy it in its entirety but we have designed it so that each chapter stands alone to allow readers to focus their reading as they wish. In this way, a range of different perspectives can be drawn depending on your particular background and interests. However, we would urge you to read at least one chapter from the book that does not fit with your immediate focus as we have found that editing the book has stimulated our thinking about new ways of addressing the underlying issues of hostile intent and counter-terrorism.

This book has taken a long time to produce, not least because of the extensive range of supporting activities upon which it is based. With the high profile and impact of counter-terrorism research, alongside established and developing university teaching around the world in security, criminology, terrorism studies and forensic psychology, there is no single resource available to researchers, students and stakeholders from a human factors and applied psychology perspective. With this book, the editors hope that it will not only expand the knowledge base of the subject area and therefore be of prime relevance to researchers in counter-terrorism, but also provide a valuable resource to security stakeholders at policy and practitioner levels.

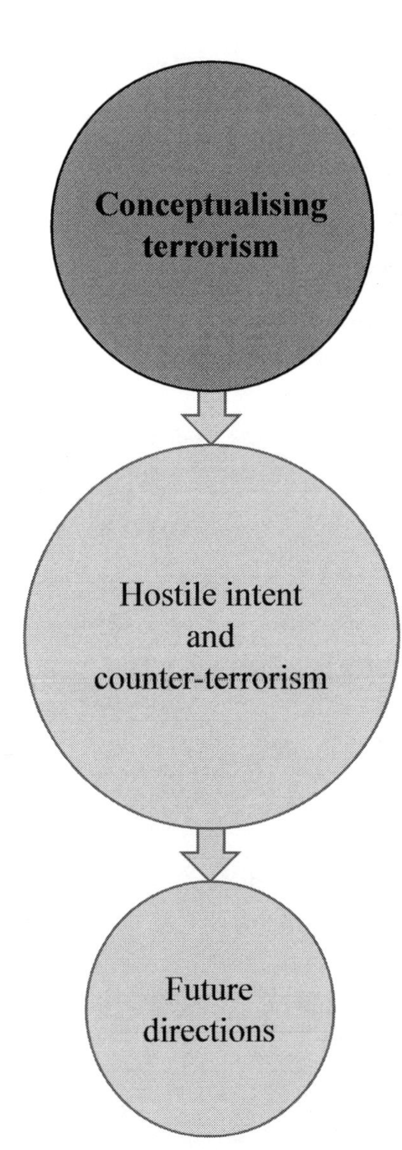

It is imperative to understand the issues surrounding terrorism, and the contexts in which it occurs in order for counter-terrorism initiatives to succeed. The chapters in Part 1, Counter-terrorism, deal with such aspects. These include the role of fear in terrorism and the emotions involved with research in this area. Ethics play an important part in our work and are addressed from both the researcher perspective and with regards to surveillance and privacy. This Part also includes a review of current approaches to counter-terrorism in light of new forms of terrorism and discusses the opportunity to merge criminology and counter-terrorism.

Chapter 2
The Role of Fear in Terrorism

Alex Braithwaite
School of Government & Public Policy, University of Arizona, USA

Introduction

Perhaps the most important and consistent goal of terrorism worldwide is to inflict psychological harm upon a public audience. The provocation of fear within a local population links motivations for the use of violence with the anticipated outcome of policy change (Crenshaw, 1998). Given that the nature of terrorist violence appears to have evolved in recent years (with a premium placed upon indiscriminate targeting and mass casualty outcomes) it follows that the effect of terrorism has also changed. In this chapter, I argue that given that the public in western democracies are more fearful of terrorism now than in previous generations, government efforts to counter terrorism would benefit from attempting to alleviate this fear. Moreover, I suggest that counter-terrorism policies that focus upon communicating the dangers associated with terrorism are likely to have the counter-productive effect of increasing levels of fear.

Building upon earlier work (Braithwaite, 2013) and that of Khalil (2006), I argue that public perceptions, rather than transportation, military, government, or private property, are the primary targets of terrorist attacks. Terrorists prioritize communication of an exaggerated sense of their ability to do harm (Hoffman, 2006). They do this by attempting to convince the target population that their government is unable to protect them. It follows, then, that any attempt at improving security policy ought to centre upon gaining a better understanding of the factors that affect public perceptions of security. Thus, there is benefit in aiding public understanding of threat and reducing their experience of fear.

States with a history of terrorist attacks commonly encourage citizens to remain vigilant for suspicious activities and behaviours. In the United Kingdom, this logic underpins the Metropolitan Police and Home Office 'Anti-Terrorist hotline' facility. I suggest that this kind of policy is unlikely to improve counter-terrorism, especially given the extremely low likelihood of terrorism, but is likely to exacerbate rather than alleviate public fear. This is highly counter-productive and it is likely that counter-terrorism and public safety would be better served by policies that improve the public's experience and enjoyment of public spaces and communities.

Fear of Terrorism: Some Empirics

There is surprisingly little systematic evidence regarding public attitudes towards terrorism (Allouche and Lind, 2010; Braithwaite, 2013). A handful of surveys and public opinion polls, as summarized in Table 2.1, illustrate an overestimation of the risk of future terrorist attacks.

Table 2.1 Fear in public attitudes towards terrorism

No.	Date (source)	Population	Key finding(s)
1	July 2005 (MORI)	London	51% think it is very likely that London will be attacked again
2	September 2005 (MORI)	London	43% think it is very likely that London will be attacked again
3	2005 (British Social Attitudes Survey)	UK	43% think threat of terrorism is not exaggerated (pre-7/7) vs. 68% think threat of terrorism is not exaggerated (post-7/7)
4	October 2007 (IPSOS MORI)	London	66% very or fairly worried about another terrorist attack in London
5	July 2010 (YouGov/ Sun)	UK	25% think threat of terrorism has increased in five years since 7/7; 17% think it has decreased
6	September 2001 (Gallup)	USA	66% believe further acts of terrorism are somewhat or very likely in the coming weeks; 56% worried that they or a member of their family will become victim of a terrorist attack
7	September 2002 (Gallup)	USA	60% believe further acts of terrorism are somewhat or very likely in the coming weeks; 38% worried that they or a member of their family will become victim of a terrorist attack
8	November 2009/ January 2010 (Gallup)	USA	39% believe further acts of terrorism are somewhat or very likely in the coming weeks; 42% worried that they or a member of their family will become victim of a terrorist attack
9	2002 (Chicago Council on Global Affairs [CCGA])	USA	91% view terrorism as a vital threat to US interests
10	2008 (CCGA)	USA	70% view terrorism as a vital threat to US interests

Source: Braithwaite (2013)

Table 2.1 presents average responses to a series of opinion polls conducted in London, the UK, and the United States of America (US) between 2001 and 2010. This list is not exhaustive but is somewhat representative of the variety of questions asked by pollsters on the subject of terrorism. The table highlights four take-home points:

- There is an observable reduction in levels of fear as time passes after an attack. The first two entries in the table reflect that the proportion of adults in London that anticipated a follow up attack after the 7/7 bombings fell from 51 to 43 percent between July and September 2005 (i.e., in the three months after the bombings).
- Perceptions that the threat of terrorism is real increased after 7/7. The third entry enables us to tap directly into the principle of exaggeration. The 2005 British Social Attitudes Survey shows that the proportion of UK residents that believed the threat of terrorism was *not* exaggerated increased sharply as a result of the bombings (from 43 percent beforehand to 68 percent shortly thereafter). This clearly demonstrates that fear correlates closely with the observation of physical violence but also endures for considerable periods after the observed violence.
- It is clear that baseline levels of concern regarding the threat of terrorism are high in countries with recent terrorism experience. The fourth entry in the table shows that two-thirds of Londoners remained worried about further terrorist attacks over two years after the 7/7 bombings. Moreover, the fifth entry reflects that a quarter of all UK residents believed the threat of terrorism had, in fact, increased in the five years after bombings.
- The slow decline in the very high levels of concern about the threat of subsequent attacks is observed in the US as well as in the UK. Entries 6, 7 and 8 in the table (summarizing Gallup polls from September 2001, September 2002, and winter 2009/10) show that the proportion of Americans anticipating further imminent attacks declined from 66 to 60 to 39 percent. In addition, entries 9 and 10 show a marginal decline in the belief that terrorism represented a serious threat to the US from 91 percent in 2002 to 70 percent in 2008.

Fear of Terrorism: Two Psychological Models

Two competing paradigms from psychology and neuroscience offer accounts of fear and anxiety that are pertinent to our understanding of the fear of terrorism (Figure 2.1).

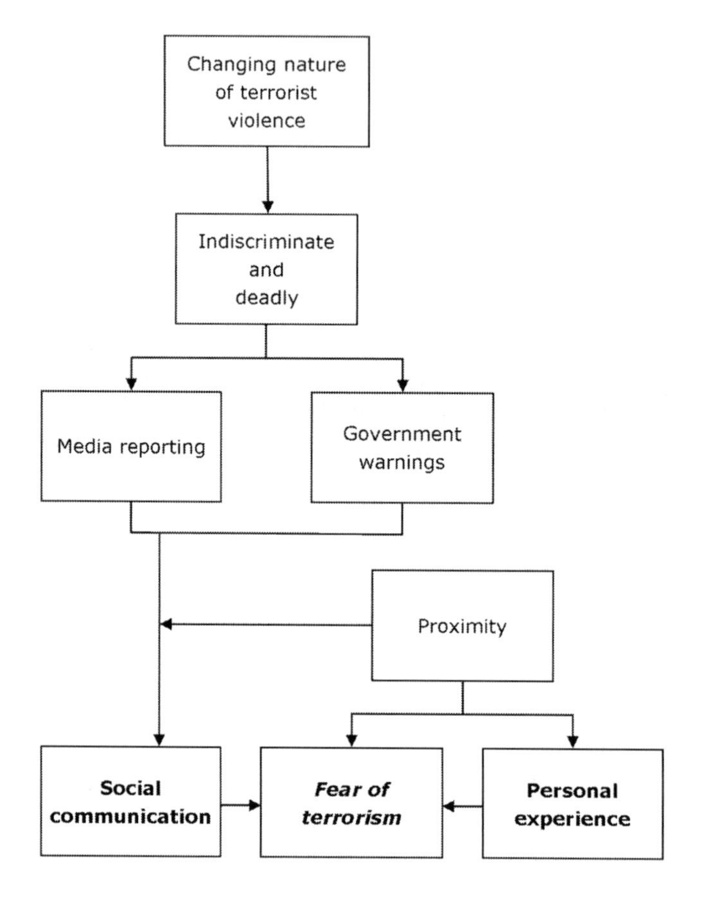

Figure 2.1 Sources of the fear of terrorism

Figure 2.1 details these alternative logics. In both cases, fear and anxiety are considered greatest towards threats that are uncontrollable or unavoidable (Ohman, 2000). These accounts differ, however, in terms of the mechanisms via which individuals are exposed to the threat in question. First, traditional models of fear conditioning centre upon the importance of personal experience of, and exposure to, aversive events (Bandura, 1977). The chance that any individual is likely to become embroiled in, or directly impacted by, a terrorist attack is incredibly low. Terrorism is an extremely rare event and the number of casualties associated with terrorist violence is also very small. As the figure suggests, it is likely that only those that reside in close proximity to the locus of an attack can reasonably be considered to have had a personal experience of terrorism. As a demonstration of this proximity effect, Fischoff et al. (2003) show that fear and anxiety of terrorism was greatest in areas in close proximity to the sites attacked on 9/11.

A more plausible alternative to the personal experience account of fear is the logic that centres upon the importance of social communication. "Our sociocultural environment provides other, indirect, means of attaining fear-relevant information, such as social observation and verbal communications which are more efficient and associated with fewer risks than learning through direct aversive experiences" (Olsson et al., 2007, p. 1). According to this logic, individuals do not need to be directly (i.e., physically or mentally) affected by an individual terrorist attack. Rather, preemption of terrorist events, government warnings and media reporting all provide fear-relevant information. Figure 2.1 demonstrates that two factors are likely to exacerbate this social communication effect. First, proximity of attacks is deemed important because this helps to guarantee extra media coverage of events. Second, the changing nature of terrorism, with a dramatic increase in the deadliness of individual attacks, has the effect of increasing both media coverage and government warnings regarding the future threat of terrorism.

Fear of Terrorism: A Political Model

At the heart of the act of terrorism is the use of violence. This violence is typically indiscriminate in application and deadly in intent. Violence targeting the public is designed to coerce them to demand that their government appease the terrorists by giving in to their demands (Crenshaw, 1986; Long, 1990; Wardlaw, 1982).

Bueno de Mesquita (2005, 2007) offers a compelling model of the general strategic interactions between terrorists and governments in competition over the support of the public. Kydd and Walter (2006) specify five strategies of terrorism that reflect the ways in which terrorists seek to win or coerce public support. Of particular note is the strategy they refer to as 'intimidation'. Terrorists compete with governments over the support of the population with two goals in mind. First, they attempt to convince the population that they (the terrorist organization) can inflict damage upon the public at will. Second, they seek to demonstrate that the government is unable to defend itself and its citizens against this threat.

Figure 2.2 (from Braithwaite, 2013) details the difference between regular political opposition to a government and the logic of intimidation within the process of terrorism. In the legal version of the logic of politics, the opposition participates in the political process as defined by the government of the day. In the context of democracy, we commonly think of such participation as including political parties trying to mobilize supporters and competing in elections. The same logic of participation holds in non-democratic settings, though in such instances, opportunities to engage with the public, to run in elections, or generally to debate against the status quo authority will be restricted. In the alternative, illegal perspective (as depicted via the left-hand path in Figure 2.2), the opposition chooses instead to threaten or employ the use of violence to spread fear within the public and intimidate them into affecting change to a policy position.

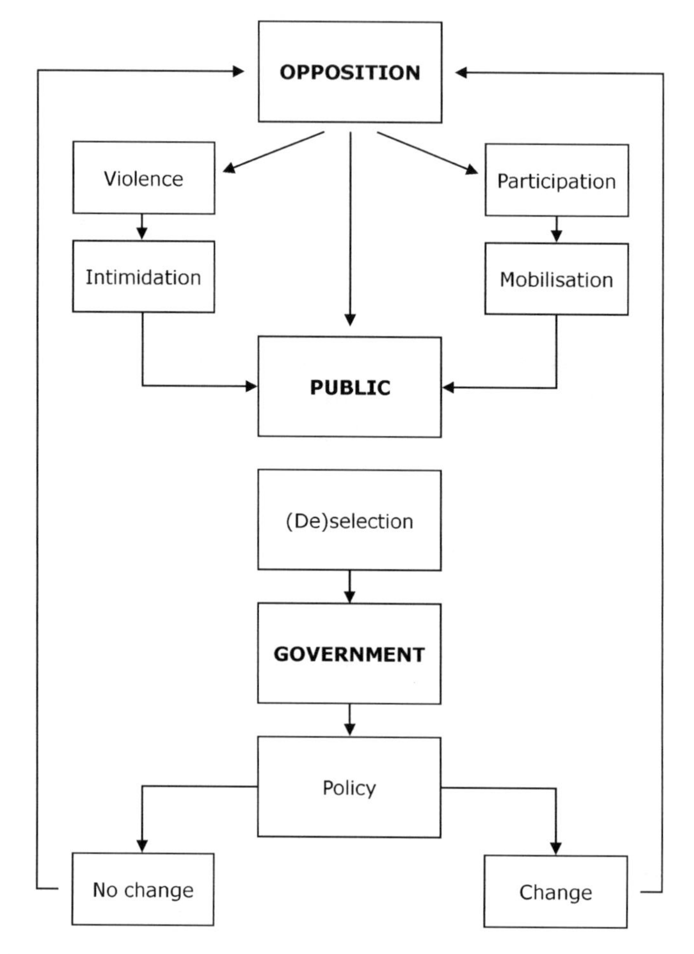

Figure 2.2 A simple model of fear, intimidation, and policy change in the process of terrorism (Braithwaite, 2013)

According to both the legal and illegal versions of reality, the opposition is attempting to affect the decisions that the public makes regarding the selection or de-selection of representatives in government. Given their low resource wealth relative to the governments they oppose, terrorists rely upon convincing or coercing disaffected or vulnerable citizens to support their cause. Success in this process centres upon the ability to convert the fear that their attacks cause within the public into demands for policy change that would help bring about changes in line with the terrorists' goals (Crenshaw, 1998).

This agenda-setting quality of terrorism is central to Crenshaw's logic of terrorism where she argues that by attracting attention, terrorism "makes the claims of the resistance a salient issue in the public mind" (Crenshaw, 1998,

p. 17). Terrorists are able, accordingly, to communicate to the public that the violence to which they are victims is likely to remain a consistent feature of the status quo political order. Thus, the successful political terror campaign coerces citizens to pressure their government to give into the demands of the terrorists in order to restore a semblance of public safety (Hoffman, 2006). Hewitt (1993) finds marginal empirical support for this claim. He demonstrates that incumbent governments (at least in 1970s and 1980s Germany and Italy) tended to be blamed for general threats to popular security.

Terrorism as psychological warfare that preys upon the fears and attitudes of individual members of the public provides the pivot point around which most political models of terrorism operate (Friedland and Merari, 1985). Terrorists "use the unreasonable fear and the resulting political disaffection it has generated among the public to intimidate governments into making political concessions in line with its political goals" (Long, 1990, p. 5). The priority of the terrorist is, therefore, to conduct violence that can be widely viewed by a public audience (Nacos, 2005). As consumers of new media, members of the public are commonly portrayed as easily manipulated (Hoffman, 2006).

Drawing upon the logic of the symbiotic relationship between the media and terrorist, Khalil (2006) argues that public perceptions are the primary target of terrorist attacks. Terrorists essentially rely upon communicating an exaggerated sense of their ability to do harm. It follows, Khalil argues, that any attempt at homeland security revolves around gaining a better understanding of the factors affect public perceptions of the government's security of the population.

A considerable literature across the social sciences and psychiatry traces the co-evolution of terrorism and individual wellbeing. In order to address this very issue, two recent studies examine the co-evolution of measures of individual-level happiness and societal-level terrorist violence. Robust patterns emerge cross-nationally. Residents in regions with higher levels of terrorism (such as Northern Ireland within the UK) are more negatively affected than those in areas with relatively little terrorist violence (Frey et al., 2009). Populations in close proximity to the attacks on 9/11 displayed significant levels of stress (Cohen Silver et al., 2002; Galea et al., 2002; Schlenger et al., 2002). Along similar lines, Bleich et al. (2003) identify a moderate psychological impact from the violence at the height of the intifada in Israel.

Implications for Responding to Terrorism

This chapter highlights a common logic by which terrorism is explained. According to this logic, terrorists employ indiscriminate physical violence against a target in a bid to intimidate a broader public audience. This intimidation is designed to compel the public to demand policy change from their government. A small body of research uncovers empirical evidence consistent with this logic. In particular, this evidence appears to show that, by and large, terrorists are quite successful in instilling a sense of fear within the public. This logic has significant implications for the design and conduct of counter-terrorism policy in societies threatened by terrorist violence.

First and foremost, it seems apparent that counter-terrorism policies must focus upon educating and reassuring the public about the real risk of terrorism (see Khalil, 2006; Mueller, 2006; Glass and Schoch-Spana, 2005). Any approach that chooses, instead, to highlight a grave danger associated with terrorism is likely to have the counter-productive effect of increasing fears. Governments remain unlikely to seek to reduce public concerns, however. From a rationalist perspective, elected representatives within government prioritize re-election (Bueno de Mesquita et al., 2004). In the context of the characterization of a terrorist threat, it is prudent for an elected representative to assume and communicate an elevated risk, because to communicate a lower level of risk and then be proven wrong by the occurrence of a terrorist attack would presumably result in a dramatic loss of public support. At the same time, if the representative communicates an elevated risk and the public subsequently observes no new terrorist activity, they are well placed to take credit for counter-terrorism successes. Moreover, it is perhaps not unreasonable to conclude that representatives may stand to benefit from inciting and exploiting public fear (perhaps through media coverage). This may help them, for instance, to achieve some set of related political goals (Altheide, 2006; Furedi, 2005; Mueller, 2006).

The implication of this possibility is quite stark. Returning, for instance, to the Metropolitan Police and Home Office 'Anti-Terrorist Hotline', the logic of fear implies that the typical frames of suspicion and alertness requested by Governments might prove imprudent as a means of reducing public fear. Indeed, they may even work to exacerbate concern by needlessly increasing public overestimations of the threat of terrorism. If it is true that terrorism is designed in part to instill a sense of fear then, in fact, countering terrorism might benefit from actively reducing levels of fear. Doing so could involve activities that range between simply providing clear information regarding the relatively low risks associated with terrorist violence (Mueller, 2006) and more proactively attempting to improve average participant enjoyment of public spaces.

References

Allouche, J., and Lind, J. (2010). *Public Attitudes to Global Uncertainties. A Research Synthesis Exploring the Trends and Gaps in Knowledge*. http://www.globaluncertainties.org.uk/Image/Global_Uncertainties_Research_Synthesis_-_Exec_Summary_tcm11-15030.pdf, accessed 5 May 2011.

Altheide, D. (2006). Terrorism and the politics of fear. *Cultural Studies, Critical Methodologies* 6(4), 415–439.

Bandura, A. (1977). *Social Learning Theory*. General Learning Press: New York.

Bleich, A., Gelkopf, M., and Solomon, Z. (2003). Exposure to terrorism, stress-related mental health symptoms, and coping behaviors among a nationally representative sample in Israel. *Journal of the American Medical Association* 290, 612–620.
Braithwaite, A. (2013). The logic of fear in terrorism and counterrorism. *The Journal of Police & Criminal Psychology* DOI 10.1007/s11896-013-9126-x.
Bueno de Mesquita, B., Morrow, J.D., Siverson, R.M., and Smith, A. (2004). Testing novel implications from the selectorate theory of war. *World Politics* 56(3), 363–388.
Bueno de Mesquita, E. (2005). Conciliation, commitment, and counterterrorism. *International Organization* 59, 145–176.
Bueno de Mesquita, E. (2007). The quality of terror. *American Journal of Political Science* 49(3), 515–530.
Cohen Silver, R., Holman, E.A., McIntosh, D.N., Poulin, M., and Gil-Rivas, V. (2002). Nationwide longitudinal study of psychological responses to September 11. *Journal of the American Medical Association* 288, 1235–1244.
Crenshaw, M. (1986). The psychology of political terrorism. In M.G. Hermann (Ed.), *Political Psychology*. Jossey-Bass: New York, 379–413.
Crenshaw, M. (1998). The logic of terrorism: Terrorist behavior as a product of strategic choice. In W. Reich (Ed.), *Origins of Terrorism: Psychologies, Ideologies, Theologies, States of Mind.* Woodrow Wilson Center Press: Washington, DC, 7–24.
Fischoff, B., Gonzalez, R., Small, D., and Lerner, J. (2003). Judged terror risk and proximity to the World Trade Center. *Journal of Risk and Uncertainty* 26(2–3), 137–151.
Frey, B.S., Luechinger, S., and Stutzer, A. (2009). The life satisfaction approach to valuing public goods: The case of terrorism. *Public Choice* 138, 317–345.
Friedland, N., and Merari, A. (1985). The psychological impact of terrorism: A double-edged sword. *Political Psychology* 6, 591–604.
Furedi, F. (2005). *Culture of Fear*. Continuum: London.
Galea, S., Ahern, J., Resnick, H., Kilpatrick, D., Bucuvalas, M., Gold, J., and Vlahov, D. (2002). Psychological sequelae of the September 11 terrorist attacks in New York City. *The New England Journal of Medicine* 346(13), 982–987.
Glass, T.A., and Schoch-Spana, M. (2002). Bioterrorism and the people: How to vaccinate a city against panic. *Clinical Infectious Diseases* 34(2), 217–223.
Hewitt, C. (1993). *Consequences of Political Violence*. Darthmouth: Aldershot.
Hoffman, B. (2006). *Inside Terrorism*. Columbia University Press: New York.
Khalil, L. (2006). Public perceptions and Homeland Security. In J.J.F. Forest (Ed.), *Homeland Security: Protecting America's Targets, Vol. 2: Public Spaces and Social Institutions*. Praeger Security International: Westport, CT, 303–332.
Kydd, A., and Walter, B. (2006). The strategies of terrorism. *International Security* 31(1), 49–80.
Long, E. (1990). *The Anatomy of Terrorism*. Free Press: New York.

Mueller, J. (2006). *Overblown: How Politicians and the Terrorism Industry Inflate National Security Threats, and Why We Believe Them*. Free Press: New York.

Nacos, B. (2005). *Terrorism and Counterterrorism: Understanding Threats and Responses in the Post 9/11 World*. Penguin: New York.

Ohman, A. (2000). Fear and anxiety: Evolutionary, cognitive, and clinical perspectives. In M. Lewis and J.M. Haviland-Jones (Eds), *Handbook of Emotions*. Guilford Press: New York, 573–593.

Olsson, A., Mearing, K.I., and Olsson, E.A. (2007). Learning fears by observing others: The neural systems of social fear transmission. *Social Cognitive and Affective Neuroscience* 2(1), 3–11.

Schlenger, W.E., Caddell, J.M., Ebert, L., Jordan, B.K., Rourke, K.M., Wilson, D., Thalji, L., Dennis, J.M., Fairbank, J.A., and Kulka, R.A. (2002). Psychological reactions to terrorist attacks: Findings from the national study of Americans' reactions to September 11. *Journal of the American Medical Association* 288, 581–588.

Wardlaw, G. (1982). *Political Terrorism*. Cambridge University Press: Cambridge.

Chapter 3
Understanding Terrorism through Criminology? Merging Crime Control and Counter-terrorism in the UK

Pete Fussey
Professor of Sociology, University of Essex, UK

Introduction

9/11 is often credited as a watershed moment for counter-terrorism (CT) practice, particularly in the US and UK. Unprecedented attention became focused on the motivations and activities of would-be terrorists. Suspicion and the practices of identifying those deemed to harbour hostile and violent intentions proliferated. Yet, while 9/11 undoubtedly introduced a number of significant changes to CT practice, the transformational nature of the change has been disputed. Some argue that it is important to recognise that many enduring trends have persevered: that for many elements, the form and practice of CT underwent little significant transformation (Ball and Webster, 2003; Fussey, 2013). Rather, what has changed is the sheer scale and intensity of CT operations: 9/11 itself and subsequent terrorist 'spectacles' staged in Madrid and London during 2004 and 2005 stimulated a sudden abundance of resources and political will aimed at tackling this often poorly defined and variously imagined threat. The effects of such commitment permeated security planning arenas at a range of levels, from the local community (e.g. funding grassroots organisation to counter narratives of extremism) to internationally focused mega-events (e.g. the London Olympics in 2012). Speaking of the US situation, Bellavita succinctly explains in relation to his role of managing security for the 2002 Salt Lake City Olympics, less than four months after 9/11, 'Very little in the plan had changed … [but] it [became] very easy to get money and people – two resources hard to obtain before the attacks' (Bellavita, 2007, p. 2). Much academic attention has rightly focused on novel approaches and interventions, seeking to peer over the horizon and identify emerging trends in security practice and assess their efficacy and operational, social and ethical impacts. While such exegeses have much to offer, they leave unexamined all that remains inert within security practice. Yet some old orthodoxies, while retained, have been reframed, applied in new settings, and, in the decade that has followed 9/11, have begun to adapt their own logics and operations.

This chapter identifies a number of enduring practices and underpinning principles of CT which, in turn, raise questions over their applicability to new, mutable and dynamic forms of terrorism. In doing so, the chapter first identifies the convergence of crime control and CT and considers the implications of drawing diverse agencies and organisations together into wider ensembles of security practice. The discussion then moves towards an identification and analysis of how, despite this plurality and diversity of practice, some homogenising tendencies are visible. In particular specific forms of understanding transgressive behaviour based on utilitarian notions of rational choice theory are identified. Via an analysis of points of convergence and divergence between crime and terrorism the remainder of the chapter examines the applicability of such explanations of transgression across both activities.

Converging Realms

One of the key changes occurring within CT practice has been the increasingly crowded landscape of organisations, agencies and institutions undertaking security functions. As Bigo (2000, 2001), writing before 9/11, presciently observed: formerly distinct 'internal' and 'external' security realms, once serviced by different institutions, have undergone significant convergence. Thus, distinctions between domestic and international security have become increasingly meaningless. Bigo (2001, p. 113) characterises this conflation through the metaphor of a Möbius Ribbon:

> The boundaries of security are not fixed through a clear belief of what security is (and what it is not). They don't know where the inside ends and where the outside begins. They don't know where security is beginning and insecurity is finishing. As in a Möbius ribbon, the internal and the external are intimately connected.

While Bigo viewed the response to migration as a primary driver for this convergence, the attacks in Madrid and London during 2004 and 2005 and the intensified focus on endogenous risks and domestic populations serve to validate his thesis further. As the then director general of MI5 recently stated, increased emphasis on 'home-grown' terrorism and on the internationalisation of domestic grievances during this period intensified to the extent that the government was compelled towards an unprecedented doubling of the agency's budget (Manningham-Buller, 2011). Thus, what was once national security becomes an 'everyday securitisation from the enemy within' (Bigo, 2001, p. 112). This is significant because the orthodoxies and staples of crime prevention (such as situational crime preventions, rational choice theory and design against crime) become applied to CT contexts. At the same time, infrastructures and techniques originally designed and deployed for CT purposes become repurposed for more civil crime reduction applications. For example, with the use of Automatic Number Plate Recognition (ANPR) surveillance cameras, that were originally deployed in the City of London to address threats from the Provisional IRA (PIRA) (see Coaffee, 2009), the 'streetcorner criminal' and the 'foreign enemy' (Bigo, 2001, p. 93) are thus drawn together.

This post-9/11 augmentation of CT practice has led to some key processes of centripetal draw, of harmonisation and cohesion of certain features of the CT effort – particularly at the strategic and central government levels of action – while simultaneously decentralising and fragmenting others, a centrifugal push, particularly at the local and operational domains. Such processes generate questions over the extent to which crime control orthodoxies become applied and appropriate to the conceptualisation and control of terrorist action.

Centripetal Draws

There is little doubt that the period immediately following 9/11 drew a number of agencies together and pooled a diverse range of resources at the higher levels of political and strategic CT planning. In the UK, CT immediately received unprecedented levels of funding and political support and began to feature prominently in the risk registers and strategic planning of a range of government departments with little previous connection to the issue, including the various versions of the Departments for International Development (DFID), Culture, Media and Society (DCMS) and Communities and Local Government (DCLG). For departments with defined statutory responsibilities relating to security alongside other duties, such as the Cabinet Office, CT started to become a dominant concern.

In some respects, this focus on post-millennial (and, in many respects, millenarian) terrorism brought some unity of purpose and cohesion across different security agencies. The Metropolitan Police Special Branch[1] and MI5, previously in conflict over the policing of Northern Ireland, were brought closer together and cloistered within the Joint Terrorism Analysis Centre (JTAC) along with multiple other agencies including GCHQ, parts of MI6 and the Transport Security Directorate (TRANSEC)[2] element of the Department of Transport (Bamford, 2004). JTAC currently holds multiple functions relating to centralising and sharing intelligence including increasing capacity to process escalating quantities of intelligence reports, sharing information across institutional boundaries and, crucially, generating consensual guidance on intelligence analysis for the central government. While these developments focused on intelligence, processes

1 'Special Branch' has since been further rationalised and key elements have merged with other 'high policing' (Brodeur, 2007) specialist functions of the Metropolitan Police to form SO15 CT Command.

2 The Department of Transport funds a number of specialist transport organisations including the British Transport Police (all other police constabularies are controlled by the Home Office). TRANSEC specialises in CT and other major security concerns.

of coalition and cohesion also applied to other security domains. UK border security, previously overseen by Special Branch, HM Customs and Excise and the Immigration Service, was unified under the remit of the UK Borders Agency (UKBA), created in 2008. Immediately after 9/11, other domestic yet disparate terrorism-related policing roles became consolidated in the Police Intelligence Counter Terrorism Unit (PICTU), followed by later additional rationalisation and reconfiguration. Subsequent terrorist events also brought harmonisation between the domestic and international spheres, such as the creation of the European Arrest Warrant following the March 2004 attacks at Atocha station in Madrid. This mechanism was used to repatriate failed 21 July bomber Hussein Osman from Italy to the UK. Such processes of integration are set to continue and draw new areas of CT capabilities together, as evidenced in the conclusions of the 2011 inquest into the 7 July London bombings.

Despite such lateral integration at strategic and policy levels, it is important to note a further two related features. First, the efforts to integrate within central government were not necessarily replicated at the more devolved operational level. Second, significant decentralising and fragmenting processes have also occurred, notably the devolving of responsibility for CT onto a diversity of local agencies and actors at different levels, ranging from large municipal manifestations of the local state to individual actors.

Centrifugal Forces

Decentralisation was a keystone of the UK's overarching CT Strategy: 'CONTEST' (COuNter TErrorism STrategy). Formulated in response to 9/11, the CONTEST strategy represents a significant deviation from the US Homeland Security model of a centralised co-ordinating institution. Instead, a number of CT roles and responsibilities were devolved and decentralised to both the local state and non-state actors and agencies. In doing so, the CONTEST is divided into four main areas of activity:

- *Prevent* (the prevention of violent extremism through measures such as providing legitimate challenges to extremist discourse),
- *Pursue* (the pursuit of those suspected of involvement in terrorist activity through intelligence and police work),
- *Protect* (the hardening of the physical environment to mitigate the impact of a terrorist attack), and
- *Prepare* (instituting anticipatory measures such as the stockpiling of vaccines or the finessing of emergency planning procedures).

As the architect of the CONTEST strategy explains, there was a conscious attempt to induce greater interdependence across different actors and institutions in order to effect 'reduced state dominance of the security environment' (Omand, 2010, p. 14). Such sentiments are replicated in the Institute for Public Policy Research's major review of UK national security in the twenty-first century, where reduction of state dominance of security policy and provision is portrayed as generating a greater reliance on 'partnership working … with the private sector, with community groups and with local government and citizens as individuals' (Institute for Public Policy Research, 2009, p. 4). One concrete manifestation of these ambitions is the Home Office's Protection of Crowded Places initiative (e.g. Home Office, 2008) promoting the local delivery of CT via multi-agency working and integrating local practices with broader policy agendas in addition to advice on frameworks to assess risks of, and vulnerabilities to, terrorism.

While there are apparent issues of legitimacy at play – as Omand recognised in Rousseauian tones, 'security in cities is not just a matter of physical protection, it rests upon civic harmony ... on consent by the governed' (Omand, 2010, p. 73) – this devolutionary approach is principally driven by reasons of operational efficacy. In particular, a widely held assumption among many CT practitioners interviewed in research for this chapter is that local and, often, non-state actors and agencies are more effectively placed to perform a number of CT functions. These include identifying suspicious behaviour, challenging extremist narratives, embedding the built environment with target-hardening and blast-resistant features and limiting the impact of a terrorist attack through contingency and emergency continuity planning. More recently, this process of devolved responsibilisation has been posed in terms of the increasingly ubiquitous (and increasingly nebulous) discourse of resilience.

Merging Crime Control and CT

Yet for all the decentralisation and fragmentation of CT practice, some homogenising tendencies remain. The proliferation of actors and agencies have lead to both a babel of discourses around how terrorism is conceived and controlled yet, paradoxically, it has seen some deep furrows of existing practice – particularly criminologically-inflected variants – become recognised as established approaches. As de Goede and Randalls (2009) have argued, some security orthodoxies – such as risk-based anticipatory approaches – have come to occupy a 'post-political' space where consensus transcends organisational and political differences. Particularly notable here has been the increasing adoption of applied criminological approaches into the practice of CT.

In this sense many post-9/11 CT practices have been shaped in the slipstream of mainstream adoptions of activities in the more general realms of policing and social control. In this more general realm, since the 1980s criminologists have identified a shift towards approaches that privilege pre-emptive methods of crime control. This is predicated on practical expedience, a response to a (controversial) perception that rehabilitative strategies were ineffective (Martinson, 1974) and conviction that 'changing people is difficult and expensive' (Simon, 1988, p. 773). Thus, as Cohen (1985) and others identified, the 1980s saw an abandonment of perpetrator-centred causation-focused perspectives of transgression. In their place have grown a number of strategies that seek to manage the distribution of risks through techniques such as target hardening, surveillance, defensible space or actuarial categorisations of suspicion (inter alios Feeley and Simon, 1994). Inherent within these approaches is an emphasis on efficiency over less tangible, more costly and labour-intensive rehabilitative social control strategies. Thus 'social control' has become less and less 'social' in nature and focus (Fussey, 2013). Moreover, a coherent theme running through many of these practices is the assumption of highly rationalised choices, often reducible to simple utilitarian cost-benefit analyses, on the part of the potential offender. Thus, despite the proliferation of agencies and actors afforded CT responsibilities, numerous core criminological assumptions of perpetrator action are in evidence.

In addition, a key component of many contemporary security practices is an emphasis on pre-emptive anticipatory logics, something afforded greater importance given the scale and ambiguous nature of the immediate post-9/11 threat (Omand, 2011). Such shifts again mirror more enduring trends in criminology, such as the growth in actuarialism and anticipation across policing (Ericson and Haggerty, 1997), surveillance (Lyon, 2003) and criminal justice more generally (Feeley and Simon, 1994) since the 1980s. While providing a broad shared conceptual approach to the identification of risk, such practices place a heavy emphasis on knowledge, specifically putative knowledge of potential transgression. Momentarily leaving aside the unequivocal failure of attempts to accurately predict offending behaviour that litter the histories of criminology and penology, these approaches therefore place a premium on the distillation and crystallisation of 'suspicion'. As the following examples demonstrate, devolved CT practices heavily accent the establishment of localised normative settings to identify risk. This occurs via a number of processes, two of which are illustrated here.

Othering Suspects and Owning Suspicion

First, recent police CT training schemes, and also public information campaigns, have placed considerable emphasis on encouraging security professionals and the general public to take ownership of, and act upon, feelings of suspicion. Most visibly, untargeted high-profile publicity campaigns such as the Metropolitan Police's 'If you suspect it, report it' and Transport for London's 'It's All Up To Us' and 'If anything suspicious catches your eye' campaigns. More targeted and formalised measures include government-funded training of security professionals and other public-facing staff working in crowded places (such as Argus training for those working in shopping malls, stadia, hospitality and night-time economy roles (Source: author attendance at training schemes November 2009, February 2010 and January 2012) in a range of security aspects including identifying and managing suspicious behaviour and materials (National Counter Terrorism Security Office, 2011). Such advice and training place heavy emphasis on local practitioners' expert readings of what is deemed normal for their particular sphere of operations.

To some extent these initiatives represent a reintroduction rather than innovation of practice. Publicity campaigns warning of the dangers of unattended packages were a common feature of Britain's urban landscape throughout the Provisional IRA's campaign on the British mainland between 1973 and 1996. Yet this emphasis on 'owning' suspicion has been further driven by specific terrorist events. Particularly influential events here include Brixton market traders' movement of David Copeland's inconveniently placed pre-detonated bomb: a significant driver for the encouragement of the public to act upon their suspicions (Source: interview with senior police CT officers December 2010 and January 2011).

One particularly significant precursor event that still influences contemporary CT thinking is the Red Army Faction (RAF) murder of Alfred Herrhausen, then Chairman of Deutsche Bank in 1989 (Source: interviews with separate senior London-based private-sector security CT agents, May 2010, August 2011 and February 2012). The device that killed him was concealed in the pannier of a bicycle, located on his normal commuting route, that had been placed there by RAF members posing as construction workers six weeks before the attack and left conspicuously less than 100 metres from permanent cycle racks, in order to both normalise its presence and to ascertain the likelihood that the device would not be interfered with. Key lessons include issues of the proximity of perpetrators and victims prior to attacks, the vulnerabilities of routine, the exposure of 'choke points' and, crucially the identification of 'matter out of place' in the known or normative environment. Baselining what is 'normal' for specific environments, and identifying matter that deviates, has since acquired axiomatic status within CT practice.

Criminological Orthodoxies and CT Practice

One of the most influential elements of applied criminology among CT practice has been a number of specific interpretations of rational choice theory (RCT). At their heart, these 'rational choice' approaches are predicated on a tripartite model which attributes offending behaviour to a motivated offender, an opportunity to commit the crime and the absence of a 'capable guardian' such as beat officers or closed-circuit television (CCTV) surveillance (inter alios Clarke and Cornish, 1985). Over time, aspects of the theory have been developed to explain (actual and potential) offenders' decision-making processes through reference to notions of the rewards and costs of the offence (inter alios Felson and Clarke, 1998). In doing so, the latter two components of RCT (opportunity reduction and capable guardianship) have been accented and subsequently translated into policy while the first element, that of the 'motivated offender', has received notably less emphasis (beyond the belief that manipulating the other features will reduce an individual's motivation to offend). Taking the example of CCTV surveillance, for instance, such theorising has established a maxim amongst many practitioners that CCTV increases the level of capable guardianship in the spaces it surveys and, as such, affects the objective rational decisions of potential offenders by increasing the perceived risk of the offence – the classic utilitarian model of deterrence.

Perhaps most explicitly, these ideas have permeated the hardening and reconfiguration of urban spaces to reduce their vulnerability from terrorist attacks – an approach explicitly outlined in the 'Protect' strand of the UK government's CONTEST strategy. The intention of such measures is to manipulate the built environment to reduce the attractiveness and physical opportunities to access targets while increasing the likelihood that perpetrators would be apprehended and, thus, the chances that an individual will deem the offence too risky. Manifest strategies include securing properties with strong locks and bolts, heavy investment in technological surveillance to monitor public spaces and deter would-be offenders, controlling access routes, installing street furniture in strategic locations to prevent vehicle-borne explosives, reducing the number of entrances to buildings and restricting traffic flows.

Perhaps the best exemplar of how such approaches have been deployed in the CT context is in the development of the so-called 'Ring of Steel' encircling important locations within London's financial district. Following the PIRA bombing of London's Baltic Exchange in St Mary Axe, 1992, and Bishopsgate, 1993, the attendant spiralling insurance premiums (Walker and McGuiness, 2002), diminished financial confidence and heightened Health and Safety obligations led to commercial and political demands for a more protected urban space from which to conduct business. As one authoritative study details, these calls were answered by a progressive fortification of the immediate environment via enhanced electronic surveillance measures (including the pioneering of Automatic Number Plate Recognition *ANPR* surveillance technology); strict controls on the access, routing and parking of traffic; and alteration to the physical design of buildings and streets (Coaffee, 2009). This marked a process of increasing physical and technological demarcation and separation of the City of London from the wider urban environment (Coaffee, 2009). What is of further interest here is the way in which such approaches have since become a template for protecting redeveloped financial districts in other countries (Marcuse, 2004) and, more broadly, across urban spaces (Coaffee, 2009; Centre for the Protection of National Infrastructure, 2008; Fussey, 2007).

Reconsidering Criminological Assumptions

This chapter does not suggest that urban target-hardening strategies have no place in countering terrorism, but raises questions over the totality and wider utility of the ideas that underpin and shape such approaches. A central aim here is to generate further debate and reflection on the assumptions of terrorist decision-making that underpin such strategies. In particular, it challenges the notion that perpetrators engage in largely objective and value-free strategic and operational choices. While an emphasis on the conjunction of motivation, opportunity and capable guardianship endow obvious practical relevance to the theory, it also ignores the expressive and emotive aspects of violence (inter alios Katz, 1988) and the unpredictable, less tangible, non-uniform and 'messy' assertions of human agency. As Drake (1998, p. 163) noted, 'when examining the actions of terrorists, one must bear in mind that one is dealing with people, with all their imperfections and unpredictability, rather than some hypothetical, hyper-rational beings'. As Adey (2008) elegantly put it, people are not rational billiard ball-like objects that react in solely logical ways. Instead, people have diverse and, for Adey, emotional and affective relationships with their environments. Thus, at the basic level, we move from binaries of cause and effect, to trialectics of an affecting process, the affected person and their capacity to be affected (Adey, 2008). Moreover, and of more direct relevance to transgressive behaviour, RCT as a whole is limited in its ability to explain why individuals perform more 'expressive crimes' (Hayward, 2007) of which acts of terrorism are arguably more closely related.

Further challenges to this approach come from criticism that, in addition to questioning their wider applicability, rational choice measures encounter difficulties when explaining the types of transgressive behaviour they were originally designed to address. In particular, there is paucity of empirical support from the perspective of perpetrators. Among those studies that do exist are challenges to the idea that increasing the 'costs' of crime necessarily discourages offending behaviour. For example, some research suggests that, at best, increasing the certainty of apprehension and severity of punishment has only a small to moderate impact on criminal behaviour (see Akers, 1999). Such findings accord with volumes of criminological and penological research contesting the impact of utilitarian notions of 'deterrence' upon crime reduction (see inter alios Cavadino and Dignan, 2007).

Crime and Terrorism as Divergent Activity

Debates over the strengths and weaknesses of utilitarian rational choice theories in explaining transgressive behaviour have been well-rehearsed and often generate significant heat, if little light. Hayward (2007) and Farrell (2010) offer a robust and entertaining exchange on this topic; and Fussey (2011) provides an overview in relation to terrorism. However, the convergence of the realms of crime control and CT has generated significant questions over how applicable such criminological ideas are to the realm of CT. In response, some proponents of rational choice theories and situational prevention, for example, assert that any differences between crime and terrorism are immaterial. As Clarke and Newman (2006, pp. 5–6) argue, '[t]here is often a high degree of overlap between the motives of "ordinary" criminals and terrorists ... the supposed differences rarely stand close scrutiny'. By implication, this assertion is used to justify the applicability of those authors' other work on more conventional crimes, such as burglary, to the domain of CT. This chapter, however, challenges such claims and argues that, with scrutiny and attention to the activities of perpetrators, some clear differences between criminal and terrorist-related activities are visible.

One useful starting point here is the comparison of what ostensibly look like similar offences committed for different criminal and terrorist-related ends. Drawing on Lyons and Harbinson's study of murder in Northern Ireland during the 1980s, for example, Drake (1998) identifies how 85 per cent of non-political murders occurred in the home (a common location for murders more generally). When related to politically motivated murders, only 6 per cent took place in the home. Reasons for this disparity included the different levels of expectation of risk among victims and, as a corollary, the extent of precautionary measures taken (accordingly, senior PIRA members installed steel internal doors and shatterproof glass in their homes). The point here, then, is that although the outcome of the offence may be similar, the divergent means of their perpetration suggests that different anticipatory and amelioratory strategies are required.

Other factors also differentiate the two activities. Perhaps most important is the role of ideology in shaping targets. Here, debate exists between those stressing the immateriality of ideology – for example, 'understanding the ideologies of terrorist groups will therefore give little insight into their selection of targets and tactics' (Clarke and Newman, 2006, p. 70) – and volumes of scholarship that stress the counter argument. As Hoffman notes in relation to the latter:

> the tactics and targets of various terrorist movements, as well as the weapons they favor are therefore ineluctably shaped by a group's ideological, its internal organisational dynamics and the personalities of its key members, as well as a variety of internal and external stimuli. (Hoffman, 2006, p. 229)

Violent right-wing extremism and religiously motivated terrorism generally exercise fewer constraints in relation to target preferences when, for example, compared to left-wing groups (Hoffman, 2006). Thus crucial elements of terrorist activities may be considered highly contingent on ideological processes. These two types of activity may be further distinguished by the intended audience of terrorism and issues of temporality. These are briefly discussed in turn.

Terrorism is often perpetrated with the aim of appealing to far wider constituencies than more conventional forms of crime. Intended audiences of the former regularly supersede actual and potential victims to include diverse domestic constituencies, patronage organisations or rival groups (see inter alios Bloom, 2005). Additionally, issues of extensive media coverage have consistently influenced the selection of terrorist targets (inter alios Schbley, 2004; Libicki et al., 2007) and tactics (inter alios Makarenko, 2007). Indeed, the now famous and oft-repeated images of Michael Adebolajo, addressing witnesses' amateur smartphone recordings, and the wider virtual social networks they access, following his murder of Lee Rigby in Woolwich during May 2013 underlines this point. Together, these external audiences, while important to terrorist action, are likely to be absent in more materialistically oriented criminality. Moreover, such appeals to wider audiences potentially undermine the role of 'capable guardianship' – essentially a form of deterrence through surveillance – which comprises one of the three core elements of situational crime prevention.

Over the last 70 years considerable debate has existed within criminology regarding the degree of planning that accompanies criminal offences. For example, advocates of the 'crime-as-planned' thesis are challenged by theories that foreground the emotive, visceral and impulsive drivers of transgressive behaviour (inter alios Katz, 1988; Presdee, 2000); a debate characterised by Hayward (2007) as rational choice theory versus the 'culture of now'. Amid such debate, however, is some consensus that degrees of pre-offence planning vary over different types of crime. Some research argues that sexual offences, for example, are often perpetrated after 'compressed' (short) decision-making processes (Ward and Hudson, 2000). By contrast, research data from interviews with over 2,000 convicted bank robbers showed that planning is a more central consideration in that activity (Kube, 1988). However, much of this planning is oriented around escape routes and perpetrators often erroneously overstated the potential yields of the offence.

For terrorism, greater consensus exists over the role of more protracted and elaborate planning processes that precede the event. Among the extant research on terrorist activity, such antecedent activities (along with other features of terrorism) have been perhaps best captured by Horgan's (2005) notion of the 'process' of terrorism, where a realised attack is but one manifestation of a deeper and more enduring set of circumstances. Given such extended processes, it can be argued that situational measures that harden the resilience of public spaces are heavily weighted to only addressing the single, final, aspect of terrorist action. Attention to these extended temporal frames of action also serves to illustrate the myriad of processes that underpin and precede the execution of a terrorist attack. Moreover, not only do more protracted campaigns raise the likelihood of mutability and adaptation within such groups, such as the replacement of old members with others harbouring different strategic outlooks – a process that has been described as also having occurred over decades during the broader shift from 'political Islam' to more nihilistic contemporary manifestations of violent Jihadi extremism (Burke, 2007) – they also indicate the potential for a wide range of influences to shape such activity.

Conclusion

The events of 9/11 drove a number of changes through the landscape of UK CT practice. Particularly prominent among these was the convergence of internal and external security, catalysing a closer merger of crime control and CT. Focusing on the UK, this chapter has argued that this union paradoxically drew a diversity of agencies into cognate areas of security practice yet, at the same time, elevated a number of understandings of criminal behaviour into to realm of CT. These included anticipatory risk-based approaches to security but, most notably, rational actor theories and notions of situational crime prevention that seek to understand transgressive behaviour through utilitarian calculi relating to opportunity, motivation and the absence of surveillance. While points of convergence between crime and terrorism clearly exist, there are a number of key differences which, in turn, generates questions over the wider applicability of such understandings of transgressive behaviour. Most importantly, the complexity and mutability of terrorist actions, along with its intrinsic infusion with values and, generally, underpinned a sense of (real, imagined or vicarious) grievance (inter alios Horgan, 2005), a number of difficulties prevent viewing such activities in utilitarian value-free terms.

References

Adey, P. (2008). Airports, mobility, and the calculative architecture of affective control. *Geoforum*, 1, 438–451.

Akers, R. (1999). Rational choice, deterrence, and social learning theory in criminology: The path not taken. *Journal of Criminal Law and Criminology* 81(3), 653–676.

Ball, K. and F. Webster. (2003). The intensification of surveillance. In K. Ball and F. Webster (Eds), *The Intensification of Surveillance: Crime, Terrorism and Warfare in the Information Age*. London: Pluto, 1–15.

Bamford, B. (2004). The United Kingdom's war against terrorism. *Terrorism and Political Violence* 16(4): 737–756.

Bellavita, C. (2007). Changing homeland security: a strategic logic of special event security. *Homeland Security Affairs* 3, 1–23.

Bigo, D. (2000). When two become one: Internal and external securitisations in Europe. In M. Kelstrup and M. Williams (Eds), *International Relations Theory and The Politics of European Integration. Power, Security and Community*. London: Routledge, 320–360.

Bigo, D. (2001). The Möbius Ribbon of internal and external security(ies). In M. Albert, D. Jacobson and Y. Lapid (Eds), *Identities, Borders, Orders: Rethinking International Relations Theory*. Minneapolis, MN: Minnesota University Press, 91–116.

Bloom, M. (2005). *Dying to Kill: The Allure of Suicide Terror*. New York: Columbia University Press.

Brodeur, J. (2007). High policing and low policing in post 9/11 times. *Policing: A Journal of Policy and Practice* 1(1), 25–37.

Burke, J. (2007). *Al Qaeda: The True Story of Radical Islam*. London: Penguin.

Cavadino, M. and Dignan, J. (2007). *The Penal System: An Introduction*. London: Sage.

Centre for the Protection of National Infrastructure (2008). *Protecting Your Assets*. Available from http://www.cpni.gov.uk/protectingassets.aspx, accessed 27 March 2008.

Clarke, R., and Cornish, D. (1985). Modeling offenders' decisions: A framework for research and policy. In M. Tonry and N. Morris (Eds), *Crime and Justice: An Annual Review of Research Volume 6*. Chicago, IL: University of Chicago Press, 147–185.

Clarke, R. and Newman, G. (2006). *Outsmarting the Terrorists*. London: Praeger.

Coaffee, J. (2009). *Terrorism, Risk and the Global City: Towards Urban Resilience*. Farnham: Ashgate.

Cohen, S. (1985). *Visions of Social Control*. Cambridge: Polity.

Drake, C. (1998). *Terrorists' Target Selection*. London: Macmillan.

Ericson, R.V. and Haggerty, K.D. (1997). *Policing the Risk Society*. Oxford: Clarendon Press.

Farrell, G. (2010). Situational crime prevention and its discontents: Rational choice and harm reduction versus 'cultural criminology'. *Social Policy & Administration* 44(1), 40–66.

Feeley, M. and Simon, J. (1994). Actuarial justice: The emerging new criminal law. In D. Nelken (Ed.), *Futures of Criminology*. London: Sage, 173–201.

Felson, M., and Clarke, R. (1998). *Opportunity Makes the Thief: Practical Theory for Crime Prevention. Police Research Series Paper 98*. London: Home Office.

Fussey, P. (2007). Observing potentiality in the global city: Surveillance and counterterrorism in London. *International Criminal Justice Review* 17(3), 171–192.

Fussey, P. (2011). An economy of choice? Terrorist decision-making and criminological rational choice theories reconsidered. *Security Journal* 24(1), 85–99.

Fussey, P. (2013). Contested topologies of UK counter-terrorist surveillance: The rise and fall of Project Champion. *Critical Studies on Terrorism* 6(3), 351–370.

de Goede, M., and Randalls, S. (2009). Precaution, preemption: Arts and technologies of the actionable future. *Environment and Planning D: Society and Space* 27(5), 859–878.

Hayward, K. (2007). Situational crime prevention and its discontents: Rational choice theory versus the 'culture of now'. *Social Policy & Administration* 41(3), 232–250.

Hoffman, B. (2006). *Inside Terrorism*. New York: Columbia University Press.

Home Office. (2008). *Working Together to Protect Crowded Places*. London: HMSO.

Horgan, J. (2005). *The Psychology of Terrorism*. London: Routledge.

Institute for Public Policy Research. (2009). *Shared Responsibilities. A National Security Strategy for the UK*. London: Institute for Public Policy Research.

Katz, J. (1988). *The Seductions of Crime: Moral and Sensual Attractions in Doing Evil*. New York: Basic Books.

Kube, E. (1988). Preventing bank robbery: Lessons from interviewing robbers. *Journal of Security Administration*, 11(2), 78–83.

Libicki, M., Chalk, P. and Sisson, M. (2007). *Exploring Terrorist Targeting Preferences*.Santa Monica, CA: RAND Corporation.

Lyon, D. (2003). *Surveillance After September 11th*. London: Blackwell.

Makarenko, T. (2007). International terrorism and the UK: Assessing the threat. In P. Wilkinson (Ed.), *Homeland Security in the UK: Future Preparedness for Terrorist Attack Since 9/11*. London: Routledge, 37–56.

Manningham-Buller, E. (2011). *Terror*. BBC Reith Lectures 2011. Available from http://www.bbc.co.uk/programmes/b0145x77, accessed 11 August 2014.

Marcuse, P. (2004). The 'War on Terrorism' and life in cities after September 11, 2001. In S. Graham (Ed.), *Cities, War and Terrorism*. Oxford: Blackwell, 263–275.

Martinson, R. (1974). What works? Questions and answers about prison reform. *Public Interest* 55(Spring), 22–54.

National Counter Terrorism Security Office (NaCTSO) (2011). *Counter Terrorism. Protective Security Advice for Major Events*. Available from https://vsat.nactso.gov.uk/SiteCollectionDocuments/AreasOfRisk/Major%20Events%202011.pdf, accessed 12 October 2012.

Omand, D. (2010). *Securing the State*. London: Hurst & Co.

Omand, D. (2011). What are the limits of Western CT policy? Paper presented at the *9/11: Ten Years On* Symposium, British Academy, London, 2 September, 2011.

Presdee, M. (2000). *Cultural Criminology and the Carnival of Crime*. London: Routledge.

Schbley, A. (2004). Religious terrorism, the media, and international Islamization terrorism: Justifying the unjustifiable. *Studies in Conflict & Terrorism* 27(3), 207–233.

Simon, J. (1988). The ideological effects of actuarial practices. *Law and Society Review* 22(4), 771–800.

Walker, C. and McGuiness, M. (2002). Risk, political violence and policing the City of London. In A. Crawford (Ed.), *Crime, Insecurity, Safety and the New Governance*. Cullompton: Willan, 234–259.

Ward, T. and Hudson, S. (2000). Sexual offenders' implicit planning: A conceptual model. *Sexual Abuse: A Journal of Research and Treatment* 12(3), 189–202.

Chapter 4

Analysing the Terrorist Brain: Neurobiological Advances, Ethical Concerns and Social Implications

Valentina Bartolucci
Centro Interdipartimentale Scienze per la Pace, Università di Pisa, Italy

Introduction

In the last few years, the scientific community and, more broadly, society at large have become increasingly interested in exploring the deeper causes of terrorism. More than a decade after the September 11, 2001 attacks in the U.S., terrorism is still perceived as the one of the most existential threats humanity is facing, despite the fact that the actual danger is negligible (Bartolucci, 2010). Terrorism also continues to be the subject of numerous academic publications even though it is not an easy subject of enquiry and the serious researcher faces several issues when addressing it, such as the confusion around its conceptualisation and the scarcity of primary data (Silke, 2001). Furthermore, mainstream literature on terrorism studies has always seen terrorism in moral terms, a tendency that has been accentuated since 2001 (Jackson, 2005). A related issue is that terrorism is an emotive subject and a morally detached analysis is difficult, if not impossible, to achieve.

This chapter investigates some of the most pressing ethical issues that may arise from the application of neuroscience advances devoted to detecting deception or hidden intentions by humans in security settings, and especially within counter-terrorism. The first section of the chapter presents a general overview of the field of terrorism psychology and introduces recent claims about the possibility for neuroscience to 'read' a terrorist's brain. The second section describes five neuroscience-based experiments in which investigators claim to be able, in various ways, to 'detect' lying or the 'hidden intentions' of humans. The final part of the chapter deals with the ethical dilemmas and social implications of these and similar prospective advances in neuroscience.

The Emerging Importance of Ethics in Terrorism Research

The persistent vilification of terrorism has serious implications on a practical level. For instance, 'talking with "terrorists"... becomes taboo, unless in the context of interrogation' (Gunning, 2007, p. 372). Moreover, if terrorists are depicted as 'irrational' and 'mad', the public is more likely to advocate military action against them rather than diplomacy (Pronin et al., 2006). Similarly, ethical concerns regarding alleged terrorists are more likely to be marginalised, given the assumed exceptionality of the phenomenon and the urgency to defeat it even to the detriment of basic human rights (see Bartolucci, 2012, 2013; Fekete, 2004; Jackson, 2005; and Chapters 4, 5 and 7 of this volume). This was illustrated following the September 11, 2001 attacks, when noted legal personalities advocated the official use of torture in dealing with Muslims for their potential involvement in terrorism (Ahmed, 2003). In this context, the constant warning by governments and media of incoming terrorist attacks and the increased levels of violence encountered in American society have led scholars and policy-makers to seek ways of predicting violent behaviour before it happens. As such, research aimed at understanding the mechanisms of brain functions for the purposes of lie detection and predicting future psychopathology has gained renewed interest (Shamoo, 2010). According to Wolpe et al. (2005, p. 39):

> for the first time, using modern neuroscience techniques, a third party can in principle, bypass the peripheral nervous system – the usual way in which we communicate information – and gain direct access to the seat of a person's thoughts, feelings, intention, or knowledge.

In a globalised world, constantly depicted by governments and media as insecure and increasingly violent, it comes as no surprise that research devoted to investigating the terrorist's mind, in order to eradicate the threat of terrorism, is attracting widespread interest. Interest in deception detection has increased especially in situations where suspected terrorists might be captured and questioned before committing specific terrorist acts (Rosenfeld, 2011). In academia, 'reading the terrorist mind' has also attracted a lot of interest within the mainstream literature on the psychology of terrorism (Bongar et al., 2007; Hudson, 1999). Material on the subject are abundant on the Internet, and several research centres specifically focused on terrorism psychology have been established, for example the International Center for the Study of Terrorism (ICST), and the Pacific Graduate School of Psychology and the National Center on the Psychology of Terrorism (NCPT) – attracting but also reacting to a growing interest from the scientific, military and defence sectors (see Horgan, 2005). However, while opening up very interesting prospects for research and diagnosis, these advances also raise a number of ethical concerns as well as social implications. This chapter examines the wider implications of this kind of research and where it may lead ethically. Most of these findings are still in the exploratory stage, but their potential impact is already significant (Illes et al., 2003; Moreno, 2006; Wolpe, 2004). It is thus timely and necessary to reflect on what social implications and ethical dilemmas such neuroscience-based technologies may have. Until now, unfortunately, society has been slow to develop a policy dialogue to anticipate and deal with potentially important ethical and social implications.

This chapter draws on a wider research project devoted to a much-neglected analysis on the 'dual-use dilemma' in neuroscience carried out at the Bradford Disarmament Research Centre (BDRC) at the University of Bradford, UK (Dando, 2010; Dando and Bartolucci, 2013). Such a work is embedded in a wider debate that has been formalised in a new discipline – neuroethics – that frames its efforts 'in terms of four "pillars": brain science and the self, brain science and social policy, ethics and the practice of brain science, and brain science and public discourse' (Lombera and Illes, 2009, p. 60).

Analysing the Terrorist's Brain

The idea that terrorists have something wrong within their brains was taken very seriously thirty years ago. Russell and Miller (1977) was one of the earliest attempts to identify a 'terrorist personality' trait. In contrast, more recent research has found that psychopathology and personality disorders are no more likely among terrorists than non-terrorists (for a full review see Crenshaw, 2007). However, the assumption that terrorists have inner psychological drives that predispose a specific personality 'type' towards such violence is still present in the literature of terrorism psychology. Brannan et al. (2001, p. 3) underline that:

> Models based on psychological concerns hold that 'terrorist' violence is not contingent on rational agency but is the result of compulsion or psychopathology. Over the years scholars of this persuasion have suggested that 'terrorists' do what they do because of (variously and among other things) self-destructive urges, fantasies of cleanliness, disturbed emotions combined with problems with authority and the Self, and inconsistent mothering.

For decades, psychologists and scholars of various disciplines have studied terrorists' individual characteristics, looking for clues that could explain their willingness to engage in bombings, terrorist attacks and the like. A related question would be how the role of security agencies shapes the discourse and framing of research. In particular it would also be of interest to see how much of this is supported by defence/law enforcement agencies as well as how much influence does the reporting of violent acts affect the analytic literature. While the majority of researchers now agree that most terrorists are not strictly clinically 'pathological', a common perception remains that perpetrators of terrorism are abnormal in more subtle ways (Silke, 1998). This assumption can also be found in the political science literature. Taylor (1988) identified two basic psychological approaches to understanding terrorists that have been commonly adopted: the terrorist is viewed as either mentally ill or a fanatic. This perception also reflects public opinion in which terrorists must be insane, psychopathic or the personifications of evil (Pearce, 1977; Taylor, 1988).

Following the events of September 11, 2001, researchers concentrated once again on finding common characteristics and inner connotations that a terrorist might possess. The initially seductive 'terrorist profiling' based on race or ethnicity showed itself to be much more complex than initially thought, given the heterogeneity of both the concept of 'terrorism' and those who conduct terrorist activities. Hence, attention increasingly turned towards advances in neuroscience believed to be capable of effectively countering terrorism through the 'reading' and understanding (as well as controlling) of the terrorist's brain. In the last two decades, there have been numerous attempts to translate neuroscience findings into technological tools to be used in society for a variety of purposes (National Research Council, 2003; Saykin, 2007; Shamoo, 2010). Wolpe et al. (2005) were among the first to show that recent advances in magnetic resonance imaging, electroencephalography and other analytical technologies can reliably measure changes in brain activity associated with feelings, thoughts and behaviours. This, in principle, could allow researchers to link brain activity patterns directly to the cognitive or affective processes or states they produce. Recent research has claimed to be able to fight 'terrorism' by 'reading' the minds of (potential) terrorists. For instance, researchers at Northwestern University (Chicago, US) argue that they are realizing the possibility of 'reading' a terrorist mind thanks to a test that could uncover nefarious plans by measuring brain waves (*The Times*, 2010). These new techniques are also interesting for the intelligence and counter-intelligence communities and for forensic sciences. However, the public should be suspicious because the motives of the user of these technologies can be equally nefarious (Shamoo, 2010). Before analysing the possible dual-use and ethical implications of such advances, the next section presents an overview of recent neuroscience experiments linked with the brain imaging of terrorists. These experiments have been chosen for their scholarly impact and the interest they attracted in the media circles, especially in the United States.

Neurobiological Research into Deception

The first experiment here discussed is known as the 'Meixner and Rosenfeld test' (2011) and was conducted at Northwestern University. This test attracted widespread interest beyond academia for its supposed ability to uncover terrorist plots before they happen through the analysis of the brain waves of 'potential terrorists' (Rosenfeld, 2011). The team used electro-electroencephalogram (EEG) techniques to measure the brainwaves of 24 participants. If a discrete stimulus occurred during the recording, the EEG would break into a series of larger peaks and troughs (waves) signalling the arrival in the cortex of neural activity. Typically, such waves last up to two seconds after an initial stimulus. The 'P300' is a special Event-related Potential (ERP) component (i.e. a type of cortical activity) that produces 'results whenever a meaningful piece of information is rarely presented among a random series of more frequently presented, non meaningful stimuli often of the same category as the meaningful stimulus' (Rosenfeld, 2011, p. 64). During the test, while 'the rare, recognized, meaningful items elicit P300, the other items do not' (Ganis and Rosenfeld, 2011, p. 105).

In the experiment, the participants were divided into two groups. Participants in the 'guilty' condition (n = 12) were given a briefing document prepared by the researchers saying that each subject had to play the role of a terrorist and plan a mock terrorist attack in a major US city. The briefing detailed several options the participant could choose regarding how to carry out the attack and described four types of bombs that could be used, four locations that could be targeted, and four dates in which the attack could have taken place. With the relative merits of each potential course of action elucidated, the participants were asked to identify their preferred type of bomb, location and date for an attack that they then detailed in a letter to a fictitious head of the terrorist organisation whilst also encoding the information in their memory. Participants in a second 'innocent' condition (n = 12) were asked a similar task but planned a vacation instead of a terrorist attack. At this point, the researchers conducted EEG scans on the participants to measure their P300 brainwaves. The ability of P300 to signal the involuntary recognition of meaningful information makes it possible to identify recognised 'guilty knowledge' known only to a person familiar with crime details, such as guilty perpetrators, accomplices, witnesses and police investigators (Rosenfeld, 2011). The amplitude of the P300 is very large when the person under examination sees an object that is meaningful to them.

In one test, the participants were shown the names of various cities and the researchers noticed that the amplitude of the P300 was very large when the name of the city where attacks were planned was presented. In these cases with the researchers knowing details about the mock terrorist attack, they claimed to be able to correlate the P300 to 'guilty knowledge' with accuracy close to 100 per cent. In more detail, they claimed to be able to identify 10 out of 12 'terrorists' and to match 21 out of 30 terrorism-related details, such as specific places, times, types of arms and explosives. They said they were also able to identify concealed information even in those cases in which they had no advance knowledge. The team then concluded that the protocol 'appears to hold promise for the anti-terrorist challenge' (Rosenfeld, 2011, p. 85). Even if in the real world the test has currently limited application, the researchers still claimed that the protocol can be used to predict concealed knowledge and identify future activity (Boyle, 2010).

Another experiment, conducted at the University of Pennsylvania, employed Functional Magnetic Resonance Imaging (fMRI) which has been proposed as one of the most promising advances in neuroscience-based technology in detecting behaviour, especially for detecting lying/truth telling (Poldrack, 2008). It is widely used to obtain measurements of cerebral blood flow in individuals engaged in deception and there are serious attempts to commercialise the product for non-medical deception purposes. For example, two commercial enterprises, Cephos Corporation and No Lie MRI Inc., were launched with the goal of bringing fMRI to the public for use in legal proceedings, employment screening and national security investigations. The No Lie MRI Inc. website claims to be able to offer the first and only direct measure of truth verification and lie detection in human history. Worryingly, results of fMRI scans have been already offered as legal evidence in a court in California as well as in Iowa (Shamoo, 2010; see also Illes and Racine, 2005; Kleiner, 2009). In the experiment discussed here, a team of researchers led by Langleben, claimed that they were able, under strict laboratory conditions, to differentiate brain activity associated with lies between 88 per cent for untrained participants and up to 99 per cent for trained participants (Langleben et al., 2002). The researchers claimed they were able, for the first time ever, to encode covert intentions in highly specific fashion and concluded that:

> These findings have important implications not only for the neural models of executive control, but also for technical and clinical applications, such as the further development of brain-computer interfaces, that might now be able to decode intentions that go beyond simple movements and extend to high-level cognitive processes. (Haynes et al., 2007, p. 326)

A further experiment was carried out by the research team led by Dr Iacoboni at Ahmanson Lovelace Brain Mapping Center at the University of California, Los Angeles. This experiment is considered to be the first attempt to use neuroscience-based technologies in the field of politics. In the study, fMRI was employed to make predictions about how swing voters would react to the candidates in the Republican and Democratic Primaries for the US Presidency. Twenty subjects (ten women, ten men) were asked to answer a list of questions about their political preferences. The researchers then measured the response of their brain activity to various photographs and videos of Hillary Clinton, Rudy Giuliani and other candidates for nearly an hour. Afterwards, the participants completed a series of questions in which they were asked to rate the candidates on a 0–10 scale ranging from 'very unfavourable' to 'very favourable'. The questionnaire responses were then compared with the brain activity data. In an opinion-editorial article published in *The New York Times* (2007), under the title 'This is your brain on politics', the team claimed to be able to determine how each participant had reacted and to deduce the acceptability of the Primary candidates to undecided voters in general –for the first time using neuroscience advances to analyse politics.

In another study, in 2005 Dr Kozel and his team at the University of Texas Southwestern Medical Center conducted the 'first study to use fMRI to detect deception at the individual level' and then further expanded their work in the following years (Kozel et al., 2005, 2009). In their first study, the researchers used fMRI to show that specific regions of the brain were reproducibly activated when subjects participating in a mock crime stealing either a ring or a watch deceived. A

Model-Building Group (MBG n. 30) was used to develop the methods of analysis that were subsequently applied to an independent Model-Testing Group (MTG n. 31). The authors claimed they were able to correctly differentiate truthful from deceptive responses and correctly identifying the object stolen for 93 per cent of the subjects in the MBG and 90 per cent of the ones in the MTG. The researchers also suggested that their method could be used in real-life settings and the protocol was then licensed by the Cephos Corporation (Simpson, 2008).

Another approach has employed one of the most publicised neuroscience techniques for detecting deception: 'Farwell's brain fingerprinting'. Farwell's technique has been promoted in various contexts including forensic investigations, counter-terrorism efforts and security testing (Wolpe et al., 2005). The principle behind brain fingerprinting is that the brain processes known and relevant information differently from unknown or irrelevant information (Farwell and Donchin, 1991). In this way it is similar to the first example of the Meixner and Rosenfeld test. The differences in processing known or unknown information (e.g. details of a crime or active deception) are revealed in specific EEG patterns (Farwell and Smith, 2001). Brain fingerprinting was originally based on the P300 brainwave response but Farwell discovered that the P300-Memory and Encoding Related Multifaceted Electroencephalographic Response (P300- MERMER) provided more accurate results and statistical confidence than the P300 alone (Farwell et al., 2012). Such results, typically report less than 1 per cent error rate in laboratory research (Farwell and Richardson, 2006) and real-life field applications (Farwell et al., 2012). Other research following similar protocols has reported similarly low error rate results (Allen and Iacono, 1997). However, the brain fingerprinting technique has proven controversial because researchers using different protocols have not yielded low error rates (Rosenfeld, Hu and Pederson, 2012). Even so, brain fingerprinting has still been ruled as admissible evidence in court (Farwell 2012) and has been applied in a number of criminal cases (e.g. serial killer J.B. Grinder) as well as helping to quash a falsely convicted murderer (Terry Harrington).

Other studies have been carried out using neuroscience advances for lie detection and predicting future behaviour. For instance, Pavlidis and Levine (2002) and Pavlidis, Eberhardt, and Levine (2002) proposed a neuroscience technique that can detect changes in temperature patterns, suggesting that it can be used for 'deception detection on the fly' to screen airline passengers. Dr Nita Farahany in an article published in *The Washington Post* (2008) invited readers to imagine a world where police officers can read the minds of potential criminals and arrest them before they commit any crimes, adding that police may soon been able to monitor suspicious brain activity from a distance as well. Indeed, neuroscience-based technologies could have wide applications in intelligence and counter-intelligence domains. Studies such as the ones using EEG scans to analyse brain waves to uncover terrorist plots before they happen or to detect deception on the fly may sound a lot like the plot of the 2002 movie *Minority Report*, based on Philip K. Dick's novel, in which 'pre-cogs' (muted humans with 'precognition' abilities) catch criminals before they commit crimes. Nevertheless, the scientific exploration of the brain and the mind is far from over and will continue to influence society in numerous ways. The implications of that and the underlying ethical issues are profound, as the next section illustrates.

Ethical Concerns and Social Implications of Analysing the Terrorist's Brain

Whether most recent neuroscience-based technologies will be effective or not outside laboratories remains to be seen, but the fact that governments and private companies are seriously investing in their potential raises several ethical questions and can have lasting implications. The focus of this section is on social and ethical issues of analysing the terrorist's brain and does not address specific legal issues. For consideration of those, see Greely and Illes (2007), and for a debate on the reliability of the findings of recent advances in neuroscience see Wolpe et al. (2005).

Common concerns in biomedical ethics, such as privacy, confidentiality, and agency are equally relevant to the most recent neuroscience-based technologies aimed at detecting deception or predicting an imminent attack. However, it is in the application of neuroscience-based technologies in the security field that some of the most serious concerns arise, and in particular 'when neuroimaging methods are naively accepted as providing an objective description of a person's brain state, particularly as it relates to predicting future behaviour' (Canli and Amin, 2002, p. 415).

As the interest in mapping and understanding the human brain constantly grows, society faces an increasing number of ethical concerns and social implications raised by these new technologies and their applications in real-life settings. The use of neuroscience-based technologies for the purpose of lie detection can have potential benefits but also pose enormous challenges in terms of privacy, informed consent, etc. Also, several companies are already selling 'lie detection services' and some neuroscientists are actively promoting their methods. As Wolpe et al. (2005) observed there is an obvious attraction of new techniques for the detection of deception in a society that is concerned with internal security and foreign threats. Furthermore, the scientific community, the political circles and the media are intrinsically linked in perpetuating the myth of an infallible science that is able to 'read' minds. As discussed below, such a myth affects the way in which science is done; the way science is reported; the way in which politics is conducted; and also has wider effects on all individuals in society with unalienable rights such as for that person not to have their brain the subject of scientific scrutiny.

The subsections below attempt to address some of the most crucial concerns and implications that recent neuroscience-based technologies can have in relation to terrorism and its impact on policy, society at large, and for science itself.

Implications for Policy

Given the interest that the intelligence and counter-intelligence communities have in using neuroscience-based technologies even though validity, effectiveness and potential misuses and abuses are well below scientific standards for public policy, an awareness of the potential implications of neuroscience is both timely and urgent (Hazlett, 2006). Indeed, the use of these technologies (e.g. P300 and fMRI) without validation in intelligence and counter-intelligence operations represents a major concern (Shamoo, 2010), and more crucially about the ethical and social implications. It has been reported that 'in the "War on Terror"', several military psychologists and psychiatrists have been designated "behavioural science consultants"' and tasked with helping interrogators develop interrogation strategies tailored for individual detainees' (Marks, 2007, p. 484).

Crucially, to approach neuroscience technologies as primarily anti-terrorism tools risks deflecting attention away from the discipline in general and its use as a wider social phenomenon. It can be argued that the over attention currently devoted to the study of brain analysis for counter-terrorism purposes results in an over simplification of what is a very complex and multifaceted phenomenon, not solvable by scientific means alone. Starting with the word 'terrorism', the problem of defining it has hindered analysis since the inception of terrorism in the 1970s (Crenshaw, 2000). Terrorism is an essentially 'contested' term (Connolly, 1993) and is politically and morally charged. Additional problems derive from the fact that even the practice of what goes under the name of terrorism is highly diverse, ranging from random mass casualty bombings to the kidnapping of an individual (see Chapters 1 and 5). It occurs within different geographical settings, political systems and at different times. It can be the 'weapon of the weak' but States may also use it against their enemies. Terrorists can originate from very different backgrounds and have very different aims. Terrorism is a social phenomenon at the crossroads of social policy, regional and international politics, economical development, public diplomacy, issues of poverty, education and health – all of which should receive more attention in academic and policy discussions. Rather, presenting terrorism as a problem solvable by scientific means precludes the exploration of alternative approaches such as public diplomacy as anti-terrorism.

Implications for Society at Large

One of the most important concerns regarding the recent advances in 'reading minds' with regard to anti-terrorism relates to the impact on civil liberties. Such advances could indeed access the most secret parts of human beings, allowing others to 'read' emotions, feelings, and thoughts. One of the primary ethical concerns relates to the right to the privacy of one's thoughts. The term 'cognitive liberty', coined by neuroethicists, refers to the 'limits of the state's right to peer into an individual's thought processes with or without his or her consent, and the proper use of such information in civil, forensic, and security settings' (Wolpe et al. 2005, pp. 39–40). As observed by Greely (2005), mind-reading threatens to invade a last inviolate area of the self; or as Farah and Wolpe (2004, p. 35) express it:

> The brain is the organ of mind and consciousness, the locus of our sense of selfhood. Interventions in the brain therefore have different ethical implications than interventions in other organs.

The implications of being able to detect when people are intentionally carrying out a deception such as lying are profound. In particular, the use of brain imaging to predict 'future psychopathology' should be the object of careful scrutiny. The proposed change of Britain's Mental Health Act that would allow for detention of individuals who have not yet committed a crime but are seen as a potential threat to public safety (on the basis of psychopathological traits) is paradigmatic of the urgency to address the ethical implications related to the application of neuroscience technologies in the security setting. In the British case, for instance, with the proposed law, individuals deemed a potential threat to society could in theory be diagnosed with 'Dangerous Severe Personality Disorder' on the basis of an EEG scan, having thus constrained their individual right to freedom. Worryingly, there is only a limited academic discussion on the applications of neuroscience-based technologies in policy and military settings (Dando and Bartolucci, 2013).

A related concern is the potential misuse of data collected by such screening technologies. Many recent high-profile lapses in data security (e.g. in Britain. intimate details for over 25 million citizens of every household with children that claims a government subsidy were lost, including names, addresses, dates of birth, insurance numbers and banking details) makes the issue of this kind of information held on each one of us particularly sensitive. For instance if an fMRI scan suggested some likelihood of 'criminal' tendencies, how should such data be used and where should it be stored?

At a more profound level, the 'opacity of truth' that science brings allows the potential of forming an opinion on the truthfulness about something without understanding all the technical details that relate to it. In this regard, Canli and Amin (2002, pp. 424–5) wrote:

> When one sees an image of a brain's activation pattern, it is difficult not to be struck by its visual persuasiveness ... one great danger lies in the abuse of neuroimaging data for presentations to untrained audiences, such as courtroom juries. What can be easily forgotten when looking at these images is that they represent statistical inferences, rather than absolute truths.

In other words, science easily baffles the non-specialist and yet its truthfulness remains unquestioned. The 'reading' of the brain has a disproportionate and persuasive impact.

Implications for the Way Science is Reported and Disseminated

The ways in which the media portrays neuroscience is also problematic. In many cases, press coverage of studies linked to the findings of neuroscience are then translated into a popular discourse that oversimplifies the complexity of the findings and the results through the use of catchy titles such as 'Fighting Crime by Reading Minds' or 'Reading Terrorist's Minds About Imminent Attack'. In other cases, press coverage of research studies in this field often includes speculation about the imminent usefulness of the technology in civil settings: a claim not made by the researchers themselves nor justified by the state of current research (see Langleben et al., 2002). There are indeed only a few journalists that are qualified to comment on the nature of the experiments themselves. For instance, the article entitled 'EEG Scans Analyze Brain Waves to Uncover Terrorist Plots Before They Happen' (Boyle, 2010) is a commentary on a recent finding in the neuroscience field. The article positively describes the research, without reporting or indicating that there are flaws in the study, thus giving the impression that there is a science available able to uncover a terrorist attack and finally reifying an assumption that has not been robustly scrutinised. Finally, the media themselves are often bolstered by excessive claims made for these methods by the very same investigators. Farwell was interviewed by the BBC in 2004 and reportedly claimed '100 per cent accuracy' for brain fingerprinting and that we now have the ability to 'scientifically' detect if certain information is stored in the brain.

The very adoption of the idea of 'reading' a terrorist's brain may create an inaccurate impression of what is possible: an impression that is not supported by scientific evidence. Nevertheless, for the commercialisation of the science, this remains a useful impression to perpetuate. In order to capitalise on the current political climate in which counter-terrorism is one of the most important security policy objectives, scientific research may be too egotistical to wait for scientifically grounded results. Indeed, the overemphasis on brain analysis as a key to addressing societal issues leads to studies that are themselves based on a misguided perspective (such as seeing 'mind reading' as an ultimate form of counter-terrorism) and fail to address pressing questions such as: who is a terrorist, or what does it mean to lie? Instead, crucial questions such as is there such a thing as a criminally violent brain, or does it make sense to talk about the neurobiology of terrorism?, are promised to be answered by neuroscientists with a straightforward 'yes', thus realising the early criminologists' dream of identifying the biological roots of criminality (Redding, 2006). The assumption that the brain is the fundamental level at which to understand terrorist conduct further assumes that terrorism is similar to a 'disease' and its study can deliver specific cures or avert further violence. This implies there is a 'normal brain' against which one could make comparisons to an individual sample, and that the complex phenomenon of terrorism arises from dysfunctions within specific locations in the brain by people 'biologically different' from the norm. This has clear implications in terms of profiling of individuals that are suspected of terrorism on the basis of a dysfunction within a specific location of the brain (for a discussion of crime as a disease, see Kirchmeier, 2004).

Discussion

This chapter has opened up a debate on the necessity of addressing pressing social implications and ethical concerns related to brain imaging technologies for potential terrorists. As the interest in mapping and understanding the human brain constantly grows, society faces an increasing number of ethical issues raised by the applications of these technologies in real-life settings, particularly in highly sensitive domains, such as security. More than a decade after September 11, 2001, counter-terrorism is still a major concern for governments all around the world. From the 1990s onwards, terrorism has been seen as a major national security threat and neurobiological advances are looked at with increasing interest by the US Defense Department for their supposed ability to defeat terrorism in a definitive way, without imposing on or harming the rest of the population. Indeed, a reliable method for detecting deception and predicting imminent psychopathology could have enormous value in national security. These advances, however, need to be properly studied and carefully examined with regard to their social and ethical implications, both for the general public, alleged terrorists and/or known terrorists. This is a pressing issue, since contrary to what is claimed, brain imaging does not tell us whether a person is lying or not, it only shows a certain state of mind, as pointed out by Abi-Rached (2008, p. 1160):

> Given the artificial environment, the statistical distribution of data, and other inherent limitations and margin of error, it seems that detecting specific mental states is not as easy as is often claimed. Consequently, we should be much more careful in applying these technologies for highly sensitive uses such as security or in the legal system.

Research within the discipline of neurobiology has not yet fully considered the implications of recent claims about the understanding of how people think and feel. As a start, the attribution of terrorism to irrationality, personality disorders and mental illness should be carefully avoided. Psychological theories of terrorism have also based their investigations on the assumption that the better we understand the roots of the terrorist mindset, the better able we should be to develop policies for effectively managing the threat (see Wardlaw, 1989). However, given the

> heterogeneity in the temperaments, ideologies, thought processes, and cognitive capacities of terrorists within political categories, hierarchical levels, and roles [it would be more sensible] to acknowledge from the outset that any effort to uncover the 'terrorist mind' will more likely result in uncovering a spectrum of terrorist minds. (Victoroff, 2005, p. 7)

Learning from past experiences of (ultimately senseless) profiling of the 'typical terrorist', the psychology of terrorism could be better positioned to offer its insights into the study of individual mentalities in a common effort (with other disciplines) to counter terrorism in a more effective way.

Crucially, we need to move away from the attribution of terrorism to personality disorders or 'irrational' thinking and face the issue of terrorism within a much wider framework. In particular, it is essential to understand the ideologies or worldviews of those who engage in violent acts on their own terms, and not to exclude them from analysis because they appear 'irrational' in a conventional sense (Crenshaw, 2000). The study of terrorism is still affected by enduring problems. In particular and very often, psychological hypotheses are based upon speculation or are derived from such a small number of cases that the findings cannot be generalised or considered reliable (Crenshaw, 2000). Moreover, the perception of particularly vivid threats may be exaggerated, especially when there is little information about them. It is also time to reflect on whether enhanced safety should be sought even when faced with a loss of liberty.

Conclusion

People all around the world have been willing to tolerate significant and enhanced security measures and greater encroachments on civil liberties since the events of 11 September, 2001. We should consider if these and similar incidents are enough to justify the use of 'pre-crime' technological screening of individuals who may not have yet committed any actual crime. The reduction of the risk of the inappropriate application of neuroscience-based technologies in non-clinical domains, requires the continued and open discussion of their societal and ethical implications (as well as of their real feasibility) within the scientific and public domains. Traditionally academics of various disciplines (e.g. sociologists, psychologists, political scientists, security experts) have had the tendency to overemphasise the importance of their own fields in providing explanations of terrorism, downplaying the insights provided by other disciplines. Terrorism nevertheless is a complex subject to analyse and multidisciplinary research is necessary to understand the phenomenon as a prelude to countering it. Psychologists study terrorism with the aim of identifying those who may engage in violence with a view to aiding in prevention. Neuroscientists research the 'terrorist brain' with the aim of preventing attacks. Political scientists, sociologists and governmental officials specialising in terrorism, all aim at studying terrorism with a view to aiding in prevention, detection or capture as a way of enhancing security. It is both timely and urgent that all these efforts finally unite into the common goal of understanding terrorism, and especially terrorists, in order to be better prepared when facing it.

Acknowledgment

I wish to thank Professor Malcolm Dando for pushing me into this fascinating topic and Professor Steve Shore for his useful comments on an earlier draft.

References

Abi-Rached, J. (2008). The implications of the new brain sciences. *EMBO Reports* 9(12), 1158–1162.

Ahmed, A. (2003). *Islam Under Siege: Living Dangerously in a Post-Honor World.* Polity Press: Cambridge.

Allen, J.J.B., and Iaono, W.G. (1997). A comparison of methods for the analysis of event-related potentials in deception detection. *Psychophysiology*, 34, 234–240.

Bartolucci, V. (2010). Analysing the elite discourse on terrorism and its implications: The case of Morocco. *Critical Studies on Terrorism*, 3(1), 119–135.

Bartolucci, V. (2012). Terrorism rhetoric under the Bush's Administration. *Journal of Language and Politics*, 11(4) Winter, 562–582.

Bartolucci, V. (2013). Security vs. liberty: Terrorism as discourse and its societal effects in the U.S. and Morocco. *Journal of Democracy and Security*. 10(1), 1–21.

Bongar, B., Brown, L.M., Beutler, L.E., Breckenridge, J.N., and Zimbardo, P.G. (2007). *Psychology of Terrorism*. Oxford University Press: Oxford.

Boyle, R. (2010). EEG scans analyze brain waves to uncover terrorist plots before they happen. *R&D (Research and Development) Magazine*, online, available at http://www.rdmag.com/News/Feeds/2010/07/manufacturing-reading-terrorists-minds-about-imminent-attack. Accessed 2 July 2013.

Brannan, D.W., Esler, P.F., and Strindberg, N.T.A. (2001). Talking to "terrorists": Towards an independent analytical framework for the study of violent substate activism. *Studies in Conflict & Terrorism*, 24(1), 3–24.

Canli, T., and Amin, Z. (2002). Neuroimaging of emotion and personality: Scientific evidence and ethical considerations. *Brain and Cognition*, 50, 414–431.

Connolly, W. (1993). *The Terms of Political Discourse*. Princeton University Press: Princeton, NJ.

Crenshaw, M. (2000). The psychology of terrorism: An agenda for the 21st century. *Political Psychology*, 21(2), 405–420.

Crenshaw, M. (2007). Explaining suicide terrorism: A review essay. *Security Studies*, 16(1), 133–162.

Dando, M. (2010). Neuroethicists are not saying enough about the problem of dual-use. *Bulletin of the Atomic Scientists*, available at http://www.thebulletin.org/neuroethicists-are-not-saying-enough-about-problem-dual-use, accessed 2 July 2013.

Dando, M. and Bartolucci, V. (2013). What does neuroethics have to say about the problem of dual-use? In B. Rappert and M. Selgelid (Eds), *On the Dual Uses of Science and Ethics: Principles, Practices, and Prospects*. ANU Press: Canberra, 29–44.
Farah, M.J. and Wolpe, P.R. (2004). Monitoring and manipulating the human brain: New neuroscience technologies and their ethical implications. *Hastings Center Reports*, 34(3), 35–45.
Farwell, L.A. (2012). Brain fingerprinting: a comprehensive tutorial review of detection of concealed information with event-related brain potentials. *Cognitive Neurodynamics*, 6, 115–154.
Farwell, L.A. and Donchin, E. (1991). The truth will out: Interrogative polygraphy ('lie detection') with event-related brain potentials. *Psychophysiology*, 28, 531–547.
Farwell, L.A. and Richardson, D.C. (2006). Brain fingerprinting in laboratory conditions. *Psychophysiology*, 43, S38.
Farwell, L.A. and Smith, S.S. (2001). Using brain MERMER testing to detect concealed knowledge despite efforts to conceal. *Journal of Forensic Sciences*, 46(1), 135–143.
Farwell, L.A., Richardson, D.C., and Richardson, G.M. (2012). Brain fingerprinting field studies comparing P300-MERMER and P300 brainwave responses in the detection of concealed information. *Cognitive Neurodynamics*, 6, 115–154.
Fekete, L. (2004). Anti-Muslim racism and the European security state. *Race and Class*, 46(1), 3–29.
Ganis, G. and Rosenfeld, J.P. (2011). Neural correlates of deception. In J. Illes and B.J. Sahakian (Eds), *The Oxford Handbook of Neuroethics*. Oxford University Press: New York, 101–118.
Greely, H.T. (2005). Premarket approval regulation for lie detections: An idea whose time may be coming. *American Journal of Bioethics*, 5(2), 50–52.
Greely, H.T. and Illes, J. (2007). Neuroscience-based lie detection: The urgent need for regulation. *American Journal of Law, Medicine & Ethics*, 33, 377–431.
Gunning, J. (2007). The case for a critical terrorism studies? *Government and Opposition*, 42(3), 363–394.
Haynes, J.D., Sakai, K., Rees, G., Gilbert, S, Frith, C. and Passingham, R.E. (2007). Reading hidden intentions in the human brain. *Current Biology*, 17, 323–328.
Hazlett, G. (2006). Research on detection of deception: What we know vs. what we think we know. Educing Information. In Swenson, R. (Ed.) *National Defence Intelligence College, Educing Information Interrogation: Science and Art. Phase I Report*. Library of Congress: Washington, DC, 45–62.
Horgan, J. (2005). *The Psychology of Terrorism*. Routledge: New York.
Hudson, R.A. (1999). *The Sociology and Psychology of Terrorism: Who Becomes a Terrorist and Why?* Library of Congress, Federal Research Division: Washington, DC.

Illes, J. and Racine, E. (2005). Imaging or imagining? A neuroethics challenge informed by genetics. *American Journal of Bioethics*, 5(2), 5–18.

Illes, J., Kirschen, M.P. and Gabrieli, J.D. (2003). From neuroimaging to neuroethics. *Nature Neuroscience*, 66(3), 201–215.

Jackson, R. (2005). *Writing the War on Terrorism: Language, Politics and Counter-Terrorism*. Manchester University Press: Manchester.

Kirchmeier, J.L. (2004). Aggravating and mitigating factors: The paradox of today's arbitrary and mandatory capital punishment scheme. *William and Mary Bill of Rights Journal*, 6(345), 397–431.

Kleiner, K. (2009). Truth machine to be used for court evidence? *Singularity Hub*, www.singularityhub.com/2009/03/16/truth-machine-to-be-used-as-court-evidence/. Accessed 14 February 2012.

Kozel, F.A., Johnson, B.E., Grenesko, E.L., Laken, S.J., Kose, S., Lu, X., Pollina, D., Ryan, A., and George, M.S. (2009). Functional MRI detection of deception after committing a mock sabotage crime. *Journal of Forensic Sciences*, 54(1), 220–231.

Kozel, F.A., Johnson, B.E., Mu, Q., Grenesko, E.L., Laken, S.J., George, M.S. (2005). Detecting deception using functional magnetic resonance imaging. *Biological Psychiatry*, 58(8), 605–613.

Langleben, D.D., Schroeder, L., Maldjian, J.A., Gur, R.C., McDonald, S., Ragland, J.D., O'Brien, C.P., and Childress, A.R. (2002). Brain activity during simulated deception: An event-related functional magnetic resonance study. *NeuroImage*, 15(3), 727–732.

Lombera, S., and Illes, J. (2009). The international dimensions of neuroethics. *Developing World Bioethics*, 9(2), 57–64.

Marks, J.H. (2007). Interrogational neuroimaging in counterterrorism: A 'no-brainer' or a human rights hazard? *American Journal of Law & Medicine*, 33(2–3), 483–500.

Meixner, J.B., and Rosenfeld, J.P. (2011). A mock terrorism application of the P300-based concealed information test. *Psychophysiology*, 48(2), 149–154.

Moreno, J.D. (2006). *Mind Wars: Brain Research and National Defense*. The Dana Foundation: Washington, DC.

National Research Council (NRC) (2003). *The Polygraph and Lie Detection*. The National Academies Press: Washington, DC.

New York Times, The (2007). This is your brain on politics. OP-ED Contribution. November 11.

Pavlidis, I., and Levine, J. (2002). Thermal image analysis for polygraph testing. *IEEE Engineering in Medicine and Biology Magazine*, 21(6), 56–64.

Pavlidis, I., Eberhardt, N.L., and Levine, J. (2002). Seeing through the face of deception. *Nature*, 415(6867), 35.

Pearce, K.I. (1977). Police negotiations. *Canadian Psychiatric Association Journal*, 22, 171–174.

Poldrack, R.A. (2008). The role of fMRI in cognitive neuroscience: Where do we stand? *Current Opinion in Neurobiology*, 18, 223–227.

Pronin, E., Kennedy, K., and Butsch S. (2006). Bombing versus negotiating: How preferences for combating terrorism are affected by perceived terrorist rationality. *Basic and Applied Social Psychology*, 28(4), 385–392.
Redding, R.E. (2006). The brain-disordered defendant: Neuroscience and legal insanity in the twenty-first century. *American University Law Review*, 56(1), 51–124.
Rosenfeld, J.P. (2011). P300 in detecting concealed information. In B. Verschuere, G. Ben Shakhar and E. Meijer (Eds), *Memory Detection: Theory and Application of the Concealed Information Test*. Cambridge University Press: Cambridge, 63–89.
Rosenfeld, J.P., Hu, X., and Pederson, K. (2012). Deception awareness improves P-300 based deception detection in concealed information tests. *International Journal of Psychophysiology*, 86(1), 114–120.
Russell, C.A. and Miller, B.H. (1977). Profile of a terrorist. *Terrorism: An International Journal*, 1(1), 17–34.
Saykin, A.J. (2007). Brain imaging and behavior: Progress and opportunities. *Brain Imaging and Behavior*, 1, 1–2.
Shamoo, A.E. (2010). Ethical and regulatory challenges in psychophysiology and neuroscience-based technology for determining behaviour. *Accountability in Research*, 17, 8–29.
Silke, A. (1998). Cheshire-cat logic: The recurring theme of terrorist abnormality in psychological research. *Psychology, Crime and Law*, 4, 51–69.
Silke, A. (2001). The devil you know: Continuing problems with research on terrorism. *Terrorism and Political Violence*, 13(4), 1–14.
Simpson, J.R. (2008). Functional MRI lie detection: Too good to be true? *Journal of the American Academy of Psychiatry Law*, 36, 491–498.
Taylor, M. (1988). *The Terrorist*. Brassey's: London.
Times, The (2010). Fighting crime by reading minds. August 7.
Victoroff, J. (2005). The mind of the terrorist: A review and critique of psychological approaches. *Journal of Conflict Resolution*, 49(1), 3–42.
Wardlaw, G. (1989). *Political Terrorism, Theory, Tactics and Countermeasures*. Cambridge University Press: Cambridge.
Washington Post, The (2008). The government is trying to wrap its mind around yours. April 13.
Wolpe, P.R. (2004). Neuroethics. In S.G. Post (Ed.), *Encyclopedia of Bioethics*, 3rd edn. Macmillan Reference: New York, NY, 1894–1898.
Wolpe, P.R., Foster, K. and Langleben, D.D. (2005). Emerging neurotechnologies for lie-detection: Promises and perils. *American Journal of Bioethics*, 5(2), 39–49.

Chapter 5
Ethical Issues in Surveillance and Privacy

Ron Iphofen
Independent Consultant and Senior Visiting Research Fellow (Open University) UK

Introduction

This chapter seeks to explore issues involved with balancing public security with individual privacy. To illustrate these issues specific reference is made to an ongoing project part-funded by the European Commission (EC) under Framework Programme 7: SECUR-ED (http://www.secur-ed.eu). This is a demonstration project aimed at supporting mass public transport security and entails security and surveillance exercises in complex travel interchanges in selected major European cities. Passengers in mass transport interchanges are notoriously vulnerable to terrorist attack (as evidenced by the bombings in subways in London, Paris and Madrid and the toxin attacks in Tokyo – see below). There is a need to find ways to protect the travelling public, but many of the underlying issues highlight problems associated with surveillance and security in a range of other complex settings.

Surveillance has a long and challenging history in penal reform, welfarism and civil society both in ideal, realised and fictional forms (cf. George Orwell, Aldous Huxley). From Jeremy Bentham's proposed Panopticon (Bentham, 1995) to Michel Foucault's concern with the 'gaze' (Foucault, 1991) both scholarly and fiction writers have toyed with notions of how individuals perceived to be 'deviant' in a society could be kept under observation and subsequently 'controlled'.

Any attempt at operationalising surveillance practices in the current regulatory climate requires comprehensive consideration of the ethical implications of proposed implementation. As with all ethical considerations this is never easy since it raises a host of dilemmas and contradictions. Writers and researchers in the field of ethics frequently refer to conceptual 'tensions' (see, for example, Bauman [1993] on moral uncertainty; Vardy and Grosch [1999] on conflicting rights; or La Follette [2002] passim). However, when it comes to security research, to refer to a tension is almost a euphemism. There are distinct conflicts between 'security' and 'privacy' – the latter lying at the core of ethical issues.

The fundamental problem of politics, indeed of civil society, has always been how to allow individuals to freely pursue their own needs and desires while at the same time maintaining public order. The Hobbesian problem was certainly one of order – and the nasty, brutish and short life Thomas Hobbes was confronting required controlling those individuals who might be responsible for the nastiness, brutishness and brevity (Hobbes, Rogers and Schuhmann, 2006). Our problem here is the pursuit of two cherished goals that may be, in certain circumstances, fundamentally opposed to each other – individual privacy and public security. Thus in the case of the SECUR-ED project, to protect the travelling public it may be necessary to intrude upon the privacy of individuals who threaten passengers and/or to intrude upon the privacy of individuals who may be threatened, for their own protection. This presents us literally with a dilemma and ethics is precisely the study of how one addresses and, hopefully, resolves such dilemmas.

One of the major funders of surveillance studies, the EC, has incorporated this dilemma in its policies. Privacy is espoused by the European Union as a human right under Article 8 of the European Convention on Human Rights (ECHR). At the same time the EC is massively funding research and development in security technologies. SECUR-ED is only one of more than 200 current projects dealing with surveillance and security (these can all be accessed via the EC's CORDIS database http://cordis.europa.eu/fp7/security/home_en.html).

Defining the Scope of Surveillance and Privacy

Careful definitions of the relevant terms would require fairly comprehensive treatises and as such, are outside the scope of this chapter. 'Surveillance' as employed here is not just generic 'people-watching'. It involves paying close and sustained attention via visual, aural or other data-gathering techniques to a particular person or group for a specific reason. The reason is the assumption that the person or group is of interest to security agencies and therefore perhaps engaged in 'suspicious behaviour' which could be inferred to be of a threatening (terrorist and/or criminal) nature. Such surveillance is usually pre-planned, rigorous and focused on an identifiable person or group. It is the 'identifiability' and the linking of data, of whatever form and however gathered, to a specific person or group that raises the challenge to privacy. Of course surveillance is not only used for public security purposes. Loyalty cards are used by retailers to examine the demographics of consumer preferences; smart cards and ticketing devices enable transport managers to control traffic flow for vehicles and/or pedestrians; closed-circuit television (CCTV) in public locations enables safe crowd control and even cookie technology facilitates and monitors our interactions with the World Wide Web. Those surveilled to some degree knowingly trade off their privacy in return for credit points, special offers, or expanded information about goods and/or services or to gain from the efficiencies of traffic or crowd control etc. The inter-corporate monitoring of credit card use can protect against identity fraud and the suspension of credit following unusual spending patterns. In such cases the individual condones the monitoring (and by virtue of this their loss of privacy) in the interests of their own self-protection. Thus consented surveillance may be regarded as less problematic as long as the limits to such surveillance are clarified in the consenting process.

Privacy is a much more contentious concept, although generally presumed to be 'valued' it cannot be assumed to be an 'unqualified good' (Wacks 2010) since its absolute application could conceal criminality or harm perpetrated on others in secret. The range of contested definitions range from claims over the ability to control information about oneself to the right to be 'let alone' (Warren and Brandeis, 1890; Westin, 1967). In general terms it entails the ability to engage in personal, intimate actions without the unwarranted and/or unwanted intrusions or interferences of others and to control the release of information about those actions. There are some international agreements on how it is to be applied, but most States either have their own specific laws on it or link it to other human rights legislation or case law (Wacks, 2010). For many people, privacy has come to be regarded as a key human right. However, it is difficult to hold as an absolute and distinct privilege, separate from other overlapping rights such as freedom (of speech and action) and autonomy, the maintenance of human dignity and resistance to oppression. Thus Article 8 of the ECHR (Council of Europe, 1950) declares:

- Everyone has the right to respect for his private and family life, his home and his correspondence.
- There shall be no interference by a public authority with the exercise of this right except such as in accordance with the law and is necessary in a democratic society in the interests of national security, public safety or the economic well-being of the country, for the prevention of disorder or crime, for the protection of health or morals, or for the protection of the rights and freedoms of others.

These 'exceptions' to the 'right' to privacy illustrate the fundamental conflict quite clearly. As is argued later, it is how these exceptions are interpreted that is key to managing the balance between privacy and security that becomes the justification for appropriate surveillance practices. Despite this status of a fundamental right within Europe (not discounting rights to privacy elsewhere around the world), the interpretation of the right varies across the legal jurisdictions involved in a project such as SECUR-ED that represents more than a dozen Member States.

The Analytic Problem of Rights and Inherent Contradictions

All ethical issues can be framed by asking the same sets of questions: who is doing what to whom, why and how? For the answers to each of these questions we would be seeking a sense of 'legitimacy'. In other words, to a specified degree it is 'acceptable' to engage in the questioned actions or there is an authoritative rationale for the answers to each of these questions. The sense of legitimacy is determined by a set of criteria that are all linked to human rights and data protection.

For the purpose of this chapter the focus is on the relevant rights and part of the analytic problem we must address is that the espoused rights can contain inherent contradictions. Thus, for example, as ECHR Article 8 makes clear, the right to privacy holds 'unless...' – a criminal act is involved; a right to safety and security holds 'unless...' – one individual's security has to be sacrificed to protect the majority. Rights cannot be absolute: they are delimited by contextual factors. The favourite phrase of lawyers and ethicists when asked to make a moral judgement is: "It depends upon...". So whether the proposed actions are legitimate depends upon certain criteria:

- *WHO* is conducting surveillance on whom? Professionals in all security-related fields and authorities at all levels comprise this 'who' element. The range of actors and agents could be enormous, but at the very least includes: the State, usually via its 'agents' (e.g. security agents, police and judicial authorities); property owners; media organisations such as newspapers and broadcasters; transport organisations including rail, road, airport, bus, tram, etc. In accordance with Article 8 of ECHR these agents are seen as legitimate surveillers if their intrusion on privacy is conducted in the interests of security (of the public, of the State) and they are only doing it to individuals and/or groups who they suspect of nefarious activity. The legitimate authority to engage in surveillance conferred upon these agents is in turn dependent upon their motives.
- *WHY* are the actors conducting surveillance (i.e. what will the information be used for)? What are their aims and intentions, and what is the purpose? If 'defence against terrorism' is the perceived threat then public security becomes a justifiable rationale. With 'adequate' surveillance the threat of terrorism can be prevented or at least minimised. Baldly stated, this is too simple since a range of 'other factors' now come into play: Who is to judge the extent of the 'threat'? What sort of behaviour constitutes a threat? And how 'adequate' does the surveillance need to be to ensure the threat is countered? To illustrate this, is political dissent an adequate reason to ensure surveillance of the dissenters? Perhaps the answer is 'yes' if there is a security threat, but who estimates that threat and how do they make the judgement?
- *WHAT* are the available and emergent technologies of surveillance? What are they and what can they 'technically' do? And if they cannot yet 'do' certain things what things might be technically feasible in the future? (Whether they might be 'allowed' to develop is another related question but this should be dealt with under the 'Why?' umbrella. For example if nuclear, biological or chemical weaponry had not yet been devised, but knowing what we now know, would we be advised to encourage or permit their development?). In a more familiar context, we have the apparently ubiquitous gaze of CCTV. It is not merely limited to public spaces, but its effectiveness is often constrained by resources, image quality, and

limitations of human–machine interaction. Radio frequency identification (RFID) devices can be located in products and attached to persons or their clothing. Bluetooth technology can be accessed by any people in public spaces using their mobile phones, rather than those intended. Online databanks such as Amazon or eBay hold enormous amounts of personal information about their users and are accessible by databank owners (and hackers alike). Social media (Facebook, Twitter, Instagram, LinkedIn) and cloud computing services (iCloud, Dropbox, GoogleDrive, Mega) all exist in an essentially public domain. Thus the unlimited storage of searchable personal data is technically feasible and access to it only partially delimited by quantity and time. All of these technologies pose threats to privacy because they are vulnerable to interception and misuse. The complication is that many of them are based on 'consented' surveillance, where individuals could/should be aware or reminded of the vulnerability of their privacy when they engage to use such media.

- *HOW* do those conducting surveillance go about collecting data? What methods are adopted for the collection and retention of data? It is here that the issue of consent is important. As we illustrated earlier, consented surveillance may act in the interests of a traveller, passenger or consumer of products and services – enhancing or facilitating their experiences. Even then, this is dependent on the limits to the surveillance as only so much interference in the individual's privacy should be allowed to secure their comfort/enjoyment. However, it is becoming increasingly feasible to accurately identify and track individuals in both the physical and digital world, perhaps without their knowledge. It is in this identification of a specific person that we are no longer dealing with impersonal movements or actions of units (crowds, groups of anonymous passengers) but enhanced knowledge about a specified person. Herein lies the threat to the cherished principle of privacy.

Many of the commitments of European institutions have fundamental contradictions within them. In some respects these are inevitable since the very concepts of privacy and security contain inherent contradictions. Necessarily there appears to be wide variation in interpretations of privacy in security practices in different jurisdictions. Many are inconsistent with statements of European institutions such as a recent Article 29 Working Group declaration on tracking: http://ec.europa.eu/justice/policies/privacy/workinggroup/wpdocs/index_en.htm

There is clearly a contradiction between the funding call SECUR-ED responded to and the principles held in value by other European institutions such as the European Parliament itself. This could be even seen as a problem within the Commission: funding and supporting surveillance and security projects that challenge the fundamental ethical principles that the Commission seeks to secure. SECUR-ED has been contracted to carry out a particular piece of research and development and it is obliged to deliver it; but we should call attention to areas where the research seems to run up against policy contrary to its funding call. Some of the deliverable written reports for the Project should be able to point out the ways it may come into conflict with existing European law or challenge cherished ethical principles.

Mass Passenger Transport Security

It is evident from the analytic criteria summarised above that the balance between surveillance, privacy and security is context-specific. The only route to fully examine the ways in which privacy and security can be balanced against each other is to apply the test of 'justifiability' to 'who is doing what, how and why' in each case. Most public spaces are inherently insecure in terms of privacy: they are often characterised as being 'crowded' and 'open'. Controlling information about ourselves is harder to accomplish in a public domain than in our own private space. This is necessarily even more the case in large transport interchanges where large numbers of people are moving between different forms of transportation in a shared, crowded space. The terrorist risk is obvious and historical experience offers adequate evidence and clearly demonstrates the extent of the risk: the Paris Metro bombings (25 July and 17 August, 1995) and the Sarin gas attack in the Tokyo subway (20 March, 1995); the Madrid commuter train bombings (11 March, 2004); the London suicide bombings in the Underground and on a double-decker bus during rush hour (7 July, 2005).

The surveillance of such public spaces constitutes normal procedure for transport authorities, and passengers have some expectation of being observed in most public transport settings (indeed, public announcements and posters often remind them of this). The physical presence of CCTV cameras confirms this (regardless of whether they are operational) and well-placed warning notices and prior notification of RFID technologies on tickets are recognised as 'adequate consent'. So when surveillance takes place in public where people expect to be watched, and announcements can be made to call the public's attention to the possibility of surveillance, there should not be too much of a privacy problem. Anyone seen 'behaving suspiciously' could legitimately be reported to security officers for them to take 'appropriate action' without any additional 'tracking' of an identifiable individual in any substantial sense being undertaken. It is this singling out of an identifiable individual that is seen as the main threat to privacy.

The real problem lies with covert surveillance, especially when it is targeted at a particular individual. This is where the criteria for selecting that individual are crucial, especially if it is based on some form of profiling. However, it is hard to imagine the assessment of 'suspicious behaviour' without establishing a 'profile' that clearly has to be ethically and legally rigorous. So far the profiling within SECUR-ED has been mindful of the standard prejudices and biases that are proscribed in most European and national legislation. The problem of consent in such a situation remains, as passengers are not informed that certain actions or behaviour they might engage in could lead to them being singled out and tracked. This could be seen as a violation of privacy. If someone is tracked without their permission do they have any recourse to law? If they are engaged in criminal activity could unpermitted tracking lead to the case being thrown out of Court? Or if a non-criminal individual discovers they have been tracked do they have any civil redress? (If no action is taken they might never find out they had been tracked if it did not interfere with their lives.)

There is also a crucial distinction between actions carried out by empowered public authorities and those conducted by private institutions, or when private institutions carry out actions on behalf of those public authorities. While police authorities do have powers to track individuals, even this tends to be subject to permissions from judicial authorities that would have to be applied for and granted in advance. Here again the legitimacy of the surveillance would have to be decided by a higher authority prior to the act of surveillance taking place.

Evidently any surveillance that targets an identified individual over time would require prior regulatory approval. If both the means and methods for certain kinds of surveillance require prior approval in specific circumstances that would be of little use in a crisis or in emergency situations. Thus in emergency situations 'targeted' surveillance might turn out to be impractical, although much depends on what is regarded as an emergency. In any case the technology that would enable spontaneous, automated individual tracking of someone engaged in suspicious behaviour in an emergency situation is still unproven. Another EC-funded project (INDECT) that proposed to do just that, produced something of a public outcry in the popular press in Europe and calls on social networking sites to stop the project (e.g. see Johnston, 2009; http://noindect.fr/ or http://pirateparty.be/stop-indect).

Once again, in the field of security, the ethical principle of 'voluntary informed consent', as a protector of privacy, becomes nonsensical if applied to the prospective terrorist or criminal. In that case non-consented, possibly covert surveillance could become vital to the protection of the public. Thus the who and the why analytical questions raise issues of particular forms of power (specifically authority) and their legitimacy, combined with an adequate rationale for the surveillance (i.e. whether or not the activity is justifiable). So it is hard to separate these two sets of questions: The who depends upon the why and the necessity of 'non-consented surveillance' becomes justified by the answers to the who and the why questions.

Assessing what Might be Regarded as 'Proportionate Action'

For many of the SECUR-ED demonstrations we need to be reassured that if, for example, a transport worker were to identify suspicious behaviour on the part of a passenger, maintaining contact (via automated tracking devices) with that individual might then be allowed to commence. Clearly police involvement, at some level, would be essential. It might be possible to allow a standardised tracking protocol, previously agreed between the police and the transport security stakeholders that might clarify the legal division of labour. And it is even possible to argue that the more automated a device the more possible it might be to protect privacy since, subject to design controls, automation might remove the potential for more personal prejudices (Macnish, 2012).

Another conceptually difficult but vital test of the ethical/legal risk is proportionality. So while there is an acceptance that privacy might (have to) be sacrificed if there is a security risk, the degree to which privacy can be sacrificed has to be proportionate. To act proportionately seems such a sound expression of principle – so measured and balanced. But when and how does one decide that a response is or can be proportionate? Who makes the decision? In whose interests do they make it? Even then, would everyone consider it proportionate? Some commentators suggest a 'threshold' model: surveillance of an area for hazards might be justified, as would surveillance of an area for violent behaviour (see: http://www.surveille.eu/). However, the threshold for acceptable intrusion will be lower when pursuing a suspected shoplifter than a suspected terrorist. In the same way covert Internet monitoring or bugging of domestic premises by authorities is not seen as justified for minor crime. Furthermore, facial recognition technologies that link potential suspects to a database, or listening in to conversations, might not be seen as appropriate for such trivial misdemeanours as shoplifting or fare evasion on public transport. In terms of proportionality it might be hard to justify the use of covert surveillance and tracking technologies if their principal purpose is merely to stop graffiti artists. Human rights lawyers could see this as an oppressive response to antisocial and/or minor criminal acts that might be regarded as inconsistent with European standards. Thus the use of most tracking technologies for low-level illegal activity could be regarded as excessive.

It might be difficult to sustain this kind of threshold judgement on anything other than resource grounds where the costs of time, energy, distractions from other more important/serious contraventions, and/or the likelihood of conviction have higher priorities. In fact there is a problem of where and how tests of proportionality should be applied, given that criminals are often experts at distraction techniques such as simulating normal behaviour to disguise their illicit intentions or actions. The maintenance of normal appearances is a classic criminal distraction technique and it would be impossible to judge from apparently normal behaviour if there was criminal intent (see Goffman, 1971 and Chapters 6 and 7 this edition). The expert criminal is one who actually does nothing to suggest that he is a criminal.

In a hypothetical example from a SECUR-ED exercise, an identifiable male traveller might be seen to act suspiciously, his behaviour recorded and monitored on CCTV, his location tracked with RFID in a ticket picked up by sensors in the environment, and various data checked at specific security checkpoints or when he passes through an explosives/chemical detector. To test whether this violates the traveller's privacy one might first ask how much the traveller knows that he is being subject to tracking techniques or could reasonably expect to be tracked. If he is not aware, and there is nothing else to suggest that the traveller has criminal intent, this kind of surveillance could be regarded as disproportionate. However, all of this depends on the initial judgement call and the rationale for suspecting the traveler of some form of criminal intent. In many ways, this runs counter to the observation that cultivating normal appearances is standard practice for seasoned criminals. If a crime takes place but is not observed directly at the time it occurs, and the criminal is not linked to it and/or neither the crime nor criminal is detected at the time; it might be only at some later date that a putative act of observation might be realised to have been not disproportionate. Thus the automated record (video from CCTV for example) becomes vital for detection of those responsible for the commission of the criminal act – the recent Boston Marathon bombing (April 15, 2013) offers a case in point, although not transport -related. Automated predictive technology might have identified the 'suspicious behaviour' of leaving a large, heavy rucksack in a crowded environment and may even have assisted prevention. As it is sight of the CCTV recording retrospectively aided the capture of the culprits.

In summary, safeguarding the public from either organised crime or terrorism is not the sole responsibility of transport security agents. In practical terms, it relates to where the police should take the lead with the co-operation of transport security, in which case proportionality can then be subject to a judicial decision. Even then, the duty of care that transport authorities have to their customers can only offer limited justification for surveillance.

Such stringencies about the uses of the technologies being developed reduce the overall usefulness of a surveillance product to the prospective client base. Devising and selling surveillance products and systems will require clear direction concerning the proportionality of the techniques employed. Counter-terrorism evidently warrants greater intrusions of privacy, but it is also clear that technology should not be developed under the label of counter-terrorism when its principal purpose is to catch graffiti artists. It is here that the notions of 'dual use/misuse' or 'function creep' become relevant and the technology is used for something other than its original specified design purpose. The problem is that catching terrorists requires the design and implementation of sophisticated surveillance technology. Once in place it becomes hard not to justify its use for the detection of less serious criminal offences.

It could then be argued that it is unreasonable to adopt high-level counter-terrorist measures indiscriminately. They could only be justified for sites of major risk and risk assessments must be rigorous to determine the level of technological sophistication appropriate for different sites. Once again it is difficult to ascertain who should make that judgement call and on what grounds they could do so. Since terrorists will expect high levels of security at high-risk sites, this could end up displacing attacks to less obvious targets. And if a technology had not been implemented merely due to some abstract risk assessment – that its use might be disproportionate – it would be hard to explain this to families who have lost a loved one through terrorist action. Subject to costs limitations it would seem viable to apply the technology wherever it can be afforded and the proportionality applied in practice to what it is used for.

Ethical Consequences of Surveillance

Ethical dilemmas are rarely easily resolved in either informed judgement calls or sophisticated regulations. They have to be implemented in practice by human actors fully cognisant of the complexities involved. Moreover the law and ethics are not necessarily analogous. In some cases it might be more ethical, and in the interests of public safety, to ignore a known law – again, for example, the remaining suspect in the Boston bombings was perhaps not read his Miranda rights to encourage disclosure of further threats to public safety; or on an individual basis a police officer might not arrest a juvenile for a transgression of the law if the officer judged that an informal reprimand might produce a more durable solution. However, that is perhaps expecting too much of public transport operatives responding to emergency situations. Their ability to make the correct decisions will depend on the relevant authorities implementing a clear, informed policy, the operatives being very well-trained in the technology and even in the interpretation of human behaviour.

Additional protective measures are being embodied within the technology itself. Thus while a Bluetooth gateway has to be 'opened' by the individual, the act of opening it, in a sense, accords access permission. Once the gateway owner switches it on it is akin to consenting to whatever the technology can do, what it may announce it can do, or what it may permit to be done. So as long as the owner is 'informed' that they 'could' be tracked, this is similar to warning them that they 'might' be tracked without specifying the exact conditions under which it might occur. That is informed consent. But proposed privacy by design policies seek to embed privacy and data protection throughout the entire life cycle of surveillance technology from initial design to their deployment, use and ultimate disposal, since vulnerability remains even with discarded materials such as disposed hard drives, old account profiles, etc. (Information Commissioner's Office, 2008).

Macnish (2011) summarises comprehensively some of the ethical consequences of the growth of surveillance. Inappropriate surveillance can lead to the 'social sorting' or categorising of people based on social stereotypes with the danger of institutionalising prejudices (Lyon, 2002). Thus CCTV operators in the absence of evident suspicious behaviour might disproportionately monitor young, male and ethnic minorities out of an expectation of untoward behaviour (Norris and Armstrong 1999). If these groups are watched more frequently than others, they are more likely to be seen as doing something suspicious, thus reinforcing the stereotype and the prejudice.

Inappropriate or unjustified use of surveillance can lead to a diminution of trust in general and trust in authority in particular. Suspicion as to a State's motives for conducting surveillance may lead to cynicism as to how the State will employ its surveillance technology in self-protection. Even if there is no evidence of wrongdoing, the State may nonetheless choose to keep records on those who they believe to pose a future threat. If the records are retained they are open to use and abuse at a later stage: 'These so-called "chilling effects" are at odds with human rights and democratic practice and can lead to behavioural uniformity and a stifling of creativity. In … dictatorial regimes this may be seen as advantageous' (Macnish, 2011, section 12).

As Macnish suggests, whether consented or not, surveillance transfers power from the surveilled to the surveiller with all the inherent potential for loss of dignity, informational control and, ultimately, responsibility for one's own life. The unnatural distancing between the observer and observed disempowers the latter in terms of the possibilities of direct negotiation in normal forms of face-to-face interaction. This loss of privacy is a fundamental challenge to the democratic foundations of a civil society: freedom of thought, speech and action. It is essentially unethical, unless deemed necessary and justified for some higher moral purpose.

Issues for Further Research

From these observations and the inherent contradictions and tensions between surveillance and privacy, research, policy and practice should be approached within a comprehensive and transparent ethical framework. Only then can individuals be reassured that their privacy will be afforded some degree of protection. The overarching question remains: how might the balance of privacy with public security be best achieved? It is evident from the preceding discussion that, even from a basis in law, there is a need for flexibility in judgement and interpretation both of regulations and contexts for surveillance. Those implementing surveillance need help in answering the following questions:

- What can legitimately be regarded as 'suspicious behaviour'?
- How can individuals engaging in such behaviour be categorised and identified?
- Who makes this judgement call? (What qualifies them to do so?)
- How far is it possible to anticipate levels of threat or risk?
- If privacy is not to be preserved at all costs, then under what conditions can it be compromised?

While these constitute viable research questions there already exist many sources of evidence-based advice drawing on previous research and experience. The EC itself has funded an ethics advisory service for surveillance research within SURVEILLE, a multidisciplinary collaborative research project covering the range of ethical and legal issues (http://www.surveille.eu/). A plethora of research-based collaborative centres seek to address these concerns. The Centre for Research into

Information, Surveillance and Privacy (CRISP) offers just one example (http://www8.open.ac.uk/researchcentres/crisp/). In addition there are many interest and pressure groups seeking to challenge unwarranted encroachments into privacy, such as Privacy International, based in the UK (https://www.privacyinternational.org/), or the Electronic Privacy Information Centre in the USA (http://epic.org/).

Whatever ethical framework is adopted there are two principles I have argued elsewhere that apply here (Iphofen, 2011). One is the need to maintain a perspective of ethical pluralism. That is, no one moral theory can apply to complex security questions and a consequentialist appeal to the greater good would suggest that the security of the community is best secured by pervasive monitoring throughout a society so that rapid responses in emergencies are possible. The deontological perspective would resist overriding the rights of the few in return for the speculative safety of the many, the loss of rights being too great a price to pay and, ultimately, one that might damage the rights of the many in pursuit of a few. The second principle I espouse is the advocacy of a form of virtue ethics. Thus the only people who can really know when ethical transgressions are occurring are those employing the surveillance: the authorities and the operators. If a culture of acceptable practice can be cultivated and sustained via their training and shared awareness, there may be some degree of public reassurance.

Within the SECUR-ED Project the existence of an Ethical Advisory Group with a range of expertise has proven invaluable to warn and guide the consortium partners in their implementation and operation of surveillance technologies and systems. Most importantly this Group has maintained an ongoing negotiation with industrial partners, public authorities and private transport operators that has highlighted and anticipated ethical risks. Ultimately, this might help with the acceptable implementation of surveillance technologies that, after all, are intended to secure the safety of the travelling public. This role of ethical advisor (whether alone, or by committee) has been consistently endorsed by the EC and offers the kind of mentoring and monitoring that can support the justified and justifiable implementation of surveillance to safeguard public crowded spaces.

References

Bauman, Z. (1993). *Postmodern Ethics*. Oxford: Blackwell.

Bentham, J. (1995). *The Panopticon Writings*. London: Verso Books.

Council of Europe (1950). *Convention for the Protection of Human Rights and Fundamental Freedoms as amended by Protocols No. 11 and No. 14*. Rome: Council of Europe.

Foucault, M. (1991). *Discipline and Punish: The Birth of the Prison*. London: Penguin.

Goffman, E. (1971). *Relations in Public: Micro-studies of the Public Order*. London: Allen Lane.

Hobbes, T., G.A.J Rogers and K. Schuhmann (2006), *Leviathan (Continuum Classic Texts)*. Bristol: Thoemmes Continuum.
Information Commissioner's Office (ICO) (2008). *Privacy by Design*. Wilmslow: ICO.
Iphofen, R. (2011). *Ethical Decision Making in Social Research: A Practical Guide*. London: Palgrave Macmillan.
Johnston, I. (2009). EU funding 'Orwellian' artificial intelligence plan to monitor public for 'abnormal behaviour'. *Daily Telegraph*, 19 September.
LaFollete, H. (2002). *Ethics in Practice*, 2nd edn. Malden, MA: Blackwell.
Lyon, D. (2002). *Surveillance as Social Sorting: Privacy, Risk and Automated Discrimination*. London: Routledge.
Macnish, K. (2011). Surveillance Ethics. *Internet Encyclopedia of Philosophy*. Available from http://www.iep.utm.edu/surv-eth/ accessed 18 March 2013.
Macnish, K. (2012). Unblinking eyes: the ethics of automating surveillance. *Ethics and Information Technology*, 14(2), 151–167.
Norris, C. and Armstrong, G. (1999). CCTV and the social structuring of surveillance. In K. Painter and N. Tilley (Eds), *Surveillance of Public Space: CCTV, Street Lighting and Crime Prevention*, Crime Prevention Studies. Monsey, New York: Criminal Justice Press, 157–178.
Vardy, P. and Grosch, P. (1999). *The Puzzle of Ethics*. London: HarperCollins.
Wacks, R. (2010). *Privacy: A Very Short Introduction*. Oxford: Oxford University Press.
Warren, S.D. and Brandeis, L.D. (1890). The right to privacy. *Harvard Law Review*, 1–19.
Westin, A.F. (1967). *Privacy and Freedom*. New York: Atheneum.

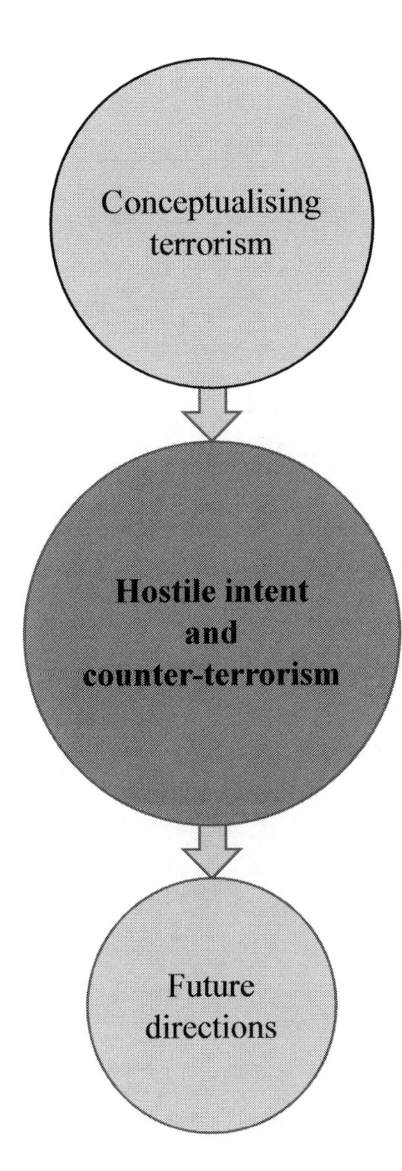

Deception and decision-making

Central to most counter-terrorism initiatives is a better understanding of hostile intent and the underlying deceptions that are employed by terrorists in order to exploit weaknesses in security. Part 2 focuses on research that can help identify cues, both verbal and non-verbal, to deception. Other challenges include understanding deception and decision-making when the research laboratory differs from the contexts in which terrorism occurs in the real world.

Chapter 6

Non-verbal Cues to Deception and their Relationship to Terrorism

Dawn L. Eubanks
Warwick Business School, University of Warwick, UK

Ke Zhang
Warwick Business School, University of Warwick, UK

Lara Frumkin
School of Psychology, University of East London, UK

Introduction

The desire to detect deception has existed as long as acts of deception, and both have been pursued across generations in an attempt to gain an advantage over another. While the precise motives and nature may differ (i.e. strategic advantage in battle or business, maintaining positive relations with a lover we have wronged, sparing a friend hurt feelings) the desire to deceive and detect this deception remains constant. For this reason, vast amounts of research time and money have been poured into this topic in order to gain a better understanding of the nature of deception (e.g. DePaulo et al., 2003; O'Brien, 2008). While we have learned a great deal, there are still many mysteries associated with deception. The main areas that have been pursued in the study of deception include verbal and nonverbal cues (including speech tone, speed, pitch, body and facial movements). The purpose of this review is to outline the progress of research related to the use of nonverbal cues in deception detection within the specific context of terrorist activities, while the following chapter deals more specifically with detecting deception from novel interview techniques. Conducting research in this area is a challenge because of the fortunate rarity of terrorist events. However, relying on our understanding of cognitive processes and extrapolating from other situations, we can begin to understand how nonverbal cues work in a terrorist domain.

In this review, we will outline four main areas:

1. Why is nonverbal behaviour important in studying deception?
2. What are the main findings about nonverbal behaviour and deception?
3. What are the challenges of studying nonverbal behaviour and deception?

4. What are the future paths for exploring research in the topic of nonverbal behaviour and deception?

The majority of interactions we have with people are truthful in nature so we have an inherent tendency to trust (DePaulo et al., 1996; DePaulo and Kashy, 1998). However, every day individuals engage in many lies that are generally harmless. Unfortunately there are some acts of deceit that can have dire consequences. Deception can be used to swindle someone out of his or her fortune, cause emotional damage or physical harm. Deception is one of the main tactics that can be used by terrorists to make attack plans 'invisible' until the actual incident (Caddell, 2004; Godson and Wirtz, 2000; Hoffman and McCormick, 2004; Jessee, 2006; Moore, 1997), and therefore requires systematic investigation from researchers. It has been contended by the Security Service (MI5) that 'terrorism presents a serious and sustained threat to the United Kingdom and UK interests abroad' (MI5, 2012). In recent years, terror attacks have become more strategic than before, and terrorists are deploying new techniques and tactics in the processes of realising their plans (Jessee, 2006; Seib and Janbek, 2011; Victoroff, 2005).

Definitions

For the purpose of this review, nonverbal behaviour includes anything that takes the form of a gesture or physical movement. Other research has included speech tone, speed or pitch as a nonverbal behaviour, but that is beyond the scope of the current review. Here we are particularly interested in what would allow an individual to detect deception purely through visual cues.

There are many ways of defining the term *deception* and many characteristics may be adapted to feature deception behaviour in relation to terrorism. Krauss (1981) and other researchers (Zuckerman, DePaulo, and Rosenthal, 1981, p. 3) have defined deception as 'an act that is intended to foster in another person a belief or understanding which the deceiver considers to be false'. Burgoon and Buller (1994, pp. 155–156) later defined deception as 'a deliberate act perpetrated by a sender to engender in a receiver beliefs contrary to what the sender believes is true to put the receiver at a disadvantage'; this emphasises the resulting disadvantage of the receiver, due to deception. More recently, Vrij (2008, p. 15) has defined deception as 'a successful or unsuccessful deliberate attempt, without forewarning, to create in another a belief which the communicator considers to be untrue'. In the context of terrorism, we would expect deception to include for example concealment, or creating an impression of being in a place for another purpose other than the true intention.

Nonverbal Behaviour, Deception and Terrorism

Before we explore deception and its relation to terrorism in greater detail, it is essential to understand what specific role nonverbal deception serves in terrorist attacks. Deception, as a deliberate activity, is employed as a tactic in terrorism activity from the pre-attack planning to the post-attack cover-up stage. It is first a crucial component of the pre-attack stage. For example, terrorists need to covertly collect and test information/material required for the attack and sometimes they use fake identities to do so (Jessee, 2006). Both covert and feigning activities are deception. Deception is not only used in hostile reconnaissance and intelligence gathering, but on the day of the execution of the attack, deception is also needed (Jessee, 2006; O'Brien, 2008). Terrorists need to use fake identities in order to pass security checks; such a deception is not easy to detect. According to the commission report of the 9/11 terrorist attacks (US National Commission on Terrorist Attacks Upon the United States, 2004), airport security workers were unable to recall anything out of the ordinary regarding the terrorists. It would appear that they looked like 'normal' passers-by. This may well be due to the fact that these terrorists, mostly trained prior to the attacks (e.g., Horgan, 2005; Shane, 2009), were skilled at using deception techniques such as concealing and falsifying their real intentions, emotions and identities on the day of the event (Porter, Juodis, ten Brinke, Klein, and Wilson, 2009). Furthermore, at the post-attack stage, deception is used if a terrorist needs to leave the scene of the attack without being identified. In the longer term, terrorists utilise deception in their signalling strategies, so as to 'enhance their freedom of action' with regard to their targets (Bowyer, 1982; Hoffman and McCormick, 2004). The above is an example of how terrorists might use deception as a mechanism at all points in planning, executing and exiting the act. Table 6.1 summarises possible deception tasks by stage of the attack.

Table 6.1 Terrorist activities that may involve deception

Stage of attack	Possible deception task
Pre-attack	Information/material collection and testing, financing, communication, travelling, locating and supporting for training.
Event execution	Passing the security operations, keeping out of the public eye, and securing weapons pre-attack.
Post-attack	Escaping from the attack scene and securing weapons post-attack if appropriate, destruction of evidence, and securing the terrorist organisation if caught.

Note. These activities that may involve deception are partially adopted from the identified crucial issues of terrorist attack proposed by (Horgan, 2005) and (Jessee, 1996).

Identifying behavioural cues and understanding their underlying cognitive or emotional processes may help to identify terrorists, from the pre-attack stage. This is considered as one of the crucial strategies that may be applied in order to detect terrorism in advance of a catastrophe, and in addition, to help people understand how terrorists may plan and carry out attacks.

In many areas where there is a potential security threat, security checkpoints will be established. This allows security officials a chance to observe behavioural and verbal cues in a systematic manner. However, frequently, if there is no checkpoint in place, for example in open spaces such as shopping centres, security officials must rely purely on visual cues to identify suspicious individuals. While there is clearly a range of expertise in observing cues based on experience and skill, it is important to understand how these individuals can assess nonverbal cues that can indicate someone may be acting in a deceptive manner – i.e. the deceiver's intentions are not what they immediately appear to be.

Behavioural Cues to Deception

Observers are often inclined to attend to nonverbal behaviour because there is an assumption that it is more difficult to control nonverbal behaviour (DePaulo, Rosenthal, Eisenstat, Rogers and Finkelstein, 1978; Hale and Stiff, 1990; Kalbfleisch, 1992; Maxwell, Cook and Burr, 1985; Stiff, Hale, Garlick and Rogan, 1990; Vrij, Dragt and Koppelaar, 1992). Therefore observers believe that it is more likely that a deceiver will 'leak' the information they are trying to hide through their nonverbal cues (Vrij, 2000).

In some instances there may not be both verbal and nonverbal cues to rely upon, but the nonverbal behaviour and speech content are discrepant. When this happens, individuals tend to emphasise nonverbal cues. For example, a job applicant that appears reserved but claims to be excited about the job will be perceived as less enthusiastic about the job than he or she reports (DePaulo et al., 1978; Hale and Stiff, 1990; Zuckerman, Driver, and Koestner, 1982; Zuckerman, Speigel, DePaulo, and Rosenthal, 1982). In sum, we are used to drawing inferences about a person based on nonverbal behaviour (Vrij et al., 2010a). This is particularly troublesome because in a recent deception detection experiment, researchers found that nonverbal cues, particularly visual ones, decrease the accuracy rate of deception detectors compared to those relying on only audio information (Burgoon, Blair, and Strom, 2008).

DePaulo and Kirkendol (1989) and Vrij (1996) identify at least four reasons why we rely more on nonverbal cues than verbal cues. First, there are automatic links between emotions and nonverbal behaviours and these automatic links do not exist between emotions and speech content. These automatic links make it difficult to remove emotion from nonverbal behaviours – thus the leaks that occur when we are under stress. Second, we have more practice in using words than behaviour to convey a message. This allows us to be more adept at changing our words rather than behaviours in order to deceive. Third, because words are more important than behaviour in conveying a message, we become more aware of the words we are using than the accompanying gestures. This focus upon words rather than behaviours allows our true feelings or intentions to leak out through our behaviours. Finally, we cannot stop the nonverbal messages being transmitted. We can stop speaking, but our body language is always conveying a message. This is an advantage for those skilled in reading nonverbal communication since there is a constant monologue to attend to.

Approaches to Deception

When trying to detect deception, the idea that there is one universal behavioural cue that people exhibit when they are lying is simply untrue (DePaulo et al., 2003; Masip, Sporer, Garrido, and Herrero, 2005; Sporer and Schwandt, 2006, 2007; Vrij, 2005). In other words there is nothing like 'Pinocchio's nose'. However, we are not without hope in trying to detect deception. According to DePaulo et al. (2003), there are 23 verbal and nonverbal indicators of deception, i.e. cues to deception, which showed significant effects in a meta-analysis study. However, many conflicting results indicate that no single cue can be confidently used to detect deception, due to the inconsistent pattern that cues present in deceivers (Vrij, 2008). There are some behaviours that are more likely to occur when an individual is being deceptive versus not deceptive (Vrij, Granhag, and Porter, 2010). It is here that we focus our attention. There are three main processes involved in conducting deception: the emotional approach, the cognitive effort approach and the attempted behavioural control approach as described in The Multi-Factor Model (Zuckerman et al., 1981) and later described by Vrij (2000, 2008). These processes have been tested by many others, and researchers have subsequently suggested supporting evidence for these processes which will be introduced in the following three sections together with discussion of how such processes relate to terrorism activity.

The Emotional Approach

There are three types of emotion associated with deception: fear, guilt and excitement (Ekman, 1985). These common aspects of emotion may occur simultaneously or separately, and contribute to the complexity of deception behaviour. Ekman and Frank (1993) have proposed a series of determinants that influence the feeling of fear during deception: the first is the liar's beliefs about the target's skill in detecting lies. Thus, liars facing lay people might feel less fear than when facing security officers, since they know that security officers are likely to be trained to detect deception. The second determinant is concerned with the liar's deception skills and preparation to deceive. When a liar is skilled or well prepared, they might feel less fear when lying, compared to those who are not skilled or well prepared. Many terrorists have experience conducting attacks (Pedahzur, Perliger, and Weinberg, 2003) or at least, involvement in related training (Bowyer, 1982; Hoffman and McCormick, 2004). This might explain why they experience less fear than other liars. Thirdly, liars' feelings about punishment are related to fear. Also individuals may feel more fear when the stakes are high (Vrij, 2000). However, even though the stakes are high for terrorists, it is rare to find ones who fear punishment (Kippenberg and Seidensticker, 2006). Nonetheless, some terrorists do fear the unsuccessful execution of their mission, and this relates to their failure of fulfilling their strong political, religious or psychological purposes (e.g., Hoffman, 2006; Victoroff and Kruglanski, 2009; Whittaker, 2007). These factors that influence the emotion of fear could, at least to some degree, influence terrorists' behaviour while engaging in deception.

Guilt is another emotion that relates to deception. In line with Ekman's (1985) proposition about guilt, perhaps deceivers who hold different social values to that of the target, for example Al-Qaeda who see the western world as the enemy, are inclined to show less guilt when being deceptive in attacks. In addition, liars feel less guilty when their targets are anonymous than when they are known (Ekman and Frank, 1993), perhaps making it easier for liars (terrorists) to operate in situations where the targets (the public) are generally anonymous. Moreover, the strength of guilt is related to the difference between the liar's gain and the target's loss (Ekman and Frank, 1993). In terms of deception in general, liars normally feel a stronger sense of guilt when a stronger disadvantage to the target is perceived. However, terrorists are inclined to regard innocent people as 'justified' targets according to their own definition of 'enemy' (Hoffman, 2006), and thus are not disturbed by the loss of their 'enemy'. These characteristics might explain why terrorists lack the presence of guilt (Hare and Cox, 1978; Post, 2007).

In spite of having negative feelings, liars may feel positive emotions such as excitement when lying. They may also feel relief and pride in their sense of achievement afterwards (Ekman, 1985; Ekman and Frank, 1993). This feeling of delight may be enhanced when liars feel challenged, or when there are audiences for their acts (Ekman, 1985; Ekman and Frank, 1993). Correspondingly, deception during terrorist attacks is a significant challenge and accomplishment for terrorists; committing the attack fulfils deception that involves a large 'audience' (the public). This might result in positive feelings while conducting terrorism-related deception.

Although there is no evidence to show that terrorism-related deception will involve fewer emotions, terrorists who are normally trained before their mission might be aware of the need to conceal signs of emotions (Horgan, 2005; Kippenberg and Seidensticker, 2006). In addition, the strength of emotional feelings and their effect on building up deception behaviour may also be moderated by other individual and contextual factors.

The Cognitive Effort Approach

Lying sometimes requires extra mental and cognitive effort (Vrij, 2008). Indeed, since liars need to formulate lies they may be preoccupied by the need to remember to play their role and they must also pay particular attention to their behaviour whilst monitoring the reaction of their targets. In addition, deliberate efforts to balance the conflict between lies and truth in an individual's mind places cognitive demands upon liars (e.g., Walczyk, Roper, Seemann, and Humphrey, 2003; Walczyk et al., 2005). This has been demonstrated in several studies (Vrij et al., 2009; Vrij et al., 2008). Specifically, cognitive complexity can lead to fewer hand and arm movements and increased gaze aversion (Ekman, 1997; Ekman and Friesen, 1972). Cognitive load is related to neglect of body language and eye contact with a conversation partner and can act as a distracter when concentrating on a cognitively demanding task. These phenomena are supported by evidence given by neuroimaging studies (e.g., Carrión, Keenan, and Sebanz, 2010; Kozel et al., 2005; also see the review in Abe, 2011).

Vrij et al. (2010b) outlined six reasons for why lying can be more cognitively demanding than truth-telling. To start, thinking of a convincing lie may be cognitively demanding. Second, because liars do not take their credibility for granted, they must expend effort in acting 'believable' (DePaulo et al., 2003; Kassin, 2005; Kassin, Appleby, and Perillo, 2010; Kassin and Gudjonsson, 2004; Kassin and Norwick, 2004). Third, because liars are concerned with appearing credible, they must monitor the reactions of those with whom they are interacting (Buller and Burgoon, 1996; Schweitzer, Brodt, and Croson, 2002). Fourth, liars have to remind themselves to maintain a façade that requires cognitive effort (DePaulo et al., 2003). Fifth, not only do liars need to generate a lie and maintain it, they must also suppress the truth from emerging (Spence et al., 2001). Finally, activating a lie requires mental effort due to the associated intentionality, while the truth generally flows forward automatically (Gilbert, 1991; Walczyk et al., 2003, 2005).

A study conducted by Carrión et al. (2010) where participants lied and told the truth with or without deceptive intentions found that the most difficult part of telling a convincing lie was in handling the cognitive conflict resulting from the need to keep alert to others' mental states and reactions while deceiving them. While most terrorism-related lies are relatively straightforward to formulate and terrorists are normally trained in advance (Jessee, 2006), a high cognitive demand on terrorists arises from the need to monitor reactions from targets, paying special attention to their behaviour, and overcoming the conflict between truth in their mind and the deception they are performing. In addition, although information from intelligence, surveillance and reconnaissance (ISR) (O'Brien, 2008) is gathered prior to the attack, terrorists may still experience cognitive load when recalling information and engaging in the attack.

A crucial factor that influences deception behaviour is that of stakes, insofar as it refers to the perceived consequences of successful and unsuccessful attempts at deception (Ekman and Frank, 1993). The fact that the stakes are high can also have an influence on cognitive load, thus influencing emotion, cognitive effort and attempted behavioural control. Although high-stakes deception happens in everyday life, the consequence of terrorists' deception is usually more severe. For example, terrorists planning threats to public security are faced with skilled lie-catchers, and they will normally perceive the stake of severe consequences if caught. For example, the high risk to religious terror groups if detected would prevent them from realising their self-sacrifice (Post, Sprinzak, and Denny, 2003) and could result in harm to their, or their family's, social status in their group (e.g., Kuznar and Lutz, 2007; Silke, 2003).

The Attempted Behavioural Control Approach

Those engaging in deception continually monitor their audience and adjust their behaviours accordingly (Burgoon and Buller, 1994). Early theories such as Interpersonal Deception Theory (IDT) (Buller and Burgoon, 1996; Burgoon and Buller, 1994) proposed that liars might adjust their behaviour by monitoring the reaction of their targets (Burgoon et al., 2008; Burgoon, Buller, and Floyd, 2001) or by monitoring their targets' suspicions. Before this point, deception was not viewed as an interaction but as an individual act by one person. It is likely that such behavioural adjustment happens more when interpersonal interaction occurs (Burgoon, Buller, Ebesu, White, and Rockwell, 1996). For this reason, dialogue rather than monologue is advantageous when the goal is to deceive. In terms of terrorist attacks, the occurrence of interpersonal interaction depends on the specific task of the deception and the stage of the attack. For example, in some cases terrorists only need to walk in the public space and make sure they are not doubted by the public or security officers (no interaction is involved), whereas sometimes they need to talk to security officers while passing the security check (interaction is involved). Moreover, the IDT perspective provides an explanation of deceivers' increase in cognitive load and how self-monitoring leads to attempted behavioural control.

Perceiving, monitoring and communicating with targets helps liars deceive successfully (e.g., Burgoon et al., 2008; Burgoon et al., 2001). Notably, in order to appear honest or normal, liars may attempt to control their behaviour during deception. Some evidence shows that liars may try to exhibit behaviours they believe are credible, such as trying to behave positively, and in a friendly fashion, to convince their targets of their honesty (DePaulo et al., 2003). However, such deliberate self-regulation sometimes makes liars seem over-controlled (Vrij, 2008). The attempted behavioural control is an essential strategy for terrorism-related deception, as terrorists normally try their best to 'blend in', so as to reduce attention from the public or security officers (Shane, 2009). There is also information regarding behavioural control that states that individuals should not show signs of confusion or tension, but to smile, be bright and confident, as seen in the Attackers' Spiritual Manual (Kippenberg and Seidensticker, 2006). It can be inferred that terrorists are seeking to give a positive impression to the receivers of their deception.

This idea of regulating behaviour was discussed in DePaulo's (1992) self-presentation theory which emphasises the possibility that in some cases, a truth-teller may experience similar cognitive processes of deception compared with a liar and regulate his or her behaviour. For example, a truth-teller may also experience the emotion of fear when they perceive the negative consequence of not being believed (Ofshe and Leo, 1997), which might lead to the presence of behaviour similar to those who are actually deceiving. In other words, there are many reasons a person may want to regulate their nonverbal behaviour. DePaulo et al.'s (2003) proposition reveals the dilemma of detecting terrorism-related deception: some innocent people may also perform deception-like behaviour when they are nervous or the stakes are high. Yet terrorists are well trained to behave as normally as possible, in order not to be detected. This complicates the process of detecting deception in the counter-terrorism field.

Through attempts to control behaviour, there have been studies that have identified reduction in hand movements in particular (Vrij, Akehurst, and Morris, 1997; Vrij and Winkel, 1991). One explanation given for this is that there is an increase in cognitive load during deception due to the fact that liars not only have to deceive the other but also have to try to promote a credible impression. A second explanation is that they realise they are under increased scrutiny, and in order not to be caught they reduce their body movements and focus on their script (Vrij, Akehurst, and Morris, 1997).

Findings and Challenges

Out of all the nonverbal cues that have been identified, only a handful consistently emerged across studies. In fact, in many instances there are blatant contradictions in what nonverbal cues indicate deception (Miller and Stiff, 1992). This may be because each individual differs in how they express themself when lying or anxious, for example (Miller and Stiff, 1992). It is important to state that there are behaviours that *may* indicate that an individual is being deceptive, but they do not necessarily mean that they are lying. It may mean that an individual realises that they look suspicious and this creates a degree of discomfort and unnaturalness about them (Vrij, 2000). Also, many of the cues identified as being related to deception will only be displayed if an individual is experiencing fear, guilt or excitement. If the lie is easy to fabricate then the behavioural cues to deception are unlikely to be displayed. It is for these reasons that detecting deception is incredibly difficult.

Having said that, there is some hope in identifying deceptive individuals. A meta-analysis including 120 studies and 158 cues to deception demonstrated that most behaviours are only weakly related to deception, if at all (DePaulo et al., 2003; see also DePaulo and Morris, 2004). Rather than looking for specific cues to deception, general impressions of a person are stronger predictors of accurate deception judgement (Hartwig and Bond, 2011). In general, compared to truth-tellers, deceivers appear to be more tense, having a higher pitched voice, presenting increased pauses and longer latency periods while lying, and decreases in hand and finger movements, leg and foot movements, and illustrators (hand and arm movements supporting speech) (Vrij, 2008). Deceivers may also respond faster and in a less straightforward way to the content of their statements, and may also provide more negative statements (Vrij, 2008).

Training to detect deceptive behaviour appears to be more effective when it focuses upon more global impressions such as looking for more tension (specifically more fidgeting) than when it does not. Similarly, looking for impressions of 'less sense' (less logical, more discrepant); 'less fluent' (more speech disturbances/word repetitions); and 'less forthcoming' (shorter speaking duration, fewer details) improved training effectiveness, although these differences are not significant. However, there is little difference in training effectiveness for the cue less engaged (i.e. fewer illustrators). Taking the sum of these findings, this may mean that overall perceptions are more accurate predictors of deception than a specific nonverbal behaviour (Driskell, 2011).

Some evidence has demonstrated differences in emotional displays shown by liars and truth-tellers (Ekman and O'Sullivan, 2006). For example, differences in micro-expressions have been investigated to distinguish liars and truth-tellers through these 'leaked emotions'. These micro-expressions are difficult to detect and often need to rely on slowed-down visual displays in order to identify them.

As mentioned above, the occurrence of deception cues depends on the experience of cognitive processes: thus the context in which a liar deceives will directly influence their behaviour. Cues to deception are more likely to be present when individuals are motivated to deceive than when there is little or no consequence attached to their performance (DePaulo et al., 2003; Driskell, 2011; Levine, Feeley, McCornack, Hughes, and Harms, 2005). Accordingly, terrorism-related deceptions might show verbal and nonverbal cues in terms of stress or cognitive effort and might also feature more behavioural control when compared with deception in general. In addition, terrorists are trained to behave in a 'normal' way, so as to arouse less attention. This may also result in more controlled behaviour. Detecting terrorists is not as easy as spotting suspicious individuals from the general public by following a checklist of cues to deception. Similar behaviour from both liars and truth-tellers further increases the difficulty of spotting liars, and terrorists are even harder to detect on account of their apparent normality.

Knowing that a reliable cue to deception does not exist presents us with a much more complex puzzle embedded in individual differences and contextual nuances. However, it is still important to improve our ability to detect deception. The question is if there are no reliable cues to distinguish between truth-tellers and liars, what is the purpose of this training? In fact, some authors have pointed out that deception detection training may be misleading and counter-productive if it is focused upon cues that do not actually distinguish between liars and truth tellers (Kassin and Fong, 1999). Because of the inherent weaknesses in the predictive nature of behavioural cues to deception, it may be more pertinent to focus upon intuitive notions about deception. It may be that increasing the behavioural differences between liars and truth-tellers is a better way to arm security officials to detect deception, rather than looking for specific cues to deception (Hartwig and Bond, 2011).

Stereotypes

As Vrij et al. (2010b) state in their review article, there are many stereotypes that people hold about the nonverbal cues that a liar will exhibit. Some common beliefs are that liars will act nervously, exhibit gaze aversion ('liars look away') and display grooming gestures ('liars fidget') (Strömwall, Granhag, and Hartwig, 2004; Taylor and Hick, 2007; The Global Deception Team, 2006; Vrij, 2008; Vrij, Akehurst, and Knight, 2006). Some evidence suggests that gaze aversion may be related to a way to manage cognitive load. For example, an experiment with 8-year-old children found that they did avert their gaze when engaged in pedagogical question and answer sessions. This may have been related to social factors. However, the main function found for averting gaze is in cognitive load management when processing information (Doherty-Sneddon and Phelps, 2005). The challenge here is to identify when a person is averting gaze because they are concentrating on something honest versus engaging in deceit. A problem with using cognitive load as a proxy for identifying deception is that not every person engaging in deceit will experience cognitive load in the same way. Research has shown that these cues commonly thought to be associated with lying are in fact not related. In particular, liars do not seem to exhibit the behavioural cues of gaze aversion and fidgeting (Vrij and Mann, 2001). For more details about people's beliefs about behavioural cues to deception see DePaulo (1992); DePaulo, Stone, and Lassiter (1985); Vrij, (1998); Zuckerman Driver (1985) and Zuckerman, Driver, and Guadagno (1985).

It appears that multiple sources of information can assist in identifying when an individual is engaging in deception. In fact, the best-validated cues to deception are verbal and nonverbal cues to be considered together that include illustrators, blink and pause rate, speech rate, vague descriptions, repeated details, contextual embedding, reproduction of conversations, and emotional 'leakage' in the face (Porter and ten Brinke, 2010). In a study by Blair, Levine, and Shaw (2010), they found that the mean accuracy of groups receiving contextual information was 21 per cent higher than the accuracies reported in the Bond and DePaulo (2006) meta-analysis. This clearly demonstrates the importance of context when trying to determine whether a person is acting in a deceptive manner. Perhaps once we clearly move away from the search for a reliable behavioural cue we will begin to open our eyes to the complexity involved in detecting deception.

It is a combination of science and intuitions grounded in experience and general impressions that may help us get closer to detecting a deceptive individual.

Conclusion

In conclusion, identifying nonverbal cues to deception is clearly an important but challenging task. While we understand that there is no one reliable cue indicating deception, we need to delve into the complex task of identifying pairings of cues (verbal and nonverbal). We have come a long way in understanding what drives a person to display certain cues when being deceptive, however this is unreliable because these cues are also present in other situations (e.g. when someone is anxious, nervous, lost). Future studies need to continue down the path of exploring combinations of cues to see if we can pinpoint a deceptive 'look' more generally rather than looking for one particular cue. When we begin to understand what drives these intuitions about a person being deceptive or not, we can once again begin to put science back into deception detection rather than appearing to simply rely on one's gut.

Acknowledgment

This research is conducted as part of the project 'Shades of Grey – Towards a Science of Interventions for Eliciting and Detecting Notable Behaviours' project (EPSRC reference: EP/H02302X/1).

References

Abe, N. (2011). How the brain shapes deception. *The Neuroscientist*, 17(5), 560–574.

Blair, J.P., Levine, T.R., and Shaw, A.S. (2010). Content in context improves deception detection accuracy. *Human Communication Research*, 36(3), 423–442.

Bond, C.F., and DePaulo, B.M. (2006). Accuracy of deception judgments. *Personality and Social Psychology Review*, 10(3), 214–234.

Bowyer, J. (1982). *Cheating: Deception in War and Magic, Games and Sports, Sex and Religion, Business and Congames, Politics and Espionage, Art and Science*. New York: St Martin Press.

Buller, D.B., and Burgoon, J.K. (1996). Interpersonal deception theory. *Communication Theory*, 6, 203–242.

Burgoon, J.K., and Buller, D.B. (1994). Interpersonal deception: III. Effects of deceit on perceived communication and nonverbal dynamics. *Journal of Nonverbal Behavior*, 18, 155–184.

Burgoon, J.K., Blair, J.P., and Strom, R.E. (2008). Cognitive biases and nonverbal cue availability in detecting deception. *Human Communication Research*, 34, 572–599.

Burgoon, J.K., Buller, D.B., and Floyd, K. (2001). Does participation affect deception success? *Human Communication Research*, 27(4), 503–534.

Burgoon, J.K., Buller, D.B., Ebesu, A.S., White, C.H., and Rockwell, P.A. (1996). Testing interpersonal deception theory: Effects of suspicion on communication behaviors and perceptions. *Communication Theory*, 6(3), 243–267.

Caddell, J.W. (2004). *Deception 101: Primer on Deception.* Carlisle Barracks, PA: Strategic Studies Institute, U.S. Army War College.

Carrión, R.E., Keenan, J.P., and Sebanz, N. (2010). A truth that's told with bad intent: An ERP study of deception. *Cognition*, 114(1), 105–110.

DePaulo, B.M. (1992). Nonverbal behavior and self-presentation. *Psychological Bulletin*, 111(2), 203–243.

DePaulo, B.M., and Kashy, D.A. (1998). Everyday lies in close and casual relationships. *Journal of Personality and Social Psychology*, 74(1), 63–79.

DePaulo, B.M., and Kirkendol, S.E. (1989). The motivational impairment effect in the communication of deception. In J.C. Yuille (Ed.), *Credibility Assessment.* Dordrecht: Kluwer, 51–70.

DePaulo, B.M., and Morris, W.L. (2004). Discerning lies from truths: Behavioural cues to deception and the indirect pathway of intuition. In P.A. Granhag and L.A. Strömwall (Eds), *Deception Detection in Forensic Contexts*. Cambridge: Cambridge University Press, 15–40.

DePaulo, B.M., Kashy, D.A., Kirkendol, S.E., Wyer, M.M., and Epstein, J.A. (1996). Lying in everyday life. *Journal of Personality and Social Psychology*, 70(5), 979–979.

DePaulo, B.M., Lindsay, J.J., Malone, B.E., Muhlenbruck, L., Charlton, K., and Cooper, H. (2003). Cues to deception. *Psychological Bulletin*, 129(1), 74–118.

DePaulo, B.M., Rosenthal R., Eisenstadt R.A., Rogers P.L., and Finkelstein, S. (1978). Decoding discrepant nonverbal cues. *Journal of Personality and Social Psychology*, 36, 313–323.

DePaulo, B.M., Stone, J.I., and Lassiter, G.D. (1985). Telling ingratiating lies: Effects of target sex and target attractiveness on verbal and nonverbal deceptive success. *Journal of Personality and Social Psychology*, 48, 1191–1203.

Doherty-Sneddon, G., and Phelps, F.G. (2005). Gaze aversion: A response to cognitive or social difficulty? *Memory and Cognition*, 33(4), 727–733.

Driskell, J.E. (2011). Effectiveness of deception detection training: A meta-analysis. *Psychology, Crime and Law*, 18(8), 1–19.

Ekman, P. (1985). *Telling Lies: Clues to Deceit in the Marketplace, Politics, and Marriage*. New York: Norton.

Ekman, P. (1997). Deception, lying, and demeanor. In D.F. Halpern and A.E. Voiskounsky (Eds) *States of mind: American and post-soviet perspectives on contemporary issues in psychology*. New York, NJ: Oxford University Press, 93–105.

Ekman, P., and Frank, M. (1993). Lies that fail. In M. Lewis and C. Saarni (Eds), *Lying and Deception in Everyday Life*. New York: The Guilford Press, 184–200.

Ekman, P., and Friesen, W.V. (1972). Hand movements. *Journal of Communication*, 22, 353–374.

Ekman, P., and O'Sullivan, M. (2006). From flawed self-assessment to blatant whoppers: The utility of voluntary and involuntary behavior in detecting deception. *Behavioral Sciences and the Law*, 24, 673–686.

Global Deception Team, The (2006). A world of lies. *Journal of Cross-Cultural Psychology*, 37(1), 60–74.

Godson, R., and Wirtz, J.J. (2000). Strategic denial and deception. *International Journal of Intelligence and Counter Intelligence*, 13(4), 424–437.

Guilbert, D.T. (1991). How mental systems believe. *American Psychologist*, 46, 107–119.

Hale, J.L., and Stiff, J.B. (1990). Nonverbal primacy in veracity judgments. *Communication Reports*, 3, 75–83.

Hare, R.D., and Cox, D.N. (1978). Clinical and empirical conceptions of psychopathy, and the selection of subjects for research. In R.D. Hare and D. Schalling (Eds), *Psychopathic Behaviour: Approaches to Research*. Chichester: Wiley, 1–21.

Hartwig, M., and Bond Jr, C.F. (2011). Why do lie-catchers fail? A lens model meta-analysis of human lie judgments. *Psychological Bulletin*, 137(4), 643–659.

Hoffman, B. (2006). *Inside Terrorism*, Revised and expanded edn. New York: Columbia University Press.

Hoffman, B., and McCormick, G.H. (2004). Terrorism, signaling, and suicide attack. *Studies in Conflict and Terrorism*, 27(4), 243–281.

Horgan, J. (2005). *The Psychology of Terrorism*. London: Routledge.

Jessee, D. (2006). Tactical means, strategic ends: Al Qaeda's use of denial and deception. *Terrorism and Political Violence*, 18(3), 367–388.

Kalbfleisch, P.J. (1992). Deceit, distrust and the social milieu: Application of deception research in a troubled world. *Journal of Applied Communication Research*, 20(3), 308–334.

Kassin, S.M. (2005). On the psychology of confessions: Does innocence put innocents at risk? *American Psychologist*, 60, 215–228.

Kassin, S.M., and Fong, C.T. (1999). 'I'm innocent!': Effects of training on judgments of truth and deception in the interrogation room. *Law and Human Behavior*, 23, 499–516.

Kassin, S.M., and Gudjonsson, G.H. (2004). The psychology of confessions: A review of the literature and issues. *Psychological Science in the Public Interest*, 5, 33–67.

Kassin, S.M., and Norwick, R.J. (2004). Why people waive their Miranda rights: The power of innocence. *Law and Human Behavior*, 28, 211–221.

Kassin, S.M., Appleby, S.C., and Perillo, J.T. (2010). Interviewing suspects: Practice, science, and future directions. *Legal and Criminological Psychology*, 15(1), 39–55.

Kippenberg, H.G., and Seidensticker, T. (2006). *The 9/11 Handbook: Annotated Translation and Interpretation of the Attackers' Spiritual Manual*. London: Equinox.

Kozel, F., Johnson, K., Mu, Q., Grenesko, E., Laken, S., and George, M. (2005). Detecting deception using functional magnetic resonance imaging. *Biological Psychiatry*, 58(8), 605–613.

Krauss, R.M. (1981). Impression formation, impression management, and nonverbal behaviors. In E.T. Higgins, C.P. Herman, and M.P. Zanna (Eds), *Social Cognition: The Ontario Symposium, Vol. 1*. Hillsdale, NJ: Erlbaum, 323–341.

Kuznar, L., and Lutz, J. (2007). Risk sensitivity and terrorism. *Political Studies*, 55, 341–361.

Levine, T.R., Feeley, T.H., McCornack, S.A., Hughes, M., and Harms, C.M. (2005). Testing the effects of nonverbal behaviour training on accuracy in deception detection with the inclusion of a bogus training control group. *Western Journal of Communication*, 69, 203–217.

Masip, J., Sporer, S.L., Garrido, E., and Herrero, C. (2005). The detection of deception with the reality monitoring approach: A review of the empirical evidence. *Psychology, Crime and Law*, 11(1), 99–122.

Maxwell, G.M., Cook, M.W., and Burr, R. (1985). The encoding and decoding of liking from behavioural cues in both auditory and visual channels. *Journal of Nonverbal Behavior*, 9, 239–264.

MI5. (2012). *Threat Levels*. Retrieved from https://www.mi5.gov.uk/home/the-threats/terrorism/threat-levels.html, accessed 20 August 2014.

Miller, G.R., and Stiff, J.B. (1992). Applied issues in studying deceptive communication. In R.S. Feldman (Ed.), *Applications of Nonverbal Behavioral Theories and Research*. Hillsdale, NJ: Lawrence Erlbaum Associates, 217–237.

Moore, J.N. (1997). *Deception and Deterrence in 'Wars of National Liberation': State-sponsored Terrorism and Other Forms of Secret Warfare*. Durham, NC: Carolina Academic Press.

O'Brien, K.A. (2008). Assessing hostile reconnaissance and terrorist intelligence activities. *The RUSI Journal*, 153(5), 34–39.

Ofshe, R.J., and Leo, R.A. (1997). The decision to confess falsely: Rational choice and irrational action. *Denver University Law Review*, 74, 979–1112.

Pedahzur, A., Perliger, A., and Weinberg, L. (2003). Altruism and fatalism: The characteristics of Palestinian suicide terrorists. *Deviant Behavior*, 24(4), 405–423.

Porter, S., and ten Brinke, L. (2010). The truth about lies: What works in detecting high-stakes deception? *Legal and Criminological Psychology*, 15(1), 57–75.

Porter, S., Juodis, M., Leanne, M., Klein, R., and Wilson, K. (2009). Evaluation of the effectiveness of a brief deception detection training program. *Journal of Forensic Psychiatry and Psychology*, 21(1), 1–11.

Post, J.M. (2007). *The Mind of the Terrorist: The Psychology of Terrorism from the IRA to Al-Qaeda*. Basingstoke: Palgrave Macmillan.

Post, J.M., Sprinzak, E., and Denny, L. (2003). The terrorists in their own words: Interviews with 35 incarcerated Middle Eastern terrorists. *Terrorism and Political Violence*, 15(1), 171–184.

Schweitzer, M.E., Brodt, S.E., and Croson, R.T.A. (2002). Seeing and believing: Visual access and the strategic use of deception. *International Journal of Conflict Management*, 13(3), 258–375.

Seib, P.M., and Janbek, D.M. (2011). *Global Terrorism and New Media: The post-Al Qaeda Generation*. London: Routledge.

Shane, J.M. (2009). September 11th terrorist attacks on the United States and the law enforcement response. In M.R. Haberfeld and A.V. Hassell (Eds), *A New Understanding of Terrorism: Case Studies, Trajectories and Lessons Learned*. New York: Springer, 99–142.

Silke, A.P. (2003). *Terrorists, Victims and Society: Psychological Perspectives on Terrorism and its Consequences*. Chichester: Wiley.

Spence, S.A., Farrow, T.F.D., Herford, A.E., Wilkinson, I.D., Zheng, Y., and Woodruff, P.W.R. (2001). Behavioural and functional anatomical correlates of deception in humans. *Neuroreport: For Rapid Communication of Neuroscience Research*, 12, 2849–2853.

Sporer, S.L., and Schwandt, B. (2006). Paraverbal indicators of deception: A meta-analytic synthesis. *Applied Cognitive Psychology*, 20, 421–446.

Sporer, S.L., and Schwandt, B. (2007). Moderators of nonverbal correlates of deception: A meta-analytic synthesis. *Psychology, Public Policy, and Law*, 13, 1–34.

Stiff, J.B., Hale, J.L., Garlick, R., and Rogan, R. (1990). Effect of cue incongruence and social normative influences on individual judgments of honesty and deceit. *The Southern Communication Journal*, 55, 206–229.

Strömwall, L.A., Granhag, P.A., and Hartwig, M. (2004). Practitioners' beliefs about deception. In P.A. Granhag and L.A. Strömwall (Eds), *Deception Detection in Forensic Contexts*. Cambridge: Cambridge University Press, 229–250.

Taylor, R., and Hick, R.F. (2007). Believed cues to deception: Judgements in self-generated serious and trivial situations. *Legal and Criminological Psychology*, 12, 321–332.

U.S. National Commission on Terrorist Attacks Upon the United States. (2004) *The 9/11 Commission Report: Final Report of The National Commission on Terrorist Attacks upon The United States. Authorized edn.* New York: W. W. Norton.

Victoroff, J.I. (2005). The mind of the terrorist. *Journal of Conflict Resolution*, 49, 3–42.

Victoroff, J.I., and Kruglanski, A.W. (2009). *Psychology of Terrorism: The Best Writings about the Mind of the Terrorist.* New York: Psychology Press.

Vrij, A. (1996). Police officers' aggression in confrontations with offenders as a result of physical effort. In N.K. Clark and G.M. Stephenson (Eds), *Issues in Criminological and Legal Psychology: Psychological Perspectives on Police and Custodial Culture and Organisation.* Leicester: British Psychology Society, 59–65.

Vrij, A. (1998). Nonverbal communication and credibility. In A. Memon, A. Vrij and R. Bull (Eds), *Accuracy and Perceived Credibility of Victims, Witnesses, and Suspects.*Maidenhead: McGraw-Hill, 32–53.

Vrij, A. (2005). Criteria-based content analysis: A qualitative review of the first 37 studies. *Psychology, Public Policy, and Law*, 11, 3–41.

Vrij, A. (2008). *Detecting Lies and Deceit: Pitfalls and Opportunities*, 2nd edn. Chichester: Wiley.

Vrij, A. (2000). *Detecting Lies and Deceit.* Chichester, England: Wiley

Vrij, A., and Mann, S. (2001). Telling and detecting lies in a high-stake situation: The case of a convicted murderer. *Applied Cognitive Psychology*, 15, 187–203.

Vrij, A., and Winkel, F.W. (1991). Cultural patterns in Dutch and Surinam nonverbal behavior: A analysis of simulated police/citizen encounters. *Journal of Nonverbal Behavior*, 15, 169–184.

Vrij, A., Akehurst, L., and Knight, S. (2006). Police officers', social workers', teachers' and the general public's beliefs about deception in children, adolescents and adults. *Legal and Criminological Psychology*, 11 (2), 297–312.

Vrij, A., Akehurst, L., and Morris, P. (1997). Individual differences in hand movements during deception. *Journal of Nonverbal Behavior*, 21(2), 87–102.

Vrij, A., Dragt, A., and Koppelaar, L. (1992). Interviews with ethnic interviewees: Nonverbal communication errors in impression formation. *Journal of Community and Applied Social Psychology*, 2(3), 199–208.

Vrij, A., Mann, S., Fisher, R., Leal, S., Milne, R., and Bull, R. (2008). Increasing cognitive load to facilitate lie detection: The benefit of recalling an event in reverse order. *Law and Human Behavior*, 32(3), 253–265.

Vrij, A., Leal, S., Granhag, P.A., Mann, S., Fisher, R.P., Hillman, J., and Sperry, K. (2009). Outsmarting the liars: The benefit of asking unanticipated questions. *Law and Human Behavior*, 33(2), 159–166.

Vrij, A., Leal, S., Mann, S., Warmelink, L., Granhag, P.A., and Fisher, R.P. (2010a). Drawings as an innovative and successful lie detection tool. *Applied Cognitive Psychology*, 4, 587–594.

Vrij, A., Granhag, P.A., and Porter, S.B. (2010b). Pitfalls and opportunities in nonverbal and verbal lie detection. *Psychological Science in the Public Interest*, 11, 89–121.

Walczyk, J.J., Roper, K.S., Seemann, E., and Humphrey, A.M. (2003). Cognitive mechanisms underlying lying to questions: Response time as a cue to deception. *Applied Cognitive Psychology*, 17(7), 755–774.

Walczyk, J.J., Schwartz, J.P., Clifton, R., Adams, B., Wei, M., and Zha, P. (2005). Lying person-to-person about life events: A cognitive framework for lie detection. *Personnel Psychology*, 59(1), 141–170.

Whittaker, D.J. (2007). *The Terrorism Reader*, 3rd edn. London: Routledge.

Zuckerman, M., DePaulo, B.M., and Rosenthal, R. (1981). Verbal and nonverbal communication of deception. In L. Berkowitz (Ed.), *Advances in Experimental Social Psychology, Vol. 14*. New York: Academic Press, 1–57.

Zuckerman, M., Driver, R., and Guadagno, N.S. (1985). Effects of segmentation patterns on the perception of deception. *Journal of Nonverbal Behavior*, 9, 160–168.

Zuckerman, M., Driver, R., and Koestner, R. (1982). Discrepancy as a cue to actual and perceived deception. *Journal of Nonverbal Behavior*, 7, 95–100.

Zuckerman, M., Speigel, N.H., DePaulo, B.M., and Rosenthal, R. (1982). Nonverbal strategies for decoding deception. *Journal of Nonverbal Behavior*, 6, 171–187.

Chapter 7
Deception Detection in Counter-terrorism

Aldert Vrij, Sharon Leal and Samantha Mann
Psychology Department, University of Portsmouth, UK

Introduction

One important aspect of counter-terrorism is gathering information for intelligence. Such information often comes from interviewing individuals. Loftus (2011) argued that in interviewing individuals for intelligence purposes investigators need to consider memory distortion due to poor questioning or post-event information. In addition, Loftus said, they need to avoid using oppressive interview techniques that could result in false confessions. Those two concerns can be addressed by avoiding interview techniques that are known to increase the likelihood of memory distortion (e.g., asking leading questions) or occurrence of false confessions (e.g., oppressive interview techniques). A third aspect of concern to investigators is that they may be deliberately misled and therefore lied to by interviewees (Loftus, 2011). This requires a different, more difficult, approach. Rather than *avoiding* certain interview techniques, sufficient to address memory distortion and false confessions, investigators need *to implement* specific interview techniques that increase their chances to detect deceit. In the first part of this chapter we will discuss three such techniques: imposing cognitive load, asking unanticipated questions, and using evidence in a strategic manner.

Most forensic lie detection research to date concentrates on police/suspect interviews. These interviews differ in some important aspects from counter-terrorism interviews. For example, police/suspect interviews typically concentrate on someone's past activities, whereas intelligence interviews often focus on someone's future activities (intentions) or opinions. In addition, in police/suspect interviews suspects are typically interviewed individually, but in some counter-terrorism settings it may be useful to interview them collectively. Finally, police/suspect interviews are formal and the suspects are aware that they are being interviewed. In counter-terrorism it could be useful to interview suspects without them being aware that they are interviewed (undercover interviewing) or interviewing may not even be possible, giving investigators no option other than to observe suspects. In the second part of this chapter we will discuss these aspects in more detail together with the findings of the emerging lie detection research in those areas.

In interviews investigators can detect lies in three different ways. They can (i) observe someone's behaviour, (ii) analyse their speech, or (iii) measure their physiological responses (heart rate, blood pressure, galvanic response, P300 brain wave, brain activity). Equipment is required to measure physiological responses, whereas this is not the case in nonverbal and verbal lie detection. Therefore, nonverbal and verbal lie detection is easier to apply and can be applied in more situations. In this chapter we will focus on nonverbal and verbal lie detection.

Interviewing to Detect Deception

Five decades of lie-detection research have shown that people's ability to detect deception by observing behaviour and listening to speech is limited with, on average, 54 per cent of truths and lies being correctly classified (Bond and DePaulo, 2006). The problem is that cues to deception are typically faint and unreliable (DePaulo et al., 2003). One reason is that the underlying theoretical explanations for why such cues occur – nervousness and cognitive load – also apply to truth-tellers. That is, both liars and truth-tellers can be afraid of being disbelieved and may have to think hard when providing a statement. Can interviewers ask questions that actively elicit and amplify verbal and nonverbal cues to deceit? Efforts in the past, e.g., the Behavior Analysis Interview (Inbau et al., 2001) have concentrated on eliciting and amplifying *emotions* (Vrij, 2008), but it is doubtful whether questions can be asked that will necessarily raise more anxiety in liars than in truth-tellers (National Research Council, 2003). Someone may even argue that there is hardly anything that produces more anxiety than having to address false allegations.

We will demonstrate, however, that it is possible to ask questions that raise *cognitive load* more in liars than in truth-tellers, and we will discuss three ways to achieve this: imposing cognitive load, asking unanticipated questions, and using evidence in a strategic manner.

Imposing Cognitive Load

Lying can be more cognitively demanding than truth-telling (Vrij et al., 2008). First, formulating the lie may be cognitively demanding. A liar needs to invent a story and must monitor their fabrication so that it is plausible and adheres to everything the observer(s) know or might find out. Moreover, liars must remember what they have said to whom in order to maintain consistency. Liars should also refrain from providing new leads (Vrij, 2008). Second, liars are typically less likely than truth-tellers to take their credibility for granted (Kassin, Appleby, and Torkildson-Perillo, 2010). As such, liars will be more inclined than truth-tellers to monitor and control their demeanour in order to appear honest to the investigator, and such monitoring and controlling is cognitively demanding. Third, because liars do not take credibility for granted, they may monitor the *investigator's* reactions carefully in order to assess whether they appear to be getting away with their lie (Buller and Burgoon, 1996), which requires cognitive resources. Fourth, liars may be preoccupied with the task of reminding themselves to role-play (DePaulo et al.,

2003), which requires extra cognitive effort. Fifth, liars also have to suppress the truth while they are fabricating and this is also cognitively demanding (Spence et al., 2001). Finally, while activation of the truth often happens automatically, activation of the lie is more intentional and deliberate (Walczyk et al., 2003), and thus requires mental effort.

An investigator could exploit the differential levels of cognitive load that truth-tellers and liars experience to discriminate more effectively between them. Liars who require more cognitive resources than truth-tellers will have fewer cognitive resources left over. If cognitive demand is further raised, which could be achieved by making additional requests, liars may not be as good as truth-tellers in coping with these additional requests (Vrij et al., 2011b; Vrij, Granhag, and Porter, 2010c).

One way to impose cognitive load is by asking interviewees to tell their stories in reverse order. This increases cognitive load because (a) it runs counter to the natural forward-order coding of sequentially occurring events, and (b) it disrupts reconstructing events from a schema (Gilbert and Fisher, 2006). Another way to increase cognitive load is by instructing interviewees to maintain eye contact with the interviewer. When people have to concentrate on telling their stories – for example when asked to recall what has happened – they are inclined to look away from their conversation partner (typically to a motionless point), because maintaining eye contact is distracting (Doherty-Sneddon and Phelps, 2005). In one experiment, half of the liars and truth-tellers were requested to recall their stories in reverse order (Vrij et al., 2008) and in another experiment, half were asked to maintain eye contact with the interviewer (Vrij et al., 2010a). In both experiments no instruction was given to the other half of participants. More cues to deceit emerged in the reverse order and maintaining eye contact conditions than in the control conditions. In addition, observers who watched these videotaped interviews could distinguish between truths and lies better in the reverse order condition and maintaining eye contact conditions than in the control conditions. For example, in the reverse order experiment, 42 per cent of the lies were correctly classified in the control condition, well below that typically found in verbal and nonverbal lie-detection research, suggesting that the lie-detection task was difficult. Yet, in the experimental condition, 60 per cent of the lies were correctly classified, which is more than typically found in this type of lie-detection research.

An alternative way to impose cognitive load on liars is to ensure that in a given interview setting truth-tellers will provide more information. Talkative truth-tellers raise the standard for liars, who also need to become more talkative to match truth-tellers. Liars may find it too cognitively difficult to add as many details as truth-tellers, or if they do add a sufficient amount of detail the additional information may be of lesser quality or may sound less plausible. We recently successfully tested two ways of increasing the amount of detail truth-tellers generate. In one experiment two interviewers were used (Mann et al., 2013). The second interviewer was silent but showed different demeanours during the interview. In one condition he was supportive throughout (e.g., nodding his head and smiling); in a second condition he was neutral and in a third condition he was suspicious (e.g., frowning). Being supportive during an interview facilitates talking and encourages cooperative witnesses (e.g., truth-tellers) to talk (Memon, Meissner, and Fraser, 2010; Bull, 2010; Fisher, 2010). Indeed, truth-tellers provided most detail in the supportive condition and only in that condition did they provide significantly more detail than liars (Mann et al., 2013). In a second experiment, half of the participants were primed and were asked before being interviewed to listen to an audiotape in which someone gave a detailed account of an event unrelated to the participant's interview. Participants were informed that the purpose of the priming audiotape was to give them an idea of what a detailed account actually entails. The underlying assumption of the audiotape was that if participants hear a model of a detailed answer, they are more likely to provide a more detailed answer themselves. Interviewees' expectations about how much detail is expected from them could be inadequate. Related research shows that if conversation partners do not know each other well they tend to give short answers (Fisher, 2010; Fisher, Milne and Bull, 2011). It may be that investigators can alter participants' expectations about how much detail is required by providing them with a model answer. Indeed, although truth-tellers and liars did not differ from each other in the non-primed condition, they did so in the primed condition; furthermore, primed truth-tellers gave more detailed answers that also sounded more plausible (Leal et al., 2014).

In sum, imposing cognitive load can be achieved in two different ways. First, by using interventions that increase the difficulty to recall information (reverse order and maintaining eye contact), and, second, by using interventions that make examinees more talkative.

Asking Unanticipated Questions

A consistent finding in deception research is that liars prepare themselves when anticipating an interview (Hartwig, Granhag, and Strömwall, 2007). This strategy makes sense: planning makes lying easier, and planned lies typically contain fewer cues to deceit than do spontaneous lies (DePaulo et al., 2003). However, the positive effects of planning will only emerge if liars correctly anticipate which questions will be asked. Investigators can exploit this limitation by asking questions that liars do not anticipate. Though liars can refuse to answer unanticipated questions, such 'I don't know' or 'I can't remember' responses will create suspicion and should therefore be avoided if the questions are about central (but unanticipated) aspects of the target event. To test the unanticipated questions technique, pairs of liars and truth-tellers were interviewed individually about an alleged visit to a restaurant (Vrij et al., 2009). The conventional opening questions (e.g., 'What did you do in the restaurant?') were anticipated, whereas the request to sketch the layout of the restaurant was not. (Anticipation was established with the interviewees after the interview.) Based on the overlap (similarity) in the two pair members' drawings, 80 per cent of the liars and truth-tellers were classified correctly as the drawings were less alike for the pairs of liars than pairs of truth-tellers. However, on the basis of the conventional questions the pairs were not classified above chance level. A difference in overlap between anticipated and unanticipated questions further indicated deceit. Pairs of truth-tellers showed the same amount of overlap in their answers to the anticipated and unanticipated questions whereas liars did not. The truth-tellers also showed significantly more overlap in their answers to the anticipated questions than in their answers to the unanticipated questions.

Comparing the answers to anticipated and unanticipated questions can also be used to detect deceit in individual liars, as two recent experiments demonstrated. In the first experiment truth-tellers and liars were interviewed about their alleged activities in a room (Lancaster et al., 2012). Expected questions (e.g., 'Tell me in as much detail as you can what you did in the room?') were followed by unexpected spatial and temporal questions (e.g., 'In relation to the front door, where did you and your friend sit?', 'Who finished their food first, you or your friend?'). In the second experiment truth-tellers and liars were interviewed about their alleged forthcoming trip (Warmelink et al., 2012). Expected questions about the purpose of the trip (e.g., 'What is the main purpose of your trip?'), were followed by unexpected questions about transport (e.g., 'How are you going to travel to your destination?'), planning ('What part of the trip was easiest to plan?'), and the core event ('Keep in mind an image of the most important thing you are going to do at this trip. Please describe this mental image in detail'). Liars are likely to have prepared answers to the expected questions and may therefore be able to answer them in considerable detail. Liars will not have prepared answers for the unexpected questions and may therefore struggle to generate detailed answers to them. Indeed, in both experiments, compared to truth-tellers, liars gave significantly more detail to the expected questions and significantly less detail to the unexpected questions. This resulted in a larger decline in detail between anticipated and unanticipated answers in liars than in truth-tellers.

Another effective way to use the unanticipated questions technique when assessing individuals is asking the same question twice in different formats (Leins et al., 2011). When liars have not anticipated the question, they have to fabricate an answer on the spot. A liar's memory of this fabricated answer may be more unstable than a truth-teller's actual memory of the event. Therefore, liars may contradict themselves more than truth-tellers. This approach works best if the questions are asked in different formats, as Leins et al. (2012) have demonstrated. In this experiment truthful participants had visited a room whereas deceptive participants did not. In the interview, however, all participants claimed to have visited the room. Participants were asked to verbally recall the layout of the room twice, to sketch it twice, or to verbally recall it once and to sketch it once. Liars contradicted themselves more than liars but only in the verbal recall/drawing condition. Truth-tellers have encoded the topic of investigation along more dimensions than liars. They therefore find it easier than liars to recall the event more flexibly (along more dimensions).

The Strategic Use of Evidence

Lying and truth-telling suspects enter police interviews in different mental states (Granhag and Hartwig, 2008). Guilty suspects will often have unique knowledge about the crime which, if recognised by the interviewer, makes it obvious that they are the perpetrator. Their main concern will be to ensure that the interviewer does not gain that knowledge. In contrast, innocent suspects face the opposite problem, fearing that the interviewer will not learn or believe what they did at the time of the crime. These different mental states result in different strategies for liars and truth-tellers (Hartwig et al., 2007). Guilty suspects are inclined to use avoidance strategies (e.g., in a free recall avoid mentioning where they were at a certain time) or denial strategies (e.g., deny having been at a certain place at a certain time when directly asked). In contrast, innocent suspects neither avoid nor escape but are forthcoming and 'tell the truth like it happened' (Granhag and Hartwig, 2008).

In the Strategic Use of Evidence (SUE) technique the investigator aims to detect these differential strategies via a strategic use of the available evidence, which may include possible incriminating information. The purpose of SUE is to ask open questions (e.g., 'What did you do last Sunday afternoon?') followed by specific questions (e.g., 'Did you or anyone else drive your car last Sunday afternoon?') without revealing that evidence (e.g., CCTV images of the interviewee's car driven in a specific location on that Sunday afternoon). Truth-tellers are likely to mention driving the car on that Sunday afternoon either spontaneously or after being prompted (i.e., 'tell the truth like it happened strategy'). Liars are unlikely to spontaneously mention driving the car (i.e., avoidance) or after being prompted (i.e., denial). A denial will contradict the evidence, thus indicating or confirming the guilt of the liar.

Hartwig et al. (2006) experimentally tested the SUE technique. Prior to the experiment, half of the interviewers were SUE-trained and instructed to interview the suspect using the SUE technique. The remaining interviewers were instructed to interview the suspect in the style of their own choice. The untrained interviewers obtained 56.1 per cent accuracy (similar to that typically found in nonverbal and verbal lie detection research), whereas the SUE-trained interviewers obtained 85.4 per cent accuracy. Guilty suspects contradicted the evidence more often than innocent suspects, particularly when questioned by SUE-trained interviewers. The SUE technique also raised cognitive load in liars. After the interview the participants were asked to report on the difficulty of the interview. Liars in the SUE condition reported to have experienced more cognitive load than truth-tellers, whereas no difference in cognitive load between liars and truth-tellers was found in the control condition.

Lie Detection in Counter-terrorism

Police/suspect interviews differ in certain important aspects from counter-terrorism interviews. To gain understanding in deception and lie-detection in counter-terrorism, key aspects of counter-terrorism need to be introduced and simulated in deception research. Researchers have only recently started to do this. In this section we will outline some of the key aspects of counter-terrorism together with the first research output available in this domain.

Lying About Intentions

Most forensic deception research deals with lying about *past activities*. This makes sense because most of that research focuses on police interviewing suspects about their alleged past activities. However, in counter-terrorism, being able to discriminate between true and false accounts about *future activities* (e.g., intentions) is of paramount importance, as this addresses the issue of *preventing* criminal acts from occurring, including terrorist attacks. Only a few studies about lying about intentions have been published to date (Clemens, Granhag, and Stromwall, 2011; Granhag and Knieps, 2011; Vrij et al., 2011a, b, c; Warmelink et al., 2011). An interesting pattern is emerging from these experiments (see also Granhag, 2010, for a review). In these experiments true and false intentions were detected with higher accuracy (around 70 per cent) than typically found in deception research (54 per cent, Bond and DePaulo, 2006). Of course, this may be an isolated finding and the average accuracy rate will possibly get lower when more research has been carried out. However, Vrij et al. (2011c) have demonstrated higher accuracy when examining lying about intentions than past activities in one experiment. Participants (senders) were sent on a mission and were asked to tell the truth or lie about it. They were intercepted before they started the mission and asked what they were going to do (intentions), and were again intercepted after completing their mission and asked what they had done (past activities). The interviews were transcribed and read by observers. These observers obtained higher accuracy rates for pinpointing truth-telling and lying about intentions (70 per cent) than past activities (56 per cent). The results thus suggest that detecting true and false intent was easier than detecting truthful and deceptive recall of past activities. We have several explanations for this. The intentions interviews came as a surprise to the senders whereas the past activities interviews did not. This could mean that the senders were better prepared for their past activities interviews than their intentions interviews and people are typically better liars if they have had the opportunity to prepare (DePaulo et al., 2003; Sporer and Schwandt, 2006, 2007; Vrij, 2008). In addition, perhaps the senders (and probably people in general) were more practiced in lying about their past activities than in lying about their intentions, and practice increases skill. Related to this, perhaps people have a better notion of what sounds convincing when they discuss their past activities than when they discuss their intentions (Vrij et al., 2011c).

Others have concentrated on verbal differences between telling the truth and lying about intentions rather than on accuracy rates in lie detection. Granhag and Knieps (2011) focused on Episodic Future Thought (EFT). EFT represents the ability to mentally pre-experience personal events that may occur in the future (Tulving, 1983; Schacter and Addis, 2007). EFT contains several features, and amongst them is the activation of vivid mental images (Szpunar, 2010). EFT is a core component in creating true intentions and occurs during planning of the intention (Schacter, Addis, and Buckner, 2008). Granhag and Knieps (2011) asked truth-telling participants to prepare a visit to a shopping mall to buy a gift. Lying participants were asked to plan installing a memory stick containing illegal material in the same shopping mall, but were also asked to prepare a cover story (buying a gift) in case they would be intercepted. After the planning and before executing their intentions, participants were intercepted and interviewed. Liars were hereby instructed to mask their criminal intentions and to use their cover story. Nearly all truth-tellers and significantly fewer liars reported that they created a mental image of their intention during the planning. When asked to describe these mental images, truth-tellers' answers were more detailed.

Clemens et al. (2011) employed the SUE technique in their true and false intent experiment. Participants were told that a company was starting to sell fur clothing. Truth-tellers were asked to prepare to visit the company to present a protest letter, whereas liars were asked to prepare a visit to the same company to install a virus on the company leader's computer. Liars were further instructed to give another but harmless reason for visiting the company when intercepted. During the planning phase all participants (truth-tellers and liars) (i) visited the company's website, (ii) obtained a map of the company's building, and (iii) collected a memory stick, but truth-tellers and liars carried out these activities for different reasons. During the interview the participants were asked to recall their intent (what they were going to do when visiting the company) and their planning for the intent (activities they carried out to make the visit possible). Liars were less consistent than truth-tellers when comparisons were made between (i) statements about their intent and their actual planning activities, and (ii) statements about their planning and their actual planning activities.

In sum, the emerging lying about intentions literature suggest that statements about true intentions and statements about false intentions differ from each other: Observers are to some extent able to discriminate between statements that indicate true and false intent and verbal differences have been found between these two types of statement. More verbal differences between true and false intentions emerged in undercover interviewing research which will be described below (Vrij et al., 2012b).

Lying About Opinions: The Devil's Advocate Approach

Most verbal lie-detection research (as well as police/suspect interviews) focuses on the ability to distinguish between truths and lies when people describe events that they claim to have experienced. However, people lie not only about their experiences but also about their opinions. Determining the veracity of such *conceptual* representations can be important in security settings, as demonstrated by the loss of seven CIA agents in Afghanistan on 30 December 2009. They were killed via a suicide attack by a man they believed was going to give them information about Taliban and Al-Qaeda targets in Pakistan's tribal areas. The CIA was aware that he had posted extreme anti-American views on the Internet, but believed these to be part of a cover.

The Devil's Advocate technique is designed to detect deception in expressing opinions. Interviewees are first asked an opinion eliciting question that invites them to argue in favour of their personal view ('What are your reasons for supporting the US in the war in Afghanistan?'). This is followed by a Devil's Advocate question that asks interviewees to argue against their personal view ('Playing Devil's Advocate, is there anything you can say against the involvement of the US in Afghanistan?').

People normally think more deeply about, and are more able to generate, reasons that support rather than oppose their beliefs (Ajzen, 2001). Therefore, someone could argue that truth-tellers are likely to provide more information in their responses to the true opinion eliciting question than to the Devil's Advocate question. This pattern is unlikely to occur in liars: for them the Devil's Advocate question is more compatible with their real beliefs than is the opinion-eliciting question. In effect, for liars the Devil's Advocate approach is a set-up where they first lie when answering the opinion-eliciting question and are then lured into telling the truth when answering the Devil's Advocate question. In an experiment, participants were asked to tell the truth or lie about their views regarding issues they feel strongly about, including the war in Afghanistan. Truth-tellers' opinion-eliciting answers were longer than their Devil's Advocate answers, whereas no differences emerged in liars' answers to the two types of question (Leal et al., 2010). Based on this principle, 75 per cent of truth-tellers and 78 per cent of liars could be classified correctly.

Lying in Groups (Collective Interviewing)

Counter-terrorism interviews often do not take place in formal police interview settings. Instead, investigators may wish to interview people at specific locations, such as in a shopping mall, at the entrance of a football stadium, or at a road border control. This may then result in situations in which there is only one interviewer available but a group of suspects (e.g., road border controls where cars containing several people are checked). Such situations enable investigators to examine how group members communicate with each other when lying or truth-telling. Research on truth-tellers' and liars' preferred strategies during interviews gives insight into which type of communication cues may arise in dyad communication. It has been found that, when asked to recall an event, truth-tellers reconstruct the event from memory and prefer a 'tell it all' approach, aiming to provide a detailed description of what happened. In contrast, liars do not reconstruct a story but report a prepared alibi. In terms of detail, they prefer a 'keep it simple' approach, incorporating enough detail to avoid raising suspicion, but avoid providing too much detail in case the interviewer already knows which details are accurate or could find out subsequently that the story is fabricated by checking these details (Granhag, Strömwall, and Jonsson, 2003; Hartwig, Granhag, and Strömwall, 2007; Strömwall, Hartwig, and Granhag, 2006). These contrasting approaches by truth-tellers and liars could result in differential group interactions when being interviewed simultaneously. When truth-tellers are asked about a shared experience, they will start to reconstruct the event jointly (Hollingshead, 1998; Wegner, Erber, and Raymond, 1991) and are likely to interact with each other, sharing the telling of these experiences, comparing and correcting each other's recall. As a result, they may show cues that are typical for when people jointly recall an event: they may interrupt and add information to each other's accounts while also seeking to correct or amend each other.

When liars are asked about their supposedly shared experiences a different response may emerge. A 'keep it simple' approach will likely result in fewer interactions between them. For example, one person may answer the question(s) and the other person may simply agree with what has been said. As a result, truth-tellers, more often than liars, will interrupt and add information to each other's accounts while also seeking to correct or amend each other's account. This has been found in the two collective interviewing experiments published to date (Driskell, Salas and Driskell, 2012; Vrij et al., 2012a). Of course, interacting with each other is associated with mutual gaze. Both experiments revealed that truthful dyads gazed more at their partner than deceptive dyads (Driskell et al., 2012; Jundi et al., 2013). In sum, collective interviewing has shown promise in terms of lie detection because truthful and deceptive dyads communicate differently with each other.

Interviewed Without Being Noticed (Undercover Interviewing)

In counter-terrorism it may be useful to conduct interviews without the suspect actually knowing they are being interviewed. We call this *undercover interviewing*. It may fit particularly well in determining the veracity of an individual's intentions. At the intentions stage, no crime has yet been committed, and a formal interview may therefore be inappropriate. In addition, in some investigative contexts, law enforcement and security personnel may have good reason to extract information from a suspect without them actually being aware that they are under investigation. In particular, law enforcement officers working in an undercover capacity and interacting with potential suspects in informal settings will not wish to draw attention to themselves or arouse suspicion about their motives by using direct question formats. For example, in settings where an undercover officer has become embedded within a criminal gang or is required to interact with suspects in order to collect intelligence, the ability to elicit relevant and usable information without detection is critical. In addition, the instance of police asking tourists in London to delete their photos as a precautionary counter-terrorist measure attracted significant media attention (Weaver and Dodd, 2009). Internet accounts suggest that this was not an isolated incident, with an online movement being formed to campaign for photographers' rights entitled 'I'm a Photographer, not a Terrorist!'. The issue is a global one, with a controversial terrorism prevention video being released in the USA asking the public to report photographers to the police (Terrorism Prevention Video asks Public to Report Photographers to Police, 2012). Undercover interviewing may shed light on whether an individual has criminal intentions without arousing their suspicion in such circumstances.

Vrij et al. (2012b) carried out the first undercover interviewing experiment to date. In that experiment the participants were interviewed by an 'undercover agent', acting either as a doctoral student or an amateur photographer. He approached either tourists (truth-tellers) or participants who were on a reconnaissance mission (liars) at a hovercraft terminal, and asked them questions about their forthcoming trip to a nearby island. The liars had been instructed to prepare as a cover story that they were going to visit the island as a tourist, and were given a tourist flyer about the island to prepare their cover story.

The requirement to maintain undercover determines the questions that can be asked, and has several disadvantages and advantages in terms of lie detection. Starting with the disadvantages, interview tools that have been shown to facilitate lie detection, such as imposing cognitive load and the SUE technique, cannot be employed without making the suspect suspicious about the questioner's motives. In terms of advantages, undercover interviewing creates the opportunity to ask questions that could be useful for lie detection purposes but which will not work in traditional overt interviews. For example, the undercover interviewer could invite suspects to engage in an apparently innocent activity that establishes their presence in a certain place at a certain time, such as asking whether the suspect would mind having a photograph taken that the interviewer could place on a website. Given that a plausible rationale for this request is provided (e.g. 'I've just started a new business and I'm trying to build up my reputation as a photographer') truth-tellers and liars may well respond differently to this seemingly innocuous compliance request. Guilty people do not wish to be linked to their criminal activity and tend to 'avoid and escape' when asked about it (Granhag and Hartwig, 2008), and may therefore show greater reluctance to be photographed.

Asking questions about a forthcoming trip is, in fact, asking about intentions. Research has revealed that an individual who is about to execute their intention typically has a detailed mental representation of that intention (Trope and Liberman, 2003). This representation is more detailed than those of intentions the individual plans to execute at some later time (Trope and Liberman, 2003). It also differs from the mental representations of intentions that the person has not yet decided to execute (Ferguson and Bargh, 2004; Marsh, Hicks and Bryan, 1999). Based on this, truth-tellers may be more precise when describing their intentions (e.g., more references to exact timings) and may express more certainty in what they are going to do than liars.

Of course, skilled liars may prepare a cover story about what they are going to do. In the present experiment liars were asked to say, if questioned, that they were visiting the island as a tourist and hence gave some thought to their fictitious planned activities while there. Researchers have started to investigate how people prepare for possible questioning about their whereabouts (Colwell et al., 2006; Hartwig, Granhag and Strömwall, 2007; Hartwig et al., 2010; Strömwall, Hartwig and Granhag, 2006). A popular strategy is to think of possible questions and prepare answers to those anticipated questions. People typically do not anticipate, and are thus unprepared for, spatial questions (Vrij et al., 2009). Therefore, such questions in particular should reveal deceit. In terms of visiting an island, it could be that liars identify which attractions they are supposedly going to visit (e.g., the castle) but do not check specifically where they are located. Therefore, liars may be less accurate than truth tellers in reporting exact locations of their alleged destinations on the island. In Vrij et al.'s (2012b) experiment the undercover interviewer asked the participants why they were visiting the island, at what time they were planning to come back, and to indicate on a blind map of the island the locations they were going to visit. He also asked for permission to take their photograph. Truth-tellers, compared to liars, gave more detailed answers, were more exact about at what time they were returning from the island, were more precise in indicating the exact locations of the places they were going to visit and were more willing to be photographed. When we asked participants after their encounter with the undercover agent about their experiences during the interview, they mentioned that they only found the request to have their photograph taken somewhat odd. In sum, this experiment revealed that undercover interviewing can be used as a way to detect deceit.

Passive Observations

In most deception research so far (undercover interviewing is the exception) suspects are aware that they were being observed. Although this reflects police interviewing, it may not reflect other situations, for example when alleged terrorists are secretly shadowed or crowds are observed. The question of whether observers can distinguish innocent individuals from potential wrongdoers in such circumstances has not been addressed to date. It requires an experimental set-up in which participants are allocated to a 'wrongdoing' condition, after which their behaviours will be observed when they walk around, stand or wait (e.g., Gozna and Babooran, 2004). Innocent participants, who are not aware that they are being observed, form the control group. In such a design participants will not be interviewed, so the only cues available to observers are nonverbal cues to deceit. We believe that it is more likely that nonverbal cues will reveal deceit in such situations than in formal interviews. One problem in formal interviews is that truth-tellers may be nervous as a result of being suspected of wrongdoing or out of fear of not being believed. They may thus be as nervous as liars and, as a result, cues of nervousness will not discriminate between truth-tellers and liars (Bond and Fahey, 1987; Ofshe and Leo, 1997). When truth-tellers are not aware that they are being observed, there is no reason for them to show signs of nervousness, and differences in nervousness between innocents and wrongdoers may emerge. Second, wrongdoers do not wish to be noticed or stand out, so they may attempt to avoid displaying cues they think appear suspicious (Vrij, 2008). This self-conscious behaviour may come across as rigid and over-controlled (Baumeister, 1984; Vrij, 2008). A desire to stay unnoticed may also result in showing a keen interest in security measures. This may result in scrutinising behaviours such as observing security personnel, or looking around for CCTV cameras. There is no reason for innocents to display any of those self-conscious or scrutinising behaviours as long as they are unaware that they are being observed.

Final Thoughts and Conclusion

When discussing counter-terrorism it is important to consider to what extent current deception research can be used to fight terrorism. Over the years we have come across other academics who believe that terrorists are different from the average participant in our research: however, we disagree with this view. Take for example the unanticipated questions approach. At its core is the asking of questions that a liar has not anticipated and therefore not prepared answers for. There is no theoretical reason to suggest that terrorists will struggle less with answering unanticipated questions than our participants.

A more relevant comment is that much of the current deception research does not adequately address the various settings that are relevant for terrorism. A lot of good work has been done (see also Brandon, 2011, and Loftus, 2011). The lie detection techniques that we have discussed in this chapter can be employed in various settings. SUE can be used when evidence is available, the Devil's Advocate technique can be employed when examining the veracity of beliefs and opinions, and the other techniques (e.g., imposing cognitive load, asking unanticipated questions) can be employed to determine the veracity of statements about past activities. We have also shown that the unanticipated questions technique can be employed to identify deceit in both individuals and networks (multiple liars).

However, we have also argued that much more work is required. In particular, researchers should address the questions counter-terrorism investigators deal with and should create research scenarios that accurately reflect the settings they work in. Researchers have started to carry out this important work and we have summarised their initial findings in this chapter. We hope that it will be useful to investigators and that it will encourage other researchers to carry out this important type of innovative deception research.

References

Ajzen, I. (2001). Nature and operation of attitudes. *Annual Review of Psychology*, 52, 27–58.

Baumeister, R.F. (1984). Choking under pressure: Self-consciousness and paradoxical effects of incentives on skillful performance. *Journal of Personality and Social Psychology*, 46, 610–620.

Bond, C.F., and DePaulo, B.M. (2006). Accuracy of deception judgements. *Personality and Social Psychology Review*, 10, 214–234.

Bond, C.F., and Fahey, W.E. (1987). False suspicion and the misperception of deceit. *British Journal of Social Psychology*, 26, 41–46.

Brandon, S. (2011). Impacts of psychological science on national security agencies post-9/11. *American Psychologist*, 66, 495–506.

Bull, R. (2010). The investigative interviewing of children and other vulnerable witnesses: Psychological research and working/professional practice. *Legal and Criminological Psychology*, 15, 5–24.

Buller, D.B., and Burgoon, J.K. (1996). Interpersonal deception theory. *Communication Theory*, 6, 203–242.

Clemens, F., Granhag, P.A., and Strömwall, L. (2011). Eliciting cues to false intent: A new application of strategic interviewing. *Law and Human Behavior*, 35, 512–522.

Colwell, K., Hiscock-Anisman, C., Memon, A., Woods, D., and Michlik, P.M. (2006). Strategies of impression management among deceivers and truth tellers: How liars attempt to convince. *Amercian Journal of Forensic Psychology*, 24, 31–38.

DePaulo, B.M., Lindsay, J.L., Malone, B.E., Muhlenbruck, L., Charlton, K., and Cooper, H. (2003). Cues to deception. *Psychological Bulletin*, 129, 74–118.

Doherty-Sneddon, G., and Phelps, F.G. (2005). Gaze aversion: A response to cognitive or social difficulty? *Memory and Cognition*, 33, 727–733.

Driskell, J.E., Salas, E., and Driskell, T. (2012). Social indicators of deception. *Human Factors: The Journal of the Human Factors and Ergonomics Society*, 54, 577–588.

Ferguson, M.J., and Bargh, J.A. (2004). Liking is doing: The effects of goal pursuit on automatic evaluation. *Journal of Personality and Social Psychology*, 87, 557–572.

Fisher, R.P. (2010). Interviewing cooperative witnesses. *Legal and Criminological Psychology*, 15, 25–38.

Fisher, R., Milne, R., and Bull, R. (2011). Interviewing cooperative witnesses. *Current Directions in Psychological Science*, 20, 16–19.

Gilbert, J.A.E., and Fisher, R.P. (2006). The effects of varied retrieval cues on reminiscence in eyewitness memory. *Applied Cognitive Psychology*, 20, 723–739.

Gozna, L., and Babooram, N. (2004). Non-traditional interviews: Deception in a simulated customs baggage search. In A. Czerederecka, T. Jaskiewicz-

Obydzinska, R. Roesch, and J. Wojcikiewicz (Eds), *Forensic Psychology and Law*. Krakow, Poland: Institute of Forensic Research Publishers, 153–161.

Granhag, P.A. (2010). On the psycho-legal study of true and false intentions: Dangerous waters and some stepping-stones. *The Open Criminology Journal*, 3, 37–43.

Granhag, P.A., and Hartwig, M. (2008). A new theoretical perspective on deception detection: On the psychology of instrumental mind-reading. *Psychology, Crime & Law*, 14, 189–200.

Granhag, P.A., and Knieps, M. (2011). Episodic future thought: Illuminating the trademarks of forming true and false intentions. *Applied Cognitive Psychology*, 25, 274–280.

Granhag, P.A., Strömwall, L.A., and Jonsson, A.C. (2003). Partners in crime: How liars in collusion betray themselves. *Journal of Applied Social Psychology*, 33, 848–868.

Hartwig, M., Granhag, P.A., and Strömwall, L. (2007). Guilty and innocent suspects' strategies during interrogations. *Psychology, Crime, & Law*, 13, 213–227.

Hartwig, M., Granhag, P.A., Strömwall, L., and Doering, N. (2010). Impression and information management: On the strategic self-regulation of innocent and guilty suspects. *The Open Criminology Journal*, 3, 10–16.

Hartwig, M., Granhag, P.A., Strömwall, L., and Kronkvist, O. (2006). Strategic use of evidence during police interrogations: When training to detect deception works. *Law and Human Behavior*, 30, 603–619.

Hollingshead, A.B. (1998). Retrieval processes in transactive memory systems. *Journal of Personality and Social Psychology*, 74, 659–671.

Inbau, F.E., Reid, J.E., Buckley, J.P., and Jayne, B.C. (2001). *Criminal Interrogation and Confessions*, 4th edn. Gaithersburg, MD: Aspen Publishers.

Jundi, S., Vrij, A., Mann, S., Hope, L., Hillman, J., Warmelink, L., and Gahr, E. (2013). Who should I look at? Eye contact during collective interviewing as a cue to deceit. *Psychology, Crime, & Law*, 19(8), 661–671

Kassin, S.M., Appleby, S.C., and Torkildson-Perillo, J. (2010). Interviewing suspects: Practice, science, and future directions. *Legal and Criminological Psychology*, 15, 39–56.

Lancaster, G.L.J., Vrij, A., Hope, L., and Waller, B. (2012). Sorting the liars from the truth tellers: The benefits of asking unanticipated questions. *Applied Cognitive Psychology*, 27, 107–114.

Leal, S., Vrij, A., Mann, S., and Fisher, R. (2010). Detecting true and false opinions: The devil's advocate approach as a lie detection aid. *Acta Psychologica*, 134, 323–329.

Leal, S., Vrij, A., Warmelink, L., Vernham, Z., and Fisher, R. (2014). You can't hide your telephone lies: Providing a model statement as an aid to detect deception in insurance telephone calls. *Legal and Criminological Psychology*, doi: 10.1111/lcrp.12017.

Leins, D., Fisher, R., and Vrij, A. (2012). Drawing on liars' lack of cognitive flexibility: Detecting deception through varying report modes. *Applied Cognitive Psychology*, 26, 601–607.

Leins, D., Fisher, R.P., Vrij, A., Leal, S., and Mann, S. (2011). Using sketch-drawing to induce inconsistency in liars. *Legal and Criminological Psychology*, 16, 253–265.

Loftus, E.F. (2011). Intelligence gathering post-9/11. *American Psychologist*, 66, 532–541.

Mann, S., Vrij, A., Shaw, D., Leal, S., Ewans, S., Hillman, J., Granhag, P.A., and Fisher, R.P. (2013). Two heads are better than one? How to effectively use two interviewers to elicit cues to deception. *Legal and Criminological Psychology*, 18(2), 324–340.

Marsh, R.L., Hicks, J.L., and Bryan, E.S. (1999). The activation of unrelated and cancelled intentions. *Memory & Cognition*, 27, 320–327.

Memon, A., Meissner, C.A., and Fraser, J. (2010). The cognitive interview: A meta-analytic review and study space analysis of the past 25 years. *Psychology, Public Policy, & Law*, 16, 340–372.

National Research Council. (2003). *The Polygraph and Lie Detection. Committee to Review the Scientific Evidence on the Polygraph*. Washington, DC: The National Academic Press.

Ofshe, R.J., and Leo, R.A. (1997). The decision to confess falsely: Rational choice and irrational action. *Denver University Law Review*, 74, 979–1112.

Schacter, D.L., and Addis, R.D. (2007). The cognitive neuroscience of constructive memory: Remembering the past and imagining the future. *Philosophical Transactions of the Royal Society*, 392, 773–786.

Schacter, D.L., Addis, D.R., and Buckner, R.L. (2008). Episodic simulation of future event: Concepts, data and applications. *Annals of New York Academy of Sciences*, 1124, 39–60.

Spence, S.A., Farrow, T.F.D., Herford, A.E., Wilkinson, I.D., Zheng, Y., and Woodruff, P.W.R. (2001). Behavioural and functional anatomical correlates of deception in humans. *Neuroreport: For Rapid Communication of Neuroscience Research*, 12, 2849–2853.

Sporer, S.L., and Schwandt, B. (2006). Paraverbal indicators of deception: A meta-analytic synthesis. *Applied Cognitive Psychology*, 20, 421–446.

Sporer, S.L., and Schwandt, B. (2007). Moderators of nonverbal indicators of deception: A meta-analytic synthesis. *Psychology, Public Policy, and Law*, 13, 1–34.

Strömwall, L.A., Hartwig, M., and Granhag, P.A. (2006). To act truthfully: Nonverbal behaviour and strategies during a police interrogation. *Psychology, Crime, & Law*, 12, 207–219.

Szounar, K.K. (2010). Episodic future thought: An emerging concept. *Perspectives on Psychological Science*, 5, 142–162.

Terrorism Prevention Video asks Public to Report Photographers to Police (2012) Retrieved from http://www.petapixel.com/2012/08/09/terrorism-prevention-video-asks-public-to-report-photographers-to-police/ accessed 11 August 2014.

Trope, Y., and Liberman, N. (2003). Temporal construction. *Psychological Review*, 110, 403–421.

Tulving, E. (1983). *Elements of Episodic Memory*. New York: Oxford University Press.

Vrij, A. (2008). *Detecting Lies and Deceit: Pitfalls and Opportunities*, 2nd Edn. Chichester: John Wiley and Sons.

Vrij, A., Granhag, P.A., Mann, S., and Leal, S. (2011a). Lying about flying: The first experiment to detect false intent. *Psychology, Crime, and Law*, 17, 611–620.

Vrij, A., Granhag, P.A., Mann, S. and Leal, S. (2011b). Outsmarting the liars: Towards a cognitive lie detection approach. *Current Directions in Psychological Science*, 20, 28–32.

Vrij, A., Granhag, P.A., and Porter, S.B. (2010). Pitfalls and opportunities in nonverbal and verbal lie detection. *Psychological Science in the Public Interest*, 11, 89–121.

Vrij, A., Jundi, S., Hope, L., Hillman, J., Gahr, E., Leal, S., Warmelink, L. Mann, S., Vernham, Z., and Granhag, P.A. (2012a). Collective interviewing of suspects. *Journal of Applied Research in Memory and Cognition*, 1, 41–44.

Vrij, A., Leal, S., Granhag, P.A., Mann, S., Fisher, R. P., Hillman, J., and Sperry, K. (2009). Outsmarting the liars: The benefit of asking unanticipated questions. *Law and Human Behavior*, 33, 159–166.

Vrij, A., Leal, S., Mann, S., and Granhag, P.A. (2011c). A comparison between lying about intentions and past activities: Verbal cues and detection accuracy. *Applied Cognitive Psychology*, 25, 212–218.

Vrij, A., Mann, S., Fisher, R., Leal, S., Milne, B., and Bull, R. (2008). Increasing cognitive load to facilitate lie detection: The benefit of recalling an event in reverse order. *Law and Human Behavior*, 32, 253–265.

Vrij, A., Mann, S., Jundi, S., Hope, L., and Leal, S. (2012b). Can I take your picture? Undercover interviewing to detect deception. *Psychology, Public Policy, & Law*, 18, 231–244.

Vrij, A., Mann, S., Leal, S., and Fisher, R. (2010a). 'Look into my eyes': Can an instruction to maintain eye contact facilitate lie detection? *Psychology, Crime, & Law*, 16, 327–348.

Vrij, A., Mann, S., Leal, S., and Granhag, P.A. (2010b). Getting into the minds of pairs of liars and truth tellers: An examination of their strategies. *The Open Criminology Journal*, 3, 17–22.

Walczyk, J.J., Roper, K.S., Seemann, E., and Humphrey, A.M. (2003). Cognitive mechanisms underlying lying to questions: Response time as a cue to deception. *Applied Cognitive Psychology*, 17, 755–774.

Warmelink, L., Vrij, A., Mann, S., Jundi, S., and Granhag, P.A. (2012). Have you been there before? The effect of experience and question expectedness on lying about intentions. *Acta Psychologica*, 141(2), 178–183.

Warmelink, L., Vrij, A., Mann, S., Leal, S., Forrester, D., and Fisher, R. (2011). Thermal imaging as a lie detection tool at airports. *Law & Human Behavior*, 35, 40–48.

Weaver, M., and Dodd, V. (2009). Police delete London tourists' photos 'to prevent terrorism'. *The Guardian*, http://www.theguardian.com/uk/2009/apr/16/police-delete-tourist-photos accessed 11 August 2014.

Wegner, D.M., Erber, R., and Raymond, P. (1991). Transactive memory in close relationships. *Journal of Personality and Social Psychology*, 61, 923–929.

Chapter 8

A Field Trial to Investigate Human Pheromones Associated with Hostile Intent

Peter Eachus

School of Social Work, Psychology and Public Health, University of Salford, UK

Alex Stedmon

Human Systems Integration Group, Faculty of Engineering & Computing, Coventry University, UK

Les Baillie

Welsh School of Pharmacy, University of Cardiff, UK

Introduction

Modern terrorist operations are sophisticated and to be successful they must be based on sound intelligence. This intelligence can take many forms and may be obtained using a variety of techniques and technologies (Linett, 2005). Open-source intelligence is readily available via the Internet, where it is possible to gain access to maps detailing locations of potential targets. Furthermore, potential terrorists can view targets at street level through tools such as Google Streetview. Although in the initial planning stages this level of intelligence may be useful, more detailed intelligence is often required and this can only be obtained by physical reconnaissance of the intended target (Cannon, 2004).

Most, if not all successful terrorist operations, have involved active hostile reconnaissance before the attack was launched. For example, when the World Trade Center was first attacked in 1993, as well as the bombings of the US embassies in Kenya and Tanzania a few years later, subsequent investigations revealed that they were all preceded by years of terrorist information gathering and covert surveillance (Department of Homeland Security Bulletin, 2003).

Years before the actual attack on the World Trade Center in 2001, the plotters involved had tested airline security by smuggling box cutters onto planes in their toilet bags. As we now know to our cost, this was never challenged by airport security. In the weeks leading up to the 9/11 attack, the plotters flew many dry runs, again to test security and also to familiarise themselves with the aircraft they would be hijacking.

In the United Kingdom the 2005 London underground bombers were captured on CCTV apparently conducting a dry run prior to the attack on 7 July. Similarly,

Aum Shinrikyo, who attacked the Tokyo subway system, carried out pre-operational reconnaissance before the poison gas attack in March 1995. As with the London bombing, this incident involved the co-ordination of simultaneous attacks on several lines; something that could only have been achieved with a significant degree of pre-planning. As an aside, it is known that one of the 7/7 bombers, Hasib Hussain, did not participate in the final reconnaissance before the attack and he was the one who apparently got into difficulty when his bomb failed to explode on the underground due to a faulty battery. He tried but failed to contact the other bombers, and an hour after the first bombs exploded he was caught on CCTV leaving Boots chemists where he had bought another battery. He then decided to board the number 30 bus and detonated his device at Tavistock Square. This would seem to have been an adaptation to the plan as there is no evidence that their hostile reconnaissance involved buses.

A number of research programmes have been designed to investigate the possibility of detecting hostile intent. The Department of Homeland Security has sponsored several of these, such as the Hostile Intent Detection – Automated Prototype Project that is attempting to use real-time intent detection at airports and other vulnerable parts of the transportation system (Department of Homeland Security, 2013).

Commercial companies have also developed systems for the detection of hostile intent. Suspect Detection Systems (SDS), an Israeli company, has been testing Cogito, an expert system aimed at detecting individuals who may have terrorist intentions. From the company's own promotional material, Cogito appears to be a rather sophisticated lie detection system, measuring physiological responses to a number of computer generated questions, e.g. "Are you carrying explosives?". Unfortunately, although SDS claim an 85 per cent detection rate, there is little published validation data that would substantiate this. A further problem with this system is that it requires the people being investigated to sit in a special booth for the collection of data and therefore the range of applications is somewhat limited. It might be acceptable at airports or border crossings, but it would have little value in shopping malls or other crowded public spaces.

Despite the difficulties in identifying hostile intent and hostile reconnaissance, the UK Government does recognise it as a potential problem and has implemented policies accordingly. For example, Operation Lightning aims to identify suspicious behaviours or hostile reconnaissance at, or near, prominent or potentially vulnerable structures or buildings, including the highways infrastructure. Hostile reconnaissance in this context can include taking photographs and making notes about security measures such as the location of CCTV cameras, vigilance towards unattended baggage, or security staff rosters (National Archives, 2008).

Hostile Reconnaissance and Stress

One of the basic assumptions on which this study is based is that individuals who are undertaking hostile reconnaissance will find the experience stressful because of the risks involved. This stress will manifest itself psychologically and physiologically (Seyle, 1974) and possibly at a biological level via stress pheromones.

The psychology and physiology of stress is quite well understood and instruments for the measurement of these parameters are readily available (see Chapters 4, 6 and 20 in this volume). However, our understanding of the biology of stress, in terms of the production of stress pheromones, is rather less well developed, although several studies claim to have identified these pheromones in participants who have been subjected to stressful experiences.

Pheromones and Alarm Signals in Body Odour

The ability of insects to communicate via the release of pheromones is well documented (Regnier and Law, 1968). When placed under stressful conditions (e.g. in dangerous situations) they are known to release alarm pheromones, which have been characterised in most cases as simple ketones, aldehydes and alcohols, depending on the species in question, that warn other members of the colony of an impending threat. Pheromones produced by moths can be detectable at distances in excess of five miles. In much the same way, other animal species have shown extraordinary abilities to interact via olfactory communication. Perhaps the most pertinent example of this is the ability of male dogs to locate a bitch on heat from a great distance. While in many species olfactory communication may be regarded with greater importance, for humans it is given less credence.

The chemical make-up of human body odour is sensitive to diet, health, age and emotional state (Singh and Bronstad, 2001). However, many olfactory studies have shown that individuals do indeed have their own scent. For example a study conducted in 1998 shows that mothers can recognise the smell of their newborn child after only a few hours of contact (Porter, 1998). A recent publication by Kuhn and Natsch (2009) has also shown that the body odour of sets of monozygotic twins have similar levels of certain odour-causing chemical compounds, suggesting a genetic influence to their body odour 'fingerprint'.

The notion that each individual, be it an insect, human or any other species, secretes a characteristic 'blend' of volatile chemicals, has attracted the attention of the research community (Penn, Oberzaucher and Grammer, 2007). Human body odour is a complex mixture of hydrocarbons, alcohols, ketones and aldehydes (Curran et al., 2005). Several studies have employed the Gas Chromotography-Mass Spectrometry (GC-MS) technique to elucidate the chemical make-up of sweat samples (Curran et al., 2005; Penn, Oberzaucher and Grammer, 2007). The ability of this technique to separate out complex mixtures of chemicals, coupled with its high sensitivity, makes it an ideal tool for identifying components in sweat samples.

There has been some speculation in the literature that people may secrete a stress pheromone in their body odour when put under stress (Chen, Katdare and Lucas, 2006). One such study analysed the sweat obtained from volunteers conducting their first-ever skydive as a model for stress (Mujica-Parodi et al., 2009). The study looked at differences in brain activity of the amygdala, a structure activated in humans by emotional arousal and in animals by alarm pheromones, and found that exposure to stress sweat samples (collected from the first-time skydivers) caused significantly greater stimulation than exposure to exercise sweat (collected from treadmill users). They also found that exposure to stress sweat samples vs. exercise sweat had a significant behavioural affect of threat perception in that they appeared to be incapable of communicating their physiological stress response. These results lead the researchers to conclude that sweat collected during acute emotional stress, and subsequently presented to an unrelated group of individuals, produced significant brain activation in regions responsible for emotional processing without conscious perception of the distinct odour.

However, as yet there have been no studies that have shown one absolute stress marker in body odour. One of the assumed inherent difficulties associated with the characterisation of human fear pheromones is their low concentrations in secreted media (Boeckh, Kaissling and Schneider, 1965; Butenandt, Beckman and Hecker, 1961; Grosser et al., 2000; Millar, 2002). It is also assumed that alarm pheromones in humans must be volatile to ensure greater penetration of the olfactory system (Millar, 2002) and this may lead to practical difficulties in the capture, storage and analysis of sweat samples.

Researchers have shown a correlation between stress and the level of cortisol (a known stress hormone) in salivary samples (Haussmann, Vleck and Farrar, 2007) as well as sweat (Prunty et al., 2004). The level of cortisol in the body is sensitive to many factors aside from stress, including time of day, body weight, state of health and age. Therefore, it is not possible to determine whether an individual is stressed by a single analysis. The degree of stress an individual is under is ascertained by the difference in cortisol levels from a baseline sample. However, the determination of cortisol levels in saliva is a preferred method over that in sweat, and there is limited evidence to show the diagnostic value of cortisol levels in sweat (Prunty et al., 2004). Furthermore, the idea that cortisol present in sweat is responsible for the 'smell of fear' has not as yet been validated (Ackerl, Atzmueller and Grammer, 2002). Despite a deluge of studies into the olfactory responses to, and the chemical profiles of, sweat samples produced under stressful situations, the question of whether humans are capable of secreting a fear 'marker', or detecting it, still remains.

Detection of Alarm Signals

Should an alarm pheromone be discovered and characterised from human sweat it would pave the way for the development of detection devices to determine the presence and concentration of such a compound in an individual's odour. This technology would be invaluable in many areas, including counter-terrorism and behavioural science as well as having a much broader appeal because it would enable people to gauge the stress levels of an individual at a given time.

An approach based on insect sensors is currently being pursued by a number of groups (Rains, Tomberlin and Kulasiri, 2008), including a UK company called Inscentinel (www.inscentinel.com) who with support from the Home Office have trained bees to detect the presence of explosives and hydrogen peroxide. Also, detection devices based on the chemistry detectable compounds are widely used, such as the alcohol breathalyser and several drug tests, which rely on the analyte in question to react with a testing reagent in a specific way to produce a result (such as a change in colour) that identifies the presence and level of a compound.

Regardless of the policies, procedures, or technologies that are developed to combat hostile intent, a methodology for testing the validity and reliability of these is required. The purpose of the research described in this chapter was to develop a protocol for such a methodology.

An Integrated Approach Towards a Model of Hostile Intent

This research took an integrated approach to understanding and modelling hostile intent as illustrated in Figure 8.1. The underlying premise is that an individual involved in a criminal or terrorist act (such as hostile reconnaissance) may provide data across any (or all) of the four factors. For a more detailed account of this study, see Eachus, Stedmon and Baillie (2013).

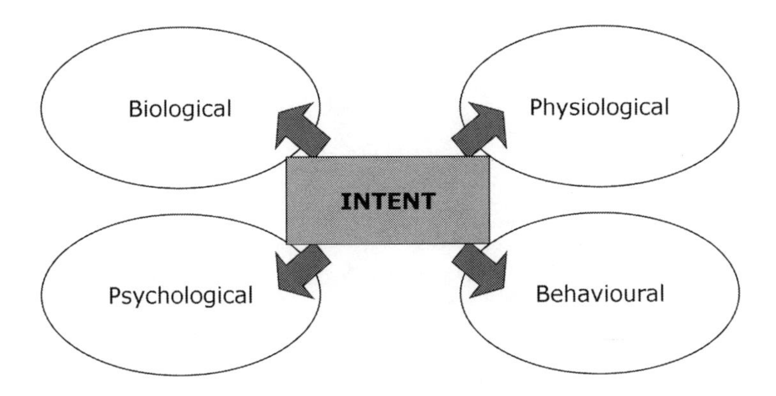

Figure 8.1 Hostile intent and stress responses

This study was based on the assumption that a person engaged in hostile reconnaissance runs a risk of being detected. This context, especially during initial hostile reconnaissance activities when a terrorist may feel particularly vulnerable to detection, would have consequences in terms of the individual's behaviour, psychological state and physiological and biological responses. Essentially, hostile reconnaissance will be experienced as stressful for the perpetrator and the purpose of this study was to determine how this stress response is manifested and what influence this might have across a range of responses for the individuals concerned.

This research comprised of two interrelated studies. The first was a field study that simulated a hostile reconnaissance activity in a crowded public space (i.e. a large shopping mall). After establishing baseline levels of stress within a laboratory situation under relaxed conditions, participants were placed in a stressful situation and their psychological, physiological and biological status was monitored (Figure 8.2). The behaviour of the participants during a 'Low Intent' and 'High Intent' condition was recorded using CCTV installed in the shopping mall. The second study was a data identification phase during which a different sample of participants who had no previous knowledge of the first phase were asked identify the levels of stress in the first sample of participants, based on CCTV recordings of their behaviour and their pheromone samples.

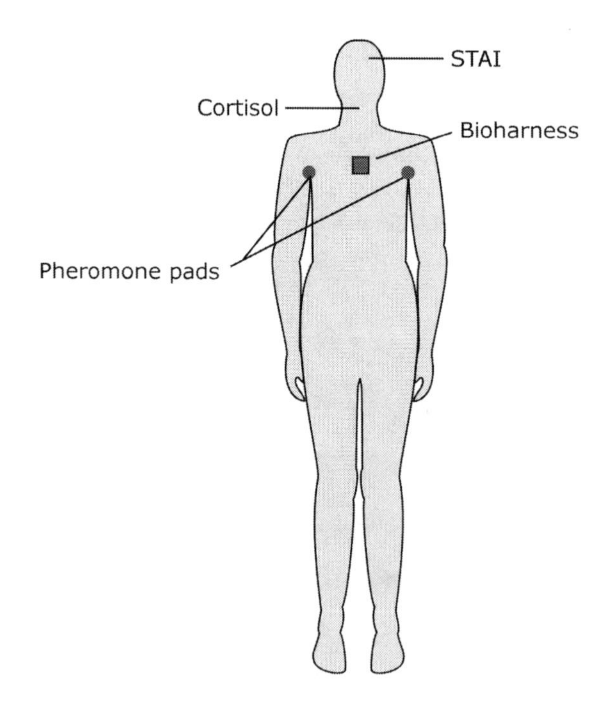

Figure 8.2 Stress data collection for the field trial

Participant Recruitment

Studies of stress pheromone secretion (Mujica-Parodi and Strey, 2006) have suggested that males tend to be better secretors, whereas females are better detectors of the stress pheromone. For this reason, 20 male participants were recruited in the first phase of the study that was advertised as an investigation into 'stress and shopping'. For the second phase of the study 15 female participants were recruited.

Apparatus and Procedural Overview

The hostile intent study progressed through three stages: pre-trial preparation, baseline data collection, and the main study that contained two traverses of the shopping mall. The baseline and main study took approximately two hours to complete.

In the pre-trial preparation, the male participants were provided with an odourless soap and asked to wash with it and instructed to avoid using perfumes, body sprays and deodorants, and to refrain from alcohol, odorous food (such as curry), cabbage, and asparagus in the 24 hours prior to taking part in the study.

Baseline data were collected with the participants seated and wearing a wireless Bioharness (produced by Biopac Systems, Zephyr Technology Ltd) along with cotton wool pads attached under their armpits. Participants completed a short form of the State Trait Anxiety Inventory (STAI) questionnaire (Marteau and Bekker, 1992). This is a six-item scale derived from the original larger scale developed by Spielberger (1983). In addition, participants provided a series of salivary cortisol samples. Participants then sat quietly for 15 minutes during which time baseline data were recorded for heart rate, respiration and body temperature. Afterwards, another salivary cortisol sample was collected and another STAI questionnaire was completed. Following this, the under arm pads were removed and all samples were placed in a freezer as soon as possible. Participants had been previously informed that for completion of this part of the experiment they would be paid £10.

In the main trial, participants were transported to a nearby shopping mall. All data were collected between 10:00 a.m. and midday when the shopping mall was reasonably busy but not overcrowded. Participants were shown a map of the shopping mall and given instructions of a route they were to follow. They were told that in order for the researchers to be able to confirm that they had taken the correct route, they were to take photographs at six specific locations. The first traverse was designated as a 'Low Intent' condition and participants were informed that if they completed the task successfully they were to be paid £20. The Bioharness and pheromone pads were re-attached, a salivary cortisol sample obtained, and the STAI questionnaire completed. The Bioharness was activated when the participant began their traverse through the shopping mall. The traverse was a circular route and on their return the Bioharness data was downloaded and the sweat samples collected. A further salivary cortisol sample was also obtained and an STAI questionnaire was completed.

The final part of the main study was designed as the 'High Intent' condition to increase stress levels in an attempt to mimic a hostile reconnaissance activity. Participants were informed that for this part of the study, their traverse would be timed and that they had to complete it more quickly than the previous traverse. Participants were also informed that CCTV and plain-clothes security staff would be looking for any unusual or suspicious behaviour. Although the security guards would not know the specific identity of the participants, each participant was told that two previous participants had been detected (this was not actually the case). Participants were also informed that if they agreed to take part in this final activity of the experiment, and were successful, they would be paid a further £70. However, if they were identified, or did not beat the timed deadline then they would not only lose the £70, but they would also lose the £20 they had been paid for the first traverse. To be sure they understood these instructions they were given a written version that required their informed consent before continuing. At the end of the second traverse, all participants were informed that they had been successful and would receive the full payment of £100 (£10 baseline payment, £20 first traverse payment and £70 second traverse payment).

Participants in both the Low Intent and High Intent conditions were recorded on CCTV installed in the shopping mall. In the second part of the study a different group of participants were presented with recordings of the two participants who had the lowest and highest STAI questionnaire scores. The CCTV recordings were presented in Low Intent and High Intent pairings in a randomised order and were rated for the perceived stress levels.

The sweat sample pads from under the participants' non-dominant arm were placed in ziplock plastic bags for refrigerated storage at –20C. These were used as the basis of a 'sniff test' to test human detection of pheromones. In this part of the study participants were instructed not to wear any perfumes, hair/body sprays that might compromise the odours being characterised. Participants undertook an initial sense of smell test using the sample of lavender oil in a jar of water and a checklist to identify the odour. Participants were asked to open the ziplock bag, place their nose in the bag and to inhale deeply. This part of the study took approximately 30 minutes to complete and participants were paid £10 for their time.

Results

A total of 19 participants completed the 3 phases of the field study. In order to demonstrate the validity of this model of hostile intent it was necessary to illustrate that it is possible to induce stress responses during the High Intent phase of the study, that were significantly higher than levels of stress obtained for the Baseline and Low Intent conditions.

In the figures below the psychological, physiological, behavioural and biological data obtained for the three phases of the field study are shown. The baseline measures were obtained under laboratory conditions with participants in a relaxed state with minimal stress. The Low Intent condition refers to the first traverse of the shopping mall, with no attempt made to induce stress. In the High Intent condition, participants completed the traverse under simulated hostile reconnaissance (i.e. with high levels of induced stress).

Psychological Responses

The STAI questionnaire was completed by participants at the start and end of each of the three conditions. Total scores ranged from 6 (minimum) to 24 (maximum). As illustrated in Figure 8.3, self-reported levels of stress increased across the three conditions.

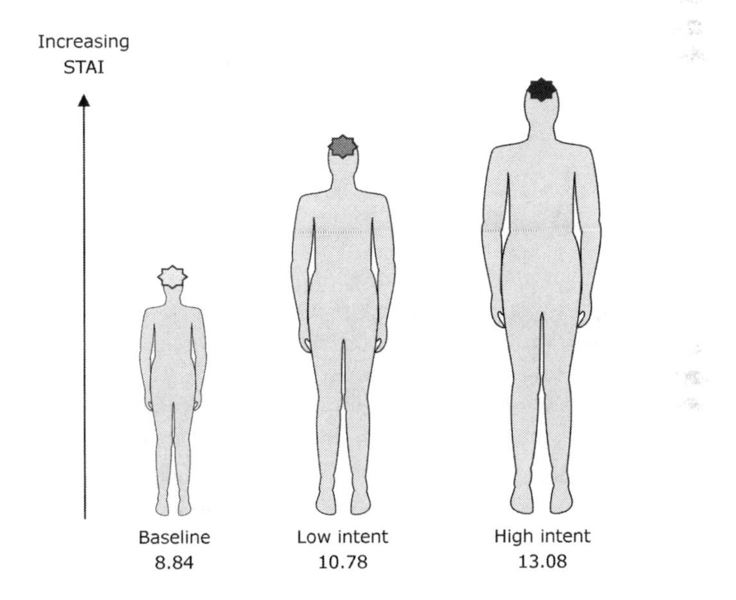

Figure 8.3 Psychological responses – STAI results (mean data)

Physiological Responses

Mean heart-rate data (beats/minute) were compared and significant statistical differences were observed between the baseline and Low Intent conditions ($p < 0.001$) and the Low Intent and High Intent conditions ($p < 0.001$) as illustrated in Figure 8.4. It should be noted that the differences between the baseline and Low Intent can be explained by the difference in activity (seated vs walking). However, in relation to the Low Intent and High Intent conditions, the task and movement involved were essentially the same. Although the High Intent condition had a time factor attributed to it and a marginal amount of the heart rate differences

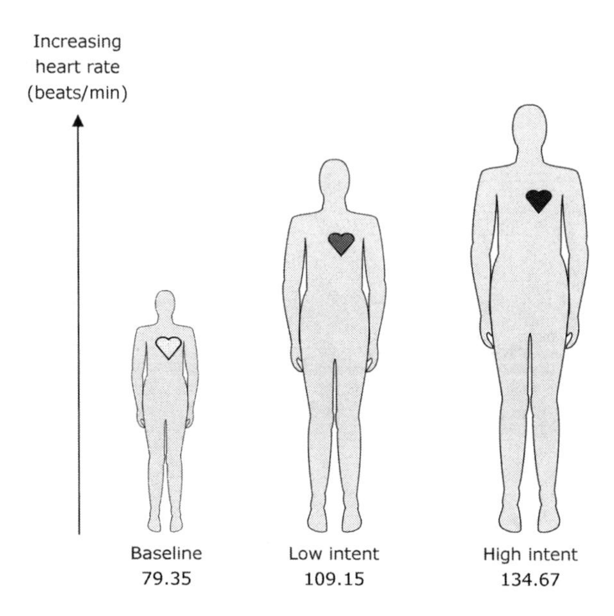

Figure 8.4 Physiological responses – heart rate results (mean data)

can attributed to increased energy expenditure, it is suggested that the major contributor to the difference observed was the level of induced stress afforded by the nature of the task.

Respiration was measured in breaths/minute and would be expected to increase in times of stress. The mean respiration data illustrated significant statistical differences between the baseline and Low Intent conditions ($p < 0.001$) and also between the Low Intent and High Intent conditions ($p < 0.01$) as represented in Figure 8.5. The caveat applied to the heart rate data must also be applied when examining the respiration data. An increase in respiration during the Low Intent condition over the baseline condition would be expected, but the increase observed between the High Intent and Low Intent conditions suggested a response to additional induced stress.

During periods of stress, when the fight or flight response is activated, blood flow is directed from the extremities to the body core in order to facilitate increases in heart rate and respiration associated with increased arousal. As a result hands and feet may feel cold or clammy, whereas body temperature should show an increase. Long, Vander and Kluger (1990) showed that increases in body temperature, hyperthermia, could be induced in rats through psychological stress. For humans, there is evidence that appears to show a relationship between body temperature and improved performance, alertness and visual attention (Wright, Hull and Czeisler, 2002); all attributes that might be expected to improve during periods of stress.

The changes in body temperature observed in participants during this study are represented in Figure 8.6.

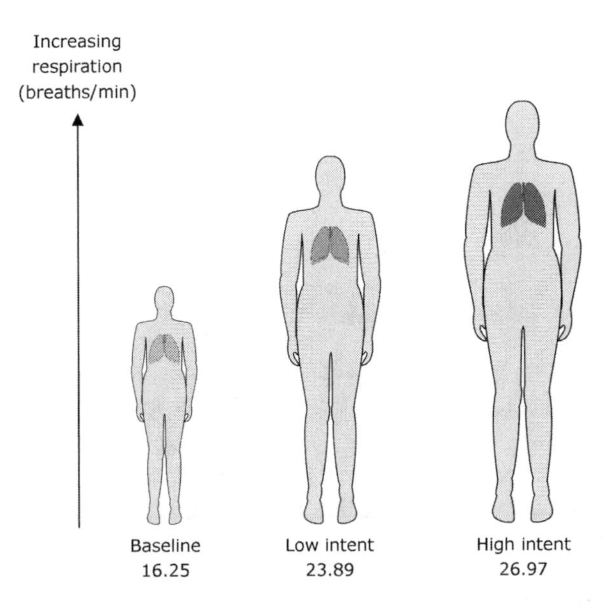

Figure 8.5 Physiological responses – respiration results (mean data)

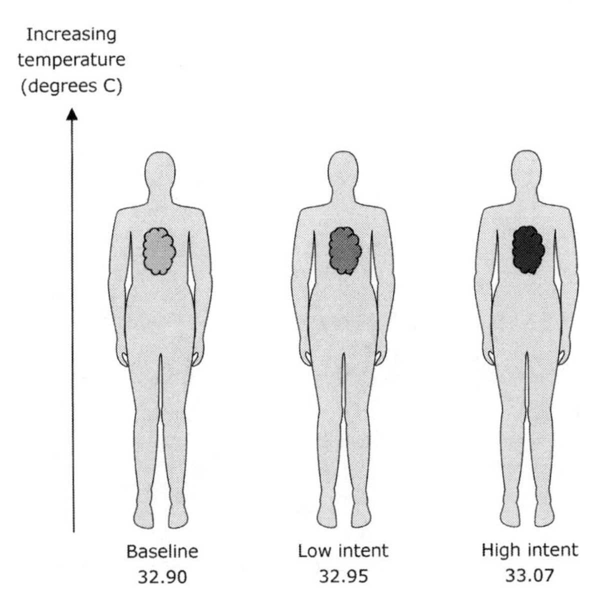

Figure 8.6 Physiological responses – body temperature (mean data)

Although the data illustrated a trend in the predicted direction with body temperature in the High Intent condition being slightly elevated relative to the other two conditions, this difference failed to reach statistical significance (p >0.05).

Biological Responses

Measures of salivary cortisol were obtained at the start and end of each of the three conditions, as described in the procedure. The interval of time between obtaining the first and second samples of cortisol varied slightly within individual participants but was approximately 15 minutes, and it was hypothesised that changes in cortisol levels during this period would be an indication of changes in arousal levels. Therefore the analysis looked for significant changes in levels of cortisol across the three conditions (Figure 8.7). The mean levels of cortisol revealed that there was no significant change between the baseline and Low Intent condition (p >0.05) however, an increase in cortisol levels during the High Intent condition was found to be statistically significant to the other conditions (p <0.05). This finding illustrates that changes in stress occurred during the High Intent condition, thus giving support to the model of hostile intent.

For the pheromone analysis, axillary sweat samples from under the participants' dominant arms were collected during the baseline, Low Intent and High Intent phases. Samples were stored in 50 ml sample tubes containing HPLC grade acetone (10ml) for analysis using GC-MS. Dominant arm pads were collected to control for any possible under arm dominance effects. The results illustrated a change in the baseline compared to the more stressed samples with the peak

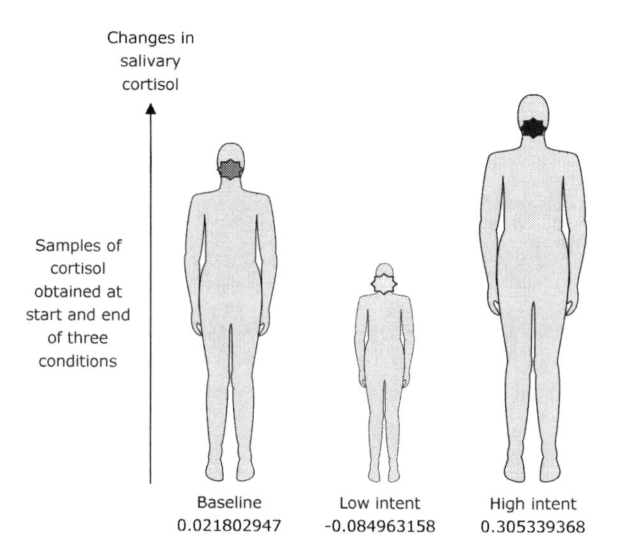

Figure 8.7 Biological responses – cortisol results (mean data)

at approximately 35.5 minutes. Mass spectrum analysis identified the compound as Sitosterol. However the presence of several fragments of higher mass in the spectrum suggest that Sitosterol is likely to be a breakdown product of a larger sterol which is fragmented by the analysis process.

The bar chart in Figure 8.8 illustrates the difference between the stressed samples and the baseline2 samples. This shows that in all cases, the level of the sterol drastically increased as a response to the test conditions.

There did not appear to be a trend in terms of any increase in the concentration of the sterol between the Low Intent and High Intent conditions. Thus, these results provide evidence of the sterol being a marker for, rather than a gauge of, the level of stress.

When the changes in sterol levels were compared with changes in cortisol levels it was found that the data shows that the concentration of the sterol increases as a function of stress regardless of the cortisol levels in the saliva samples. This finding supports the evidence for a compound that is secreted during times of stress. Further work will be required to confirm (or otherwise) that this compound is related to a stress pheromone.

Behavioural Responses

For the behavioural analysis participants were presented with Low Intent and High Intent video clips for individuals in the main study and were requested to rate the perceived stress levels.

No significant differences in the stress ratings obtained for the four CCTV clips were observed. Interestingly, the correlations between the pairs of CCTV also failed to reach significance, suggesting that stress does not have an immediately observable impact on the behaviour recorded in these CCTV clips.

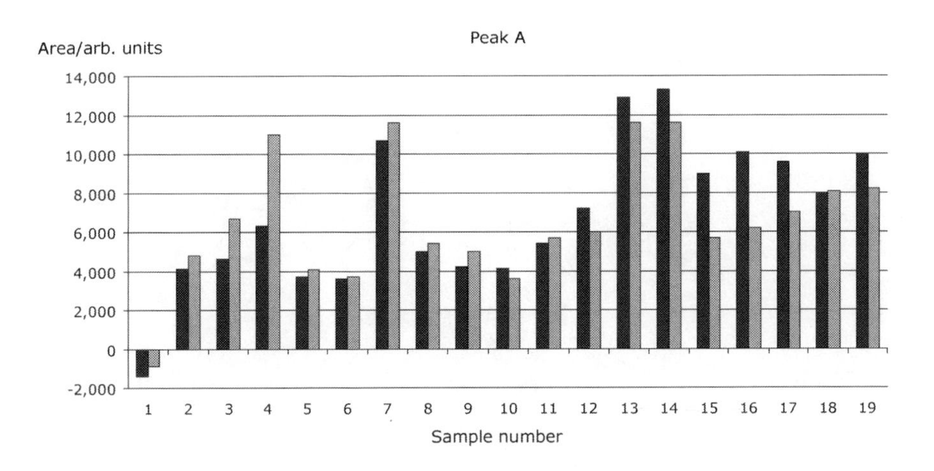

Figure 8.8 The relative concentration of Sitosterol (peak A)

Human Detection of Stress Pheromones

Olfactory analysis (the 'sniff test') was conducted on the axillary sweat samples taken from the non-dominant armpits of four of field trial participants (Figure 8.9). The samples were chosen on the basis of the STAI scores (e.g. lowest and highest during High Intent Condition) and Heart Rate (e.g. lowest and highest during High Intent Condition). This was done for each of the three conditions (e.g. baseline, Low Intent and High Intent) so that each participant in the olfactory analysis study had to sniff a total of 12 sweat samples.

The 15 female participants rated the samples across 8 characteristics: pleasantness, honesty, fear, stress, concealment, masculinity, femininity and guilt. However, the only significant result was observed for Masculinity in the Low Intent condition. A point to note is that the participants were sniffing individual pads, and it might be that the concentration of pheromone in individual sweat pads is too low to be detected by the human nose.

Discussion

The purpose of the field trial was to extend and build on previous laboratory research (Eachus, Stedmon and Baillie, 2010) to improve the ecological validity of our Model of Hostile Intent. In order to do this the field trial had to establish that it was possible to simulate the conditions that might be similar to those faced by an individual engaged in hostile reconnaissance. If successful, this simulation would induce in the participants the psychological, physiological, biological and behavioural concomitants of stress during the High Intent phase of the experiment. The results clearly show that to a large extent this has been achieved.

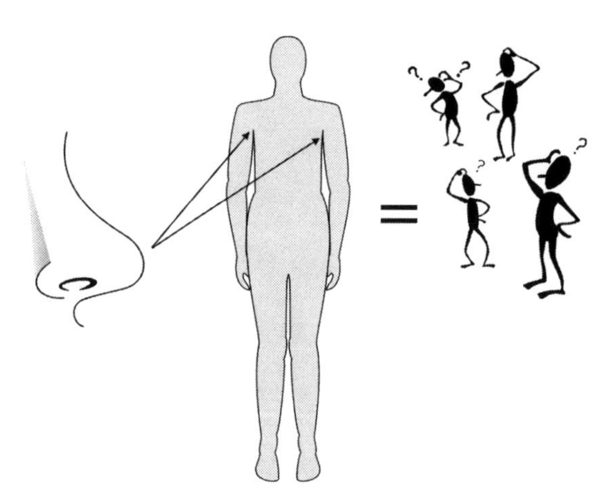

Figure 8.9 Sniff test

During the High Intent phase of this study, the task involved reconnaissance, concealment, and a degree of self-control, all of which contributed to the experience of stress, and which therefore constituted a very good approximation of hostile intent. The psychological, physiological and, to some extent, biological parameters all changed in the way that would be predicted of a person engaged in hostile intent.

The behavioural analysis has not yet revealed the anticipated results in that it was not possible to identify specific behaviours that might distinguish hostile intent. However, this aspect of the research is important in that it has provided researchers with a CCTV database of behaviours that have been produced during a simulation of hostile intent. This database can therefore be used by others to test behavioural hypotheses associated with hostile intent in the full knowledge that the behaviours being observed are those of people who are experiencing the psychological, physiological and biological changes that are likely to be felt by those involved in hostile reconnaissance.

The absence of significant behavioural differences also supports the notion that the participants were genuinely stressed and that the changes in the physiological parameters were not brought about by any changes in behaviour (e.g. walking more quickly).

All the participants in the field trial were young white males and, arguably, future terrorist activity will see more females involved (see Chapter 17). In this study the secretion and detection of stress pheromones was an important feature, and since males appear to be better secretors and females better detectors of pheromones it seemed reasonable to use this approach. However, given that gender roles in terrorist activities are likely to become more fluid, these assumptions will need to be examined by further research along with other factors such as sampling the pheromones from 'noisy' samples that might be contaminated (or actively disguised) by perfumes and other chemicals.

Conclusion

The research presented here has demonstrated that it is possible, to a large extent, to accurately simulate terrorist behaviour during hostile reconnaissance. Furthermore, we have demonstrated the feasibility of employing psychological, physiological and biological markers, such as pheromone production, to distinguish between the individual when they are simulating hostile intent and the same individual when they are not. Indeed these results suggest that individuals placed under stress secrete a volatile steroid-based marker that has the potential to be exploited as the basis of remote chemical detection. In this field trial of hostile intent the predicted changes, psychologically, physiologically and biologically have all been confirmed. It is suggested that this validation of the model will prove fruitful in further research that seeks to understand hostile intent.

Acknowledgement

This research was funded by Home Office contract: CDE12123.

References

Ackerl, K., Atzmueller, M., and Grammer, K. (2002). The scent of fear. *Neuroendocrinology Letters*, 23, 79–84.

Boeckh, J., Kaissling, K.E., and Schneider, D. (1965). Insect Olfactory Receptors. Cold Spring Harbor Symposium. *Quantitative Biology*, 30(1965), 263–280.

Butenandt, A., Beckman, R., and Hecker, E. (1961). On the sex-attractant of silk moths. I. The biological test and the isolation of the pure sex-attractant bombykol, *Hoppe-Seyler's Zeitschrift für Physiologische Chemie* (*Hoppe-Seyler's Journal of Physiological Chemistry*), 324, 71–83.

Cannon, M. (2004). *Airport surveillance and targeting threat*. Available from www.drum-cussac.comlfeatlfeat2.html, accessed 1 March 2013.

Chen, D., Katdare, A., and Lucas, N. (2006). Chemosignals of fear enhance cognitive performance in humans. *Chemical Senses*, 31(5), 415–423.

Curran, A.M., Rabin, S., Prada, P.A., and Furton, K.G. (2005). Comparison of the volatile organic compounds present in human odor using Spme-GC/MS. *Journal of Chemical Ecology*, 31(7), 1607–1619.

Department of Homeland Security Information Bulletin (2003). *Potential Indicators of Threats Involving Vehicle Borne Improvised Explosive Devices (VBIEDs); May 15, 2003*. Available from http://www.sifma.org/uploadedfiles/services/bcp/homeland_security_information_bulletin_051303.pdf, accessed 1 March 2013.

Department of Homeland Security (2013). *Science and Technology Directorate Human Factors and Behavioral Sciences Division Proj*ects. Available from: http://www.dhs.gov/files/programs/gc_1218480185439.shtm#12, accessed 1 March 2013.

Eachus, P., Stedmon, A., and Baillie, L. (2010). *A Model of Hostile Intent in Crowded Places: Final Report in Fulfillment of Home Office Contract: CDE12123*, Home Office: London.

Eachus, P., Stedmon, A., and Baillie, L. (2013). Hostile intent in public crowded spaces: A field study. *Applied Ergonomics*, 44, 703–709.

Grosser, B.I., Monti-Bloch, L., Jennings-White, C., and Berliner, D.L. (2000). Behavioral and electrophysiological effects of androstadienone, a human pheromone. *Psychoneuroendocrinology*, 25(3), 289–299.

Haussmann, M.F., Vleck, C.M., and Farrar, E.S. (2007). A laboratory exercise to illustrate increased salivary cortisol in response to three stressful conditions using competitive ELISA. *Advances in Physiology Education*, 31(1), 110–115.

Kuhn, F., and Natsch, A. (2009). Body odour of monozygotic human twins: a common pattern of odorant carboxylic acids released by a bacterial aminoacylase from axilla secretions contributing to an inherited body odour type. *Journal of the Royal Society Interface*, 6(33), 377–392.

Linett, H. (2005). *Living with Terrorism: Survival Lessons from the Streets of Jerusalem*. Paladin Press: Boulder, CO.

Long, N.C., Vander, A.J., and Kluger, M.J. (1990). Stress-induced rise of body temperature in rats is the same in warm and cool environments. *Physiology and Behavior* 47, 773–775.

Marteau, T.M., and Bekker, H. (1992). The development of a six-item short-form of the state scale of the Spielberger State-Trait Anxiety Inventory (STAI). *British Journal of Clinical Psychology*, 31, 301–306.

Millar, J.G. (2002). Sampling and sample preparation for pheromone analysis. *Comprehensive Analytical Chemistry*, 37, 669–697.

Mujica-Parodi, L., and Strey, H. (2006). *Identification and Isolation of Human Alarm Pheromones*. Report for U.S. Army Medical Research and Materiel Command, Fort Detrick, Maryland 21702–5012, The Research Foundation of SUNY: Stony Brook, New York

Mujica-Parodi, L.R., Strey, H.H., Frederick, B., Savoy, R., Cox, D., Botanov, Y., Tolkunov, D., Rubin, D., and Weber, J. (2009). Chemosensory cues to conspecific emotional stress activate amygdala in humans. *PLoS One*, 4(7):e6415. doi: 10.1371/journal.pone.0006415.

National Archives (2008). *Operation Fairway*. Available from http://webarchive.nationalarchives.gov.uk/20120810121037/http://www.highways.gov.uk/business/19767.aspx, accessed 1 March 2013.

Penn, D.J., Oberzaucher, E., and Grammer, K. (2007). Individual and gender fingerprints in human body odour. *Journal of the Royal Society Interface*, 4, 331–340.

Porter, R.H. (1998). Olfaction and human kin recognition. *Genetica*, 104, 259–263.

Prunty, H., Andrews, K., Reddy-Kolanu, G., Quinlan, P., and Wood, P.J. (2004). Sweat patch cortisol: A new screen for Cushing syndrome. *Endocrine Abstracts*, 7, 202.

Rains, G.C., Tomberlin, J.K., and Kulasiri, D. (2008). Using insect sniffing devices for detection. *Trends in Bioltechnology*, 26, 288–294.

Regnier, F.E., and Law, J.H. (1968). Insect pheromones. *Journal of Lipid Research*. 9, 541–551.

Seyle, H. (1974). *The Stress of Life*, 2nd edn. New York: McGraw-Hill.

Singh, D., and Bronstad, P.M. (2001). Female body odour is a potential cue to ovulation. *Proceedings of the Royal Society B (Biological Sciences)*, 268(1469), 797–801.

Spielberger, C.D. (1983). *State-Trait Anxiety Inventory for Adults*.Mind Garden: Palo Alto, CA.

Wright, K.P. Jr., Hull, J.T., and Czeisler, C.A. (2002). Relationship between alertness, performance, and body temperature in humans. *American Journal of Physiology: Regulatory, Integrative and Comparative Physiology*, 283(6), 1370–1377.

Chapter 9

On the Trail of the Terrorist: A Research Environment to Simulate Criminal Investigations

Alexandra L. Sandham
SCORPIO Centre, Psychology Department Lancaster University, UK

Thomas C. Ormerod
School of Psychology, University of Surrey, UK

Coral J. Dando
Department of Psychology, University of Wolverhampton, UK

Tarek Menacere
SCORPIO Centre, Psychology Department Lancaster University, UK

Introduction

At a time of raised terrorist threat, research that assists the security and police services to protect people and infrastructure is timely. At the same time, high-pressure security environments can result in inappropriate surveillance, misguided and ill-timed arrests and time-consuming and unnecessary interviews with arrested suspects. These problems have significant negative political, public relations and intelligence ramifications. Yet there is little to give confidence to those tasked with managing the terrorist threat that anti-terrorist operations can yield high-quality evidence or can conclusively demonstrate that a deceptive activity was in the process of being planned or executed. It is our contention that the sensitive application of human-centred design principles to technology development can play an important role in counter-terrorism. Here, we outline the DScent project (Detecting Scent trails), which aimed to explore a number of processes and tools to facilitate effective evidence gathering and interpretation of high-pressure, time-sensitive environments that are typical of many counter-terrorism investigations and interventions.

Scent Trails for Evidence-gathering and Use

The thesis of the DScent project was that, during an investigation into terrorism, communication and movement behaviours are likely to provide indicators of malicious intent prior to an attack. Furthermore, if these movement and communication indicators of mal-intent can be recognised and harvested prior to a terrorist event, they can support a number of positive intervention strategies. For example, those tasked with policing and protecting society would be better positioned to target limited surveillance resources, reduce the number of false positive arrests of innocent suspects, and carry out disruption-based tactics.

The concept of a scent trail is based on a rich and integrated time-, interaction- and location-based history of movements, communications and behaviours of individuals, working either alone or as a team. One of the DScent project's research challenges was to determine whether scent trails could be analysed to reveal anomalous behaviours that differentiate between innocent activities unrelated to serious crime and deceptive activities with terrorist intent. By 'scent trail' we refer to the set and trajectory of indicators left over time when an individual or group:

- visits locations revealed by positioning systems such as GPS, CCTV images or number plate recognition;
- engages in communications monitored and interpreted via signals analysis; or
- carries out transactions such as ticket purchases that leave electronic records.

By integrating and interpreting across these different indicators, a scent trail provides a detailed account of movements and activities prior to an intervention (e.g. questioning or arrest). In addition, a scent trail potentially offers predictions of trajectory for future locations and activities of the suspects. If one understands what kinds of activities and behaviours might be markers of mal-intent, scent trails could be used not only to aid monitoring of the suspects but also to provide evidence for different forms of deception for investigative purposes (see Chapters 6, 7, 10 and 19 in this volume).

One use for scent trails is in issuing 'challenges' to those suspected of wrongdoing, both in real-time while they are being monitored and during post-arrest interviews. A challenge might consist of presenting a suspect with scent trail information. For example, two individuals who deny knowledge of each other could be shown in the same unusual location at the same time. Scent trail challenges can potentially yield two benefits. First, the challenge might undermine a suspect's account during a post arrest interview (Dando and Bull, 2011, Dando et al., 2013). Second, being presented with incriminating data might change a suspect's behaviour, leading to the aborting of an attack.

Scent trails also provide an evolving set of data that can be used as evidence. This evolving disclosure of evidence is an important aspect of a criminal investigation in terms of ecological validity: rarely does an investigation team receive all the evidence for a case at once.

Using Games to Explore Scent Trails

Terrorist activities and the investigative practices used to curtail them are too sensitive to be studied directly. Yet, given the continual evolution of terrorist methods, to limit research to known event types (e.g. the 7/7 bombings) would be counter-productive. With few exceptions (e.g. cyber-terrorism) terrorist activity has a spatial or geographic context. Spatial analytical techniques to anticipate areas of crime or terrorism and to gather intelligence and intercept such activity is a growing area of research (Taylor, Bennell and Snook, 2009). Positioning data can be exploited to analyse behaviour patterns and interactions with outcomes forming location-based intelligence. From positioning data it is possible to deduce information such as where the participant was located at a particular time; the speed at which they were travelling; whether they would have had mobile phone coverage, and so on. The speed at which someone is travelling can indicate what transport they are using, and how long they spend at a location can suggest different activities that they might be engaged in (Liao et al., 2007).

The use of game environments is increasing as a psychological research tool (Washburn, 2003), providing for complex interactions between individuals and their environments to be simulated in ways that are safe and controllable. Games concerned with purposes beyond the gaming experience itself are referred to as serious games (as opposed to commercial games). Serious games have been used to simulate real-life situations that are costly or sometimes impossible to create in any other way. They provide a safe environment where valuable training or exploration can be gained at very low cost. Serious games have been used for military training (Numrich, 2008), health training (Sawyer, 2008), therapy and rehabilitation (Burke et al., 2009), education (Ardito et al., 2008), and emergency response (Chittaro and Ranon, 2009).

The rationale for using a game for research is to provide a controlled experimental environment where issues that are not amenable to real-world testing can be examined in a realistic but controlled manner. For example, games have been used to investigate the effects of different decisions made with reference to sustaining or transforming civilisations in an environment where emerging technologies can have progressive or destructive effects (Gorman, 2009). A serious game approach allowed the DScent team to explore scent trails in a non-sensitive simulation. The simulation provided an analogy to deceptive activities, and mixed minor civil deceptions with deceptions linked to a major terrorist event, thus simulating the complexity of real crime detection.

Requirements for the Research Game

The most widely used research paradigm in the field of deception detection research is a mock-crime scenario. Mock-crimes vary in complexity, from stealing a sum of money (Bell, Kircher and Bernhardt, 2008; Honts, Amato and Gordon, 2004; Vrij, 2008; Vrij, Mann and Fisher, 2006a, 2006b; Vrij et al., 2008); and buying or selling drugs (Hartwig, Granhag, Stromwall and Vrij, 2004; Strömwall, Hartwig and Granhag, 2006); to hostile reconnaissance (see Chapter 8). However, participants in these studies are typically told exactly how to carry out the mock-crime. As a consequence, the opportunity to observe realistic crime planning and the typical deceptions associated with concealing the crimes is limited, since participants are usually constrained by an experimenter's script. Moreover, the whole idea of 'deception' in such studies is questionable, since participants assigned to a deceptive condition are acting with good intent when they recount a deceptive event as required by the researcher. Thus, in this research, we set out to create an environment within which participants developed their own deceptive game play strategies and thereafter developed their own lies with respect to their game playing behaviour. The aim was to design a task that a participant in the terrorist role could conceal both during game play and afterwards in an investigative interview.

In order to test scent trail capture and use, a list of game requirements was specified to inform the game design. The overall concept was to design a game that simulated an environment in which members of the public were going about their daily lives aware that there was heightened security and that they were 'being watched' but unaware that one or more of their number was attempting to mount a terrorist attack. The game required three types of players, representing the general public, terrorists, and investigators. The game was based around a race to build sections of an Olympic site. The first player or team to complete the purchase and delivery tasks required to finish their site were the game winners.

To create scent trails containing differing types of surveillance information, players travelled across a game-playing space, making purchases from shops, carrying resources needed for tasks and delivering these resources to a specific location. This activity simulated a mix of legitimate activities (e.g. a building company developing a site) with mal-intent (e.g. a terrorist cell preparing an attack while masquerading as legitimate builders). These activities provided positional and purchasing information. Further information was provided to the investigators by a time-phased check of the stock carried in each shop. Investigators were also able to inspect a limited number of players and the items they had assembled during each round of the game. This was intended to represent an investigative activity of 'stop and search' policing tactics.

It is often necessary to investigate occurrences where individuals might withhold information for reasons other than an involvement in a crime. For example, a person may lie to police about their whereabouts, not because they were involved in criminal activity, but because they had told their employer they were sick when they were in fact attending a social engagement. To simulate the difficulty investigators face in determining what are major offences against a background of minor violations, the game provided opportunities for all players to gain advantages by a small rule violation. Scent trails resulting from cheating behaviour would introduce 'behaviour noise' in a manner indicative of that experienced by professional investigators.

In a further attempt to simulate a real investigation, game play was time phased, and scent trail information was presented to investigators in a fashion that mirrored the way an event unfolds in real life, with evidence becoming available in a sequential nature. Gaming information, for example a stock check of items sold and a list of locations visited by each player (but not which player had bought which items), was delivered to investigators at stages throughout the game, replicating the evolving nature of a terrorist event and gradual revelation of event information during the investigative process.

While the research described in this chapter concentrates on results from the detecting deception paradigm, the game design allows for a number of other research avenues to be explored. The evolving nature of the revelation of evidence will allow investigative strategies and processes to be studied, for example the development of confirmation bias. Confirmation bias appears stronger when information is presented sequentially than simultaneously (Jonas et al., 2001). During an investigation, information and evidence typically arrives sequentially. Therefore the investigator tends to make sequential decisions related to the reliability of the evidence and what further evidence is needed.

Although the game design allowed a specific set of game rules to be implemented, a number of elements remained flexible. This flexibility allowed different scenarios to be implemented according to the needs of end-user researchers. For example, the facilities for player numbers, player roles, shops, resources, cash and the number of players to be investigated were all changeable. This approach allowed the game to be changed, for example, from a scenario of investigating terrorist scent trails to one of exploring the movement and employment of migrant workers.

Design and Implementation

A web-based environment was designed to implement the game based upon the context of building contractors responsible for building part of a sports facility. A range of factors was considered during the game design, including game facilities; software environment; players aims; game aims; play; and investigations. These are expanded below.

Game facilities. The following features were automated elements of the game:

- the game-playing space or board – the current game play situation including the position of all players is displayed on each of the players laptops (but not on the investigators' laptop)

- dice roll to dictate how far the players move within each 'turn'
- virtual bank account – to allow money to be paid in and out
- resource purchase facility – showing the cost and weight of each item to be purchased
- a virtual van – used to carry the purchases
- resource 'drop-off' point
- investigation points – where evidence was revealed to the investigator/s.

Software environment. The game software was divided into four main components: a game configuration environment, a player environment, an investigator environment, and a database that stored and made available player and investigator data according to game rules. The game configuration environment provided the facility to set the game up in terms of player numbers, shops, resources, cash and investigation points. The player environment provided each player with the following information: the game board and the position of all other players, the cash in their bank account, resources purchased and stored either in the van or at the site. The player environment changed when a player entered a shop to provide a resource purchase interface. The investigator environment provided the investigators with time-phased snapshots of information including the places visited by the players and the stock purchased from each shop. This interface also allowed the investigator to choose the players to be investigated.

Players and game aims. The game had up to six players and an investigator. Players were either builders whose aim was to assemble resources in order to construct an 'event stage', or terrorists whose aim was to appear to have the same goal as the regular builders but use this as a cover for gathering resources to mount an attack. The goal of the investigator was to discover and identify which players were builders and terrorists. The investigator was given details of the builder's task, and they used this knowledge as the model for expected builder behaviour. For example, part of the builder's task required them to purchase two sticks of dynamite, therefore, if during game play, it became apparent that a player had purchased more than two sticks it would raise suspicion.

The investigator was also aware of the resources available for purchase at each shop, including the price and weight of each item. This information was useful as the nature and quantity of certain items could act as potential indicators of illegitimate activity by players. For example, while the purchase of small amounts of fertiliser late in the game was consistent with the legitimate building tasks, large amounts might be indicative of a terrorist attack preparation. Terrorist players who planned strategically would try to hide such activities (e.g. by purchasing small amounts throughout the game).

Akin to police powers of 'stop and search', the investigators were able to investigate two players at set points throughout the game. An investigation consisted of weighing the van of a suspected player and inspecting two items being carried in that van. Although a terrorist was given a different task to a builder, the tasks were similar in nature, thus ensuring that any deceptive behaviour did not stand out per se. Participants were given game-playing instructions individually via a laptop computer that included: their role as a builder (non-deceiver) or terrorist (deceiver); the game rules; an explanation of the consequence of breaking the rules (i.e. a fine which would reduce the players ability to purchases items thereby potentially slowing down their game play); that investigator/s were observing the game play and that all players would be interviewed upon completion of the game. All players were informed what investigator/s could and could not see, and told to keep the details of their task secret. Additionally, all players were encouraged to consider cheating by overloading their vans. The benefit of cheating for research was that cheating would introduce a level of minor offence so that the investigators had a more realistic investigation task to carry out rather than just looking to identify any 'bad behaviour'. Additionally if a player had cheated they were more likely to tell lies to the investigator during the interview trying to cover up their cheating, thus allowing the investigators to judge spontaneous lies and individual lies from some of the players against truth from others.

Play. Players took turns to move around a board consisting of two shops and a site (Figure 9.1), the number of steps on each move being determined by a simulated dice roll. The use of dice throws was an attempt to replicate the effect of traffic lights and/or traffic jams within a real terrorist scenario that may cause strategic and behavioural changes as a result of slow progress. A toll road between the two shops had two purposes. First, after payment of a toll, it allowed rapid movement between the two shops as a player's dice throw was doubled on the toll road. Second, players could hide from investigators during a simulated stop-and-search as a player on the toll road was not available for investigation. Each player had a laptop upon which the game software resided and each player had an initial balance of virtual money to spend on resources. A game administrator controlled the pace of the game, operating the game software and helping the players with any technical difficulties they may have had. Help was usually required for the first round of play while players got to grips with the software, but was rarely needed on subsequent rounds.

Each player started from his or her own depot (Figure 9.1). As the game progressed, players purchased resources (e.g. wiring, construction blocks and dynamite) from two shops. The shops were identified as a builder's merchant (Figure 9.1: shop with cross tools logo) and an electrical supplies store (Figure 9.1: shop with universal electric sign). These resources were carried in a virtual van between the place of purchase and the site.

Figure 9.2 illustrates the interface that players used for resource purchase while in a shop. It also illustrates the van's inventory (there is nothing currently in the van shown in Figure 9.2).

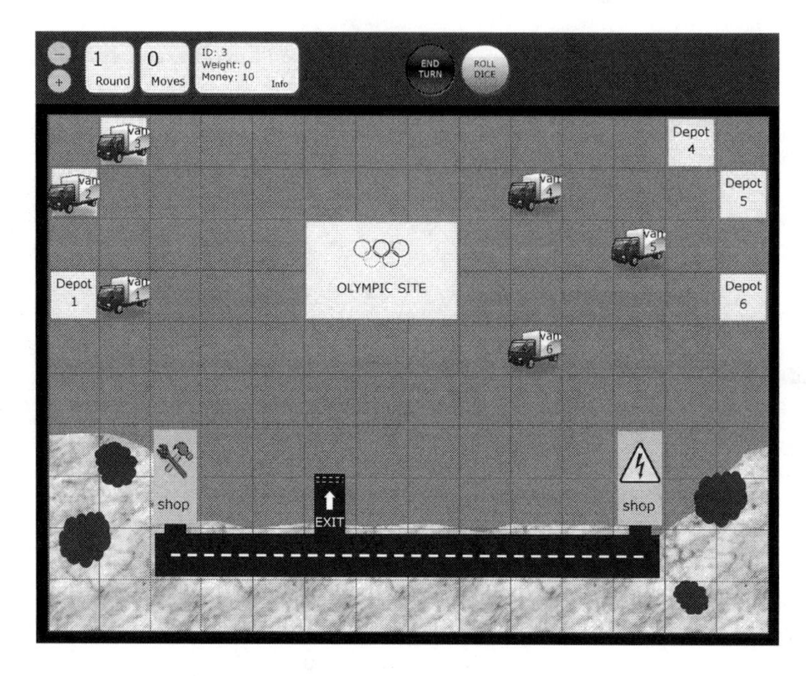

Figure 9.1 Game board. Screenshot from Cutting Corners Board Game (Dando et al., 2013)

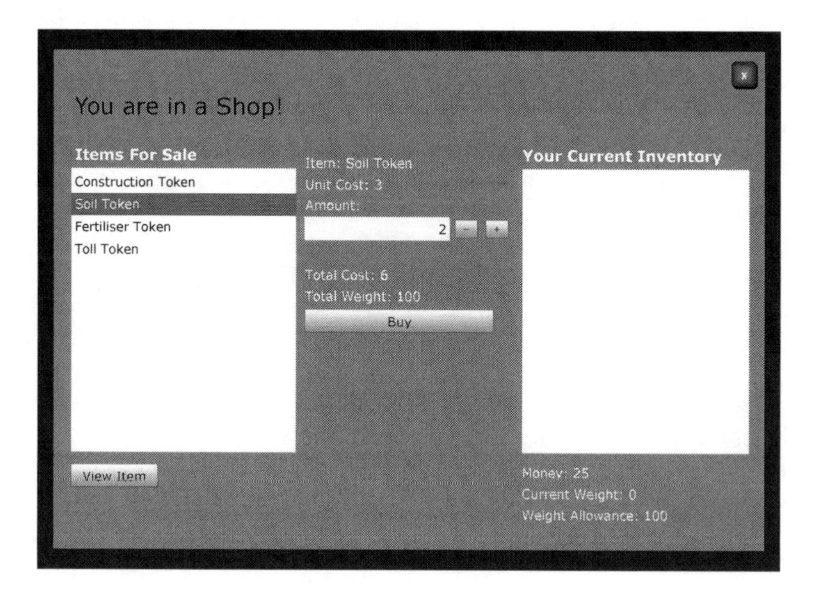

Figure 9.2 The player interface for the electrician's shop. Screenshot from Cutting Corners Board Game (Dando et al., 2013)

Investigations. At regular points throughout the game, the investigator(s) carried out an investigation. Investigators could inspect any two vans where one would be weighed and the other would have two items inspected. The investigator resided in a separate room from the players and the information was delivered to them at specific points in the game via a laptop computer. During an investigation phase of the game the investigator was able to see the locations each player had visited, but not the detail of the route taken. They could also see what stock had been purchased from each shop but were unable to see which players had purchased specific items. The scent trail provided to the investigator comprised the following information: the locations visited by each player; the stock purchased from each shop; the resources a player had shown during an investigation and the weight of each investigated van. The scent trail plus the investigator's knowledge of the builder's task along with knowledge of the weight and cost of each resource formed a basis from which deductions were made. These deductions were developed in real-time in an attempt to identify what players were doing and therefore which player might be investigated when the next opportunity arose. In addition, deductions based upon the scent trails evidence were used during post-game interviews to check the coherence of the player's own accounts. Figure 9.3 illustrates the typical scent trail information delivered to the investigator.

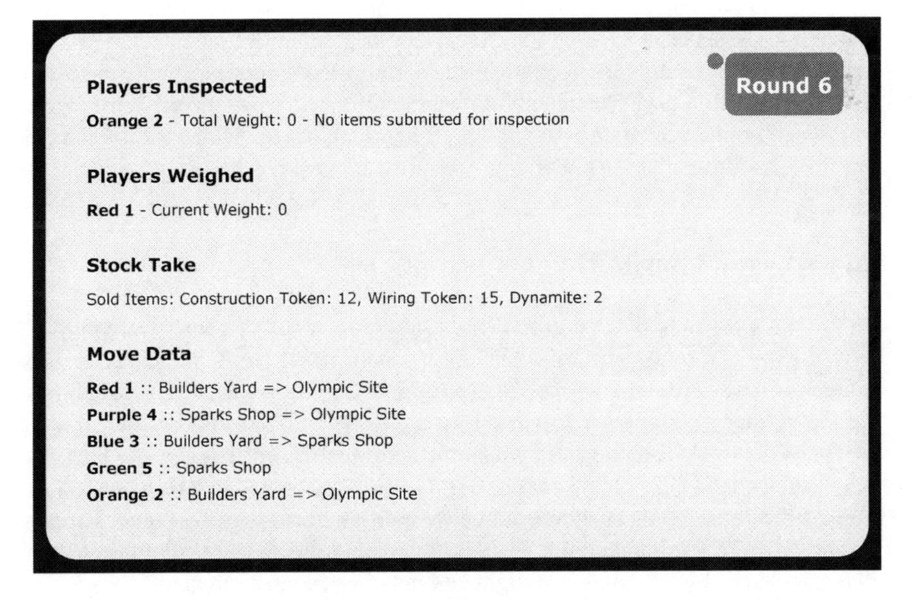

Figure 9.3 Scent trail information delivered to the investigator. Screenshot from Cutting Corners Board Game (Dando et al., 2013)

The data shows that 2 sticks of dynamite and 15 wiring tokens were purchased from the electrical supply shop and that three players (e.g. Purple 4, Blue 3 and Green 5) had all visited the electricians (labelled 'Sparks Shop'). Although this data did not immediately raise suspicions it formed the basis of future deductions. The game design allowed for between 3 and 6 participants to take part in each game. In addition there was flexibility in the number of terrorist players. Between 0 and 3 terrorists players were used in each game, thus heightening the cognitive challenge for investigators.

After an inspection all the players were awarded a small amount of virtual money to allow them to continue playing the game. However, if they had been caught cheating with their van weight over the permitted limit the payment was withheld as a fine. The winner of the game was the first player to amass all the resources needed for their task at the site. Following game play all participants were interviewed about their game playing behaviour (Dando and Bull, 2011). The interviews lasted for approximately 15 minutes each. All the participants were asked to remain 'in character', with terrorist players being reminded that throughout the interview they had to convince the investigator that their role was that of a builder.

Dando et al. (2013) compared three interview conditions: a control condition, where all incriminating evidence was disclosed at the beginning of the interview; a condition employing strategic use of evidence (SUE), where questions are asked about each piece of potentially incriminating evidence in turn and then the evidence is revealed at the end of the interview; and tactical use of evidence (TUE), where each potentially incriminating piece of evidence is questioned then revealed and challenged in turn. The results showed that trained police investigators were significantly more accurate at judging the veracity of deceivers and truth-tellers in the TUE condition than in either the SUE or the control conditions.

Application and Appraisal of the Game

In the study reported by Dando and Bull (2011), a total of 41 games were played, with between 4 and 6 different players in each game. Six experienced police officers acted as investigators, two for each game. Qualitative written feedback was collected concerning player enjoyment of the game and game comprehensibility. All players found the game initially complex during the instruction phase, but easy to understand once play had commenced. Most players stated that they would like to play the game again, as they would have a different strategy next time. Various strategies were used, including: 'I would overload my van right from the start of the game' and 'I would play it straight and not overload my van at all'.

Feedback from investigators was generally positive. Although they reported finding the game cognitively demanding, all stated that this gradually reduced as they took part in subsequent games, as the investigators better understood game play and the use of the software interface. A further piece of research will look in more detail into how the investigators' strategies changed over a number of games. For example, did they become subject to confirmation bias and stop considering all possibilities for each player, just concentrating on a few aspects of gaming behaviour (e.g. 'hiding' on the toll road) to underpin their suspicions? During a game investigators reported seeing game play evolve as the game progressed.

While detailed analysis is yet to be carried out, viewing the investigators information over a game shows that many players start off by going to a shop (and presumably making a purchase, although that cannot be verified on the investigators screen) and then moving on to the site (and presumably dropping off the purchases, although again that cannot be verified via the investigators screen). Then they start to move backwards and forwards between the shops spending time on the toll road, before making one trip to the site, presumably with a van that is considerably overweight (although that can only be seen if the investigator stops the player for a van weigh). The resulting scent trail presented to the investigator showed a decrease in frequency and number of visits to the Olympic site and an increased presence on the toll road. Thus the behaviour change from a number of the players within each game increased the number of evidence streams the investigator had to monitor, balance and address. Multiple evidence streams arising at diverse time points within the game aligned the process within the game to that within a real investigation.

It is not clear why play evolves in this way, although from the anecdotal comments from some of the players it could be for two main reasons: as a result of low dice rolls causing slow progress across the board, a player changed their playing strategy from one of staying within the van weight limit to one of overloading their van; and/or, players are watching the game play of the other players unfold and they may be (a) just copying game play of others or (b) developing strategies themselves which they think will give them an advantage over the other players and thus give them a better chance of winning the game.

To date, the game has been used as a platform upon which to investigate novel interview methods for detecting verbal deception (Dando and Bull, 2011, Dando et al., 2013) and as a tool to facilitate the systematic assessment of investigative interview training given to UK police officers (Dando and Bull, 2011). A questionnaire was given to players after they had both played the game and been interviewed about their game-playing behaviour. Results indicated that the game had provided an environment that allowed participants to be deceptive, both during the game and within the interviews that followed. Indeed, inferential statistical analysis revealed that participants playing the terrorist role had been significantly more deceptive than participants playing the builder role (Dando and Bull, 2011, Dando et al., 2013). Interviewers were taught how to use the scent trail information tactically during post-game interviews in order to maximise opportunities to detect deception (Dando and Bull, 2011). Using scent trail information in this manner proved extremely effective in terms of improving deception detection accuracy.

Generally professional lie detectors are no better than members of the public in identifying liars or truth-tellers, performing at around chance (Ekman and O'Sullivan, 1991; Vrij, 2004; Vrij and Mann, 2001). However, by teaching professional investigators how to manage the large amounts of information, delivered via the scent trails Dando and Bull (2011) increased deception detection accuracy performance to 67 per cent for liars and 74 per cent for truth-tellers. The performance of investigators in the identification of truth-tellers is as important as the performance of identifying liars, as any technique or procedure used to identify the criminals within society must also protect the innocent.

The game software as developed allows for other software applications to be used alongside it. In further research using the game (Ormerod et al., 2013), each player was given access to a messenger application running alongside the main game software. The players were told that they had an 'anonymous helper' who would be able to give them strategic direction during the game play. In reality the same person played 'helper' to each player and didn't give any direction, they just agreed with any suggestions that each player came up with. However, while they were in online conversation with each player the 'helper' worked hard to ensure that each player divulged their game-playing strategy both as initially planned and as the strategy changed throughout the game. Each player reported their playing strategy in real-time as the game progressed. During the game the rules were manipulated part way through the game to investigate the effects of 'challenges' (where the rules are changed part way through the game in order to observe any differences in deceptive and non-deceptive strategies). Results found that deceivers had more strategies than truth-tellers. Deceivers in the challenge condition were more likely to evade detection if their strategy was specifically to use the decoy of the building tasks. Additionally investigators identified 81/90 players correctly, but identification was unaffected by player role (builder or terrorist) or challenge factors (challenge or no-challenge).

Further investigations are planned to examine the effects of early, mid, late and multiple challenges. In addition, participants will replay the game a number of times in order to look at the effect of practise upon both deceptive and non-deceptive game play; the changes due to practise on the scent trails produced; and any changes to participants behaviour within an investigative interview.

The development and implementation of the board game paradigm has provided an engaging and effective environment within which players could and did choose to be deceptive. The deceptions were varied across participants, producing a multitude of diverse scent trails that have been successfully employed within investigative interviewing research and training paradigms. Although still in development this game environment has provided a research platform that has shown great promise in the behaviour that it has allowed participants to choose and display; the scent trails that were produced; and the possibilities for future research that the platform affords. For example, future research could look at the effects of co-located and disparate game play on deceptive strategies; manipulations of scent trail data like the number of dice throws (and therefore progress of game play) between investigations; difference between individual deceptive strategies and strategies employed by a team/group; the effect of learning on strategy choice, i.e. players play multiple games; and the effect of player anonymity on player strategy.

Development of a Location-based Game

The blueprint of the board game has been used as the basis for a location-based game (LBG). The LBG necessarily scaled up the game requirements from being an environment where participants worked alone in a laboratory to an outdoor environment where participants worked in teams of three travelling between various physical locations on a university campus. There were between three and four teams playing at any one time. The structure of the game was very similar to the board game, with teams representing building contractors. However, one team was briefed to undertake a terrorist attack while masquerading as regular builders. The first team to complete the task won the game. As in the board game the teams were encouraged to manipulate the rules to help them win. The game area consisted of an event site, virtual checkpoints and four shops where players purchased tokens representing items required for their tasks. The game area was laid out over an area of the university campus. A description of the LBG and figures showing the game playing space can be found in Sandham et al. (2011).

Each player was supplied with a 3G mobile phone with built in GPS and Wi-Fi running a bespoke games software interface implemented in the Android operating system. The software used the phone web browser to communicate to a main server that continually updated the main database throughout the game. It relayed surveillance signals in the form of GPS coordinates back to the main server and controlled all game-play activities. Team members communicated with each other via their phone and a central server tracked all the players' movements in real-time through their mobile device monitoring and displayed the whole game in a central control room (Sandham et al., 2011).

Detection accuracy at the end of the LBG game based upon scent trail information alone (i.e. without an interview) was 67 per cent (terrorists 64 per cent, builders 70 per cent). Thus, scent trails seem to perform reasonably well as a cue to deception detection judgements. Moreover, scent trails seem particularly to support the identification of innocent players, a bias that is not evident in other studies.

Discussion

The main project aim was to design and build an environment that provided a simulation of trails of information left by legitimate and deceptive individuals involved in activities representative of terrorism. This aim was achieved, providing an empirical environment to test methods of deception detection, in which individuals constructed both their own deceptive activities and their own verbal accounts of those activities. This allowed deception researchers to explore optimal investigative interviewing strategies within a more complex deception environment than that provided by typical mock-crime paradigms.

The games provide a realistic environment in which to examine how the planning and reactive strategies of deceptive and non-deceptive individuals and groups differ. The player interface demonstrated how location, communication and commerce (in this case purchasing information) can be integrated to support a simulation of real-time deceptive and non-deceptive activities. Challenges can be issued within game play in which game rules are changed unexpectedly, causing players to adapt their playing strategies. In addition challenges can be applied differently within a sequence of games so that some challenges allow participants time to plan their strategy change with other challenges allowing no time for planning. Likewise participants can be given differing levels of information within a series of games with some games played where there is an expectation that a strategy change may be needed and other games where the challenges are not anticipated. The differences on both deceptive and non-deceptive game play can then be analysed (Dando and Bull, 2011; Dando et al. 2013; Ormerod et al., 2013).

The investigator interface explores how different forms of information, in this case communication (for the LBG), and location and purchase information, can be integrated within a single scent trail representation to support complex investigative decision-making. While more research is needed to explore the optimum level of detail required in order to maximise deception detection without overloading the investigator's cognitive capacity, this research platform provides the information from which such analyses can be conducted. How much scent trail data and which types of scent trail data in what quantities are questions that still need to be explored within this environment.

Scent trails, both those delivered in real-time from live game play and recorded data, provide a rich research resource. For example, a planned use of the recorded data is as a resource for training investigative reasoning. Training can be given to investigators in the generation of multiple hypotheses for pieces of evidence with examples from recorded games being used both to train the investigators and to test the effectiveness of the training. Scent trails delivered in real-time from live game play can be used as a final exam. The platform allows for trainee investigators to develop their skills using fewer deceivers and fewer players, while for experienced investigators numbers of players can be potentially increased, thereby increasing the cognitive demands of the task.

This platform can be the base upon which heuristic algorithms can be layered onto scent trail representations to provide predictive inferences concerning player activities. For example, the members of a terrorist group and their individual roles within the group may be inferred from 'pattern of life' data in relation to their communications, interactions, etc. For example, identifying the pattern of incoming and outgoing telephone calls may highlight group dynamics and provide the identity of group leaders or coordinators. A leader may be identified as the initiator of short phone calls to a number of others where they may be giving instructions and the receiver of longer calls where the initiators are passing back detailed information. Once the algorithms have been created and verified they will provide a rich resource for police and security services during ongoing investigations.

Conclusion

The DScent project met its main research challenge in showing how scent trails can be analysed to reveal anomalous behaviours that differentiate between innocent and deceptive activities within a serious game environment. The development of a novel board game has allowed the identification of new forensic findings (Dando and Bull, 2011; Dando et al. 2013, Ormerod et al., 2013). A further prototype was developed in the form of a location-based game (LBG) using smartphone technology (Sandham et al., 2011). Both of these platforms provided researchers with an environment within which players could and did develop their own strategies for game play. As the game play and deceptive elements were individual to each player this provided a research environment with an important element of 'natural noise' within the data, thereby increasing the ecological validity of the research. Games such as the one described here allow the exploration of human behaviours that are either too complex or too risky to be explored in real settings. In the case of the current research, exploring terrorist strategy formation in real time would otherwise be impossible. The environment offers the potential for other applications, including training, team work, and relationship building. With changes to the game content and context, the same engine can be used in other domains within the criminal justice system (e.g., fraud investigation) and elsewhere (e.g., emergency evacuation planning), where studies in real contexts are prohibitively dangerous or costly.

Acknowledgement

EPSRC/ESRC/AHRC Countering Terrorism in Public Places Initiative, EP/F008686/1, EP/F006500/1, EP/F008562/1, EP/F008600/1, and EP/F014122/1 funded the DScent project, with co-funding from the Government Communications Planning Directorate (GCPD). The authors gratefully acknowledge support of the other DScent team members.

References

Ardito, C., Buono, P., Costabile, M.F., Lanzilotti, R., Pederson, T., and Piccinno, A. (2008). Experiencing the past through senses: An M-Learning game at archaeological parks. *IEEE Multimedia*, October–December, 76–81.

Bell, B.G., Kircher, J.C., and Bernhardt, P.C. (2008). New measures improve the accuracy of the directed-lie test when detecting deception using a mock crime. *Physiology & Behavior*, 94(3), 331–340. doi: 10.1016/j.physbeh.2008.01.022.

Burke, J.W., McNeill, M.D.J., Charles, D.K., Morrow, P.J., Crosbie, J.H., and McDonough, S.M. (2009). Serious games for upper limb rehabilitation following stroke. *Games and Virtual Worlds for Serious Applications, 2009. VS-GAMES '09 Conference*. 23–24 March, 2009, 103–110. doi: 10.1109/VS-GAMES.2009.17.

Chittaro, L., and Ranon, R. (2009) Serious games for training occupants of a building in personal fire safety skills, *Games and Virtual Worlds for Serious Applications, 2009. VS-GAMES '09 Conference*. 23–24 March, 2009, 76–83. doi: 10.1109/VS-GAMES.2009.8.

Dando, C.J., and Bull, R. (2011). Maximising opportunities to detect verbal deception: Training police officers to interview tactically. *Journal of Investigative Psychology and Offender Profiling*, 8(2), 189–202. doi: 10.1002/jip.145.

Dando, C.J., Bull, R., Ormerod, T.C., and Sandham, A. (2013) Helping to sort the liars from the truth-tellers: The gradual revelation of information during investigative interviews. *Legal & Criminological Psychology* doi:10.1111/lcrp.12016.

Ekman, P., and O'Sullivan, M. (1991). Who can catch a liar? *American Psychologist*, 46(9), 913–920. doi: 10.1037/0003-066x.46.9.913.

Gorman, M.E. (2009). Serious games, sustainable civilizations and trading zones. *Sustainable Systems and Technology, 2009. ISSST '09. IEEE International Symposium*. 18–20 May 2009, 1–3. doi: 10.1109/ISSST.2009.5156702.

Hartwig, M., Granhag, P.A., Stromwall, L.A., and Vrij, A. (2004). Police officers' lie detection accuracy: Interrogating freely versus observing video. *Police Quarterly*, 7(4), 429–456. doi: 10.1177/1098611104264748.

Honts, C.R., Amato, S., and Gordon, A. (2004). Effects of outside issues on the comparison question test. *Journal of General Psychology*, 131(1), 53–74.

Jonas, E., Schulz-Hardt, S., Frey, D., and Thelen, N. (2001). Confirmation Bias in Sequential Information Search After Preliminary Decisions: An Expansion of Dissonance Theoretical Research on Selective Exposure to Information. *Journal of Personality and Social Psychology*, 80(4), 557–571.

Liao, L., Patterson, D.J., Fox, D., and Kautz, H. (2007). Learning and inferring transportation routines. *Artificial Intelligence*, 171(5–6), 311–331. doi: 10.1016/j.artint.2007.01.006.

Numrich, S.K. (2008). Culture, models, and games: Incorporating warfare's human dimension. *IEEE Intelligent Systems*, 23(4), 58–61.

Ormerod, T.C., Dando, C.J., and Bull, R. (2013). Planning criminal acts: What does not reveal the deceiver makes them stronger. *American Psychology and Law Society 6th Annual Conference*, Portland Oregon, March 2013.

Podlesny, J.A., and Raskin, D.C. (1978). Effectiveness of techniques and physiological measures in the detection of deception. *Psychophysiology*, 15(4), 344–359. doi: 10.1111/j.1469-8986.1978.tb01391.x.

Sandham, A.L., Ormerod, T.C., Dando, C.J., Bull, R., Jackson, M., and Goulding, J. (2011). Scent trails: Countering terrorism through informed surveillance. In D. Harris (Ed.), *Engineering Psychology and Cognitive Ergonomics.* Berlin: Springer-Verlag Berlin, 452–460. doi: 10.1007/978-3-642-21741-8_48.

Sawyer, B. (2008). From cells to cell processors: The integration of health and video games. *Computer Graphics and Applications, IEEE*, 28(6), 83–85. doi: 10.1109/MCG.2008.114.

Strömwall, L., Hartwig, M., and Granhag, P.A. (2006). To act truthfully: Nonverbal behaviour and strategies during a police interrogation. *Psychology, Crime & Law*, 12(2), 207–219. doi: 10.1080/10683160512331331328.

Taylor, P.J., Bennell, C., and Snook, B. (2009). The bounds of cognitive heuristic performance on the geographic profiling task. *Applied Cognitive Psychology*, 23(3), 410–430. doi: 10.1002/acp.1469.

Vrij, A. (2004). Why professionals fail to catch liars and how they can improve. *Legal and Criminological Psychology*, 9(2), 159–181. doi: 10.1348/1355325041719356.

Vrij, A. (2008). Nonverbal dominance versus verbal accuracy in lie detection: A plea to change police practice. *Criminal Justice and Behavior*, 35(10), 1323–1336. doi: 10.1177/0093854808321530.

Vrij, A., and Mann, S. (2001). Who killed my relative? Police officers' ability to detect real-life high-stake lies. *Psychology Crime & Law*, 7(2), 119–132. doi: 10.1080/10683160108401791.

Vrij, A., Mann, S., and Fisher, R. (2006a). An empirical test of the behaviour analysis interview. *Law and Human Behavior*, 30(3), 329–345. doi: 10.1007/s10979-006-9014-3.

Vrij, A., Mann, S., and Fisher, R. (2006b). Information-gathering vs accusatory interview style: Individual differences in respondents' experiences. *Personality and Individual Differences*, 41(4), 589–599. doi: 10.1016/j.paid.2006.02.014.

Vrij, A., Mann, S., Fisher, R., Leal, S., Milne, R., and Bull, R. (2008). Increasing cognitive load to facilitate lie detection: The benefit of recalling an event in reverse order. *Law and Human Behavior*, 32(3), 253–265. doi: 10.1007/s10979-007-9103-y.

Washburn, D.A. (2003). The games psychologists play (and the data they provide). *Behavior Research Methods Instruments & Computers*, 35(2), 185–193.

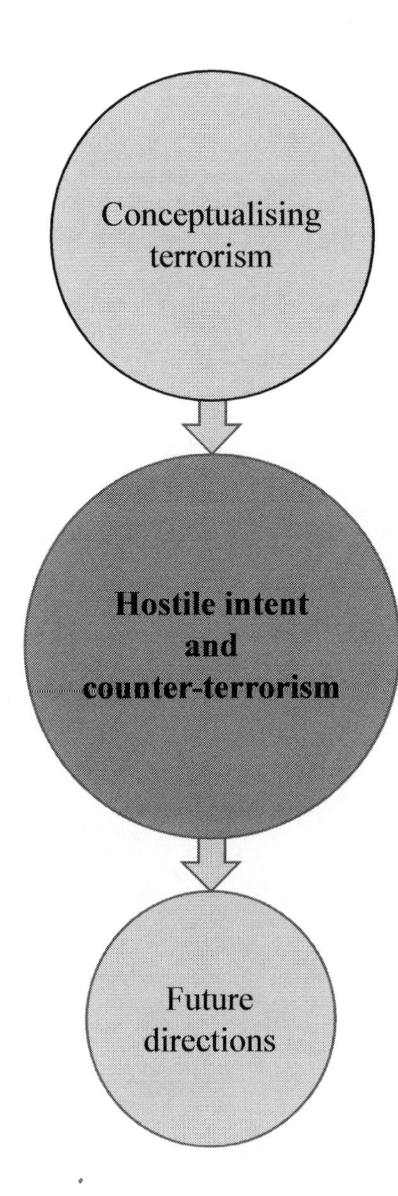

Modelling hostile intent

Part 3, Modelling hostile intent, looks at how to support counter-terrorism initiatives. It may be that we can learn from associated disciplines or application areas. Similarly, taking a multidisciplinary approach and considering the systems in which terrorism operates can bring benefits not seen when issues are considered in isolation. Further, modelling can afford tools to predict and thus proactively address terrorism.

Chapter 10
Safety and Security in Rail Systems: Drawing from the Prevention of Railway Suicide and Trespass to Inform Security Interventions

Brendan Ryan
Human Factors Research Group, Faculty of Engineering, The University of Nottingham, UK

Introduction

Each year, more than 3500 people lose their lives on the European railway system as a result of railway suicide and trespass incidents. These two different but partially overlapping types of events, account for 90 per cent of the total number of fatalities on European railways (ERA, 2013). In spite of efforts to prevent these events, the numbers of incidents are not falling (and may even be increasing). This chapter provides an overview of how these problems are being considered in the EU RESTRAIL project (REduction of Suicide and Trespass on Railway property) (www.restrail.eu), as well as examples from national prevention programmes that are implementing various interventions. The transfer of knowledge and experiences from the investigation of railway suicide and trespass to the wider arena of railway security is considered.

Suicide and Trespass-related Fatalities on the Railway

Suicides on the railway (when a person decides to take their life) can be distinguished from trespass-related fatalities (where a person who was not authorised to be on railway premises may be killed in a situation where there was no intention of self-harm). In Great Britain (GB) there are around 220 suicides per year (accounting for just under 4 per cent of the national total for all suicides) and 50 trespass-related fatalities (RSSB, 2012). The mode of death is often through contact with a moving train, with the immediate access to the railway in a number of different ways such as jumping from a platform, lying in front of an oncoming train or wandering on the track (Guggenheim and Weisman, 1972; Dinkel et al., 2011). The entry point to the railway is typically at stations or rail crossings in a largely secure rail network (as is the case in GB, although the railway is not secured in extensive parts of many European countries). There are also other types of suicide events (e.g. jumps onto the railway from bridges, suicide by other means on other parts of railway property) and deliberate or accidental contact with the electrical power rail and other equipment that also account for the fatality figures. However, not all suicidal and trespass-related events are fatal. There are thought to be more than 1600 serious injuries across Europe per year as a result of suicide and trespass. More detailed data from the British Transport Police (BTP) in GB suggest that while there may be a ratio of one injury event for every three fatalities, there may also be many other suicide-related attempts or behaviours (more than 2700 in GB per year) that do not result in injury. These include threats of suicide, situations where staff or third parties have intervened, and other reported pre-suicidal and suspicious behaviours, which (like fatality and injury events) can divert staff time and disrupt the operational railway.

Statistics from selected countries across Europe show clear differences in suicide rates. For example, in the Netherlands there are 200 rail suicide fatalities each year (1.14 per 100,000 population, 13.0 per cent of the annual suicides), whereas in Latvia there are less than 10 in total (0.09 per 100,000, 0.4 per cent of annual suicides) (Reynders et al., 2011). These figures are indicative of the relative extent of the problem across Europe, though there is need for some caution when interpreting the figures. Differences may be due to variations in the way in which suicide and accidental events are defined and classified, and details that are recorded in different databases (Reynders et al., 2011).

Railway suicide and trespass-related fatalities are influenced by a wide range of factors. They can be difficult to prevent, but are by no means inevitable. Earlier work by Rådbo et al. (2012) produced a model of railway suicide, identifying five stages of suicide events where, at each point, prevention activities can be targeted such as:

- to reduce the attractiveness of the railway and prevent ideation of suicide on the railway (such that the railway is not seen as a place for suicide)
- to restrict access to the railway (e.g. through fencing)
- to influence people in circumstances in which they are at risk of injury (e.g. encouraging people to move from a position of danger)
- to provide early warning when people are identified as being at risk (such as through surveillance activities)
- to reduce the consequences of an incident (which include design changes to the fronts of trains to reduce impact injuries).

As with other forms of hostile intent, suicide and trespass events on the railway usually involve a person (or group of people) being in a place where they should not be (e.g. unauthorised access to a part of the railway or unauthorised intent within part of the railway system). Consequently, prevention efforts for these different types of events may be similar, such as by restricting access to the railway by fencing or removing people from the track if they are spotted during surveillance activities. However, there are circumstances in which prevention strategies that are aimed at accidental events could increase the risk of suicide (e.g. warnings to the passenger of the risk of high-speed passing trains may increase the risk for some individuals who interpret this warning as information about a potential method for suicide).

Overview of the RESTRAIL Programme

RESTRAIL is a European Commission, Framework 7 funded project aiming to reduce the occurrence of suicides and trespass on railway property and the service disruption and other consequences of these events, by providing the rail industry with an analysis and identification of cost-effective prevention and mitigation measures. The project is being carried out by a consortium with 17 partners, led by Security Division at the UIC (the international professional association that represents the railway sector). The consortium members are drawn from twelve European countries, including seven UIC members, three research centres, three Universities and three industrial partners.

The project addresses a range of themes and broad areas of work (Table 10.1). The work programme focuses on the following: collating details of best practice from reviews of the knowledge base and scientific literature; in-depth consultation with the industry; assessment of available evidence to determine the likely success of known preventative measures; development of new approaches to prevention of incidents and mitigation of the impacts of events; and the field testing of some of the recommended or promising interventions (for prevention and mitigation), in a range of different contexts across Europe.

To date, the project has collated publicly available data from 19 countries and detailed incident data from 12 countries based on: the gender and age of people involved, time of events, locations of incidents, characteristic or suspicious behaviours, influencing factors (such as mental health or intoxication of people involved), weather conditions and lighting, types of trains involved, delay times and costs. Other country-specific data have been collected on investigation practices and consequences of railway suicide and trespass.

Table 10.1 Main themes of work in RESTRAIL

Components of the project	Examples of research activities
Qualitative analysis of suicide and trespass on railways properties	Literature review Liaison with stakeholders (e.g. police, infrastructure managers, safety authorities, Samaritans) and collation of national statistics, practices and processes Analysis of behaviours of people before suicide and trespass events (Ryan 2013; Ryan and Stedmon, 2012)
Assessment of measures targeted to reduce railway suicides and trespasses	Developing and applying methodology for the evaluation of known preventative measures for suicide and trespass (Ryan and Kallberg, 2013) Review and development of new or emerging approaches for prevention
Mitigation of consequences by improving procedures and decision-making	Development of methods and tools that can be integrated with existing procedures and technologies, to achieve effective and cost-efficient means of mitigating the potential impact of suicides and trespass on railway infrastructures.
Field pilot tests and evaluation	Field testing of recommended and promising measures for prevention

More than 100 preventative measures (including variations of measures) have been identified, grouped and assessed using a set of 14 criteria and related evaluation metrics that have been developed for the purpose of this preliminary evaluation within the project (Ryan and Kallberg, 2013). Sixteen measures have been recommended (or classified as promising) for the prevention of suicide (e.g. fences and barriers at specific parts of stations; increased visibility by improved lighting at railway crossings, tunnels and incident hotspots). Ten measures for the prevention of trespass have been identified (e.g. education and prevention in schools and other locations; targeted media campaigns, including shock campaigns). Seven of the measures have been assessed as having potential to prevent both suicide and trespass (e.g. fencing, surveillance, campaigns). A large amount of documentary evidence has been collated during the course of the preliminary evaluation exercises and this is being used as the basis of practical guidance for the rail industry for the planning and implementation of prevention measures and interventions. Field tests of the more promising measures for prevention and mitigation are currently in progress in seven countries across Europe (Table 10.2). Currently, there is very little available evidence on the effectiveness of different types of interventions. Results from these pilot tests will start to fill some of the gaps in knowledge in this area.

Table 10.2 Field tests of recommended and promising measures in RESTRAIL

Type of intervention	Country in which the test is being conducted
Mid-platform fencing	Great Britain
Warning signs and posters	Spain
Education campaign At schools Not in school environments	 Finland Spain
CCTV, sensors and sound warnings	Finland
Gatekeeper programme	Germany
Multiple measures at a station – gates, fencing, floor anti-trespass panels, leaflets	Turkey
Societal collaboration – collaboration between different agencies	Sweden
Incident management and information management platform	Israel
Education and cooperation of the police and legal entities	Israel

Drawing on Knowledge from RESTRAIL: Informing Security Interventions

Restricting Access

Many parts of rail networks are secured to prevent access to the railway (e.g. lineside fencing), but this is not common across all parts of Europe where extensive sections of the network (particularly in rural locations) are not fenced. Even where railways are secured (e.g. GB, and many countries in and around city areas) there are obvious locations where access to the railway is available (e.g. at stations and railway crossings). Efforts have been made in recent years to make access harder, such as using special floor panels, sensors and physical barriers at crossings and the ramps at the ends of platforms, to make it more difficult to walk onto the railway. Access directly from the platform (i.e. jumping or stepping down from the platform onto the track) is much more difficult to prevent. Platform doors or screens, as seen on some underground systems (see Law et al., 2009; Law and Yip, 2011) may be options for stations where there is a homogenous train formation (as in metro systems), but would be very costly and problematic to retrofit to stations where there is a mixed train service, typical of main line railways.

Field tests in GB as part of RESTRAIL are evaluating the potential effectiveness of a mid-platform fencing programme (preventing access to fast lines where trains are not scheduled to stop) in three separate test areas on the rail network (covering almost 50 stations). The evaluation will attempt to determine if the intervention will lead to a reduction in railway suicide at the target locations, as well as any other positive or negative operational impacts of the intervention (e.g. effects on passenger flow or perceptions of passenger safety at stations). The evaluation activities also include the collection of details of other preventative measures at the stations (e.g. support activities and resources that are provided by a major Network Rail/Samaritans programme that is running in parallel with this prevention programme) to understand potentially confounding variables in this type of complex and dynamic, real-world environment. Evaluation activities within the field test are also collecting descriptive details of the implementation of the mid-platform fencing, to understand how effectively the programme has been introduced and the range of factors that can influence the implementation of this type of programme.

Surveillance

Many of the prevention measures are linked to surveillance. This can be achieved directly, through the use of visible security staff or police at stations or other locations, who may use their experience to identify people at risk. In RESTRAIL, research activities have been carried out in four countries (GB, Finland, Germany, and Spain) to understand more about the types of behaviours that people may exhibit in the period leading up to suicide and trespass incidents. For example, the work in GB (Ryan, 2013) included analysis of various historical records of events (e.g. from event databases, descriptions of CCTV recordings, summary records of situations in which someone has intervened to prevent an incident) and used group interviews/workshops with railway staff and transport police officers to produce a preliminary classification of different observable behaviours at various time periods before these types of incidents (e.g. immediately before an incident; intermediate, in a period of deliberation at a station before an event; longer-term behaviours, such as those that might be observable in the home). This classification is being developed in ongoing analyses, with a view to enabling better interventions. In the workshops, the police and rail staff were very confident about their ability to identify people at risk. They explained how they did this using 'gut feeling' (described as 'something not quite right … using a combination of the senses') though they were not able to articulate in any more detail how they actually achieved this in practical situations. There is a risk that the experts' judgements are biased by overconfidence in their abilities (Kahneman et al., 1982). However, current evidence from the national suicide prevention programme in GB suggests that there are growing numbers of interventions at stations, by rail staff and members of the transport police. There is certainly value in more investigation of how an expert's intuitive decision-making can contribute to earlier interventions in these situations (Ryan and Stedmon, 2012).

The 'Gatekeeper approach' (Isaac et al., 2009) is one of the central components of the ongoing Network Rail/Samaritans suicide prevention programme (RSSB, 2013). Gatekeepers are people who interact with members of a community in natural settings and can be trained to recognise risk factors for suicide. This type of approach takes advantage of the presence of staff who are working in high-risk locations, and provides them with the skills to be able to approach and engage with someone who may be contemplating suicide. This approach is currently being tested in Germany, within RESTRAIL. Other security interventions that involve interactions with people include the use of novel interview techniques, as investigated in the 'Shades of Grey Project' (see Chapter 7).

Surveillance can also be carried out using CCTV, which might be monitored by staff who can initiate some form of intervention. In addition, a range of technologies and sensors can be used at stations and crossings (e.g. radar, laser, optical devices) or even along the length of the track (e.g. glass fibre cables), to identify the physical presence of unauthorised people and alert a security operator. In some circumstances, sensors can be linked to audible warnings, providing direct warnings or feedback to the person who has triggered the sensor, as currently being tested in Finland.

Modifying Behaviours

Other types of preventative measures exert their effect by trying to modify the behaviours of those with hostile intent. This is clearly an area in which there are likely to be differences in how problems of suicide and trespass are tackled. These types of interventions can be based upon education, such as advising the public of the risks associated with accessing the railway (currently being tested with schoolchildren in Finland and children at a visitor centre in Spain), or through the use of specific warning signs (currently being tested in Spain). Behaviour can also be modified through changes in the design of the railway environment, such as improving the layout at stations to modify the flow of people through space and prevent access to high-risk or secluded areas. Design changes can also improve the aesthetics of the station in an attempt to influence a person's mood or encourage a more respectful attitude to railway property (e.g. attempting to reduce vandalism or graffiti). Landscaping at lineside locations through use of embankments, hedges, and underpasses can be used to reduce the attraction to the railway or deter people from accessing vulnerable parts of the network (Silla and Luoma, 2011).

Inter-agency Collaboration

From a systems perspective, a wide range of stakeholders in the rail industry have important roles in the prevention of suicide and trespass incidents on the railway. These include the infrastructure manager (who may provide fencing for the network), train operating companies (who may have management responsibilities at stations and whose drivers suffer from the psychological effects of incidents), and railway police (who help with the prevention of crime, anti-social behaviour and unauthorised access to the railway). In GB, great emphasis has been placed on developing cooperation between stakeholders (e.g. clarifying responsibilities, data sharing) through joint initiatives on national strategic and local working groups. Recent efforts are focusing on relationships with wider stakeholders (e.g. health authorities). An interesting example of collaboration between a wide range of agencies (e.g. police, medical, fire, transport) is being tested in RESTRAIL in Sweden. This applies the precautionary principle in cases of reported trespass, with the temporary shutdown of rail traffic and the rapid response from the agencies to remove a person from the rail network.

Applying the Lessons from Prevention and Mitigation of Railway Suicide and Trespass to Broader Issues of Security

Threats to railway security are diverse. The railway is complex, with many parts, nodes and interconnections. Large numbers of people use railway stations and other parts of the infrastructure on a daily basis, with extremely high numbers of passengers passing through major stations at peak times of the day. On current projections, the demand for railway travel is likely to increase, along with the associated complexity of the system. The railway is open, with a universal operating philosophy of 'turn up and go'. For many, this is a desirable attribute of the system, which needs to be preserved and not eroded in efforts to improve safety and security of the system. However, this poses problems for protection of the railway.

Suicide and trespass are two different, but overlapping, security threats. These threats result in fatality and serious harm to those involved (directly and indirectly) and cause major disruption to train services. Each of these threats is different to other types of security threat, in part because of the different motivations of the people involved, but also in terms of the likely frequency and scale of the events (e.g. fairly frequent, single-person fatalities, accompanied by moderate to large system disruption in the cases of suicide and trespass, compared with infrequent but potentially much larger numbers of fatalities and large-scale disruption in the event of a terrorist incident).

It may be too early to suggest that there are generic lessons from the study of railway suicides and trespass that can be applied more widely in protecting the railway. However, there are a number of common issues and areas where solutions to a problem can have more widespread effects. For example, there is undoubtedly an attraction for some to the railway – for legitimate and unfortunately, illegitimate use. There is value in understanding more about the motivations of people who access the railway for a range of personal or other reasons, in order to threaten safety and operations on railway. Furthermore, there are opportunities for collection, analysis and interpretation of various data types, to give greater insight to the range of threats to safety and performance and the range of possible solutions to prevent or mitigate against these threats. This will require improvements in investigation in many circumstances and better use of available data sources, to enhance opportunities for learning and the selection and implementation of appropriate strategies for prevention. However, the often sporadic nature of many types of threats and especially the more extreme instances of hostile intent, such as terrorist attacks, will continue to present difficulties for the prediction of how, when and where these types of incidents may occur. The importance of collaboration between a wide range of organisations and stakeholders that have an interest in rail security, or who suffer the consequences of security threats, is being recognised. These efforts may be strengthened through more work to understand the complex relationships between the different agencies, by taking a sociotechnical perspective to identify the interdependencies between different parts of the dynamic railway system (e.g. different sub-systems, the operating environments, the equipment/technologies/processes that are used, and how these impact on the activities of the people that are involved).

Threats that are linked to the unauthorised access to parts of the railway or other disruptive behaviours will continue to pose problems for the railway. Experiences from the work in the prevention of railway suicide and trespass demonstrate that it is possible to make it harder for these threats to succeed. In some circumstances, successful interventions might displace the events and hostile intent to other areas (e.g. road-based suicide).

The railway is evolving. There is the potential for technological solutions for some problems of this nature (e.g. through sensors, CCTV, video analytics), though these developments are dependent upon a good understanding of suspicious behaviours. Physical changes (e.g. in the design and redevelopment of stations design) can help to restrict access, though it is difficult within the existing model of an accessible and functioning railway to design out all, or even most, of the open-access points in the system. Existing intervention strategies are helping to improve collaborative working between different agencies, but there are still challenges that need to be overcome to maintain progress and apply the emerging models of collaboration on a national scale.

Conclusion

In the short term, it may be that some of the greatest impact could be achieved through the contributions of people – the station staff and other railway staff, security, police, as well as the travelling public. All of these individuals have the combined opportunity to observe other people in places where they should not be or when they appear to behave in a suspicious manner. These people have the opportunity to do something in these circumstances. Early experience in the area of railway suicide and trespass suggests that people can be supported with some guidance to heighten their awareness of the risks, as well as empowered with the confidence and strategies to contribute and intervene in an appropriate way. This is clearly a very challenging proposition, where threats to security in the railway environment are diverse and circumstances associated with these are complex. However, events of this nature need not be inevitable.

Acknowledgements

This work has been carried out as part of the EU Seventh Framework Programme project RESTRAIL – SCP1-GA-2011-285153. The work would not have been possible without the support and contributions of colleagues from the RESTRAIL project and staff from the BTP, Network Rail, The Samaritans and RSSB.

References

Dinkel, A., Baumert, J., Erazo, N., and Ladwig, K.H. (2011). Jumping, lying, wandering: Analysis of suicidal behavior patterns in 1,004 suicidal acts on the German railway net. *Journal of Psychiatric Research*, 45, 121–125.

ERA (2013). *Intermediate Report on the Development of Railway Safety in the European Union*. European Railway Agency Safety Unit, Valenciennes, France.

Guggenheim, F.G., and Weisman, A.D. (1972). Suicide in the subway. Publicly witnessed attempts of 50 cases. *Journal of Nervous and Mental Disease*, 155, 404–409.

Isaac, M., Elias, B., Katz, L.Y., Belik, S.L., Deane, F.P., Enns, M.W., and Sareen, J. (2009). Gatekeeper training as a preventative intervention for suicide: A systematic review. *Canadian Journal of Psychiatry*, 54(4), 260–268.

Kahneman, D., Slovic, P., and Tversky, A. (Eds) (1982). *Judgment under Uncertainty: Heuristics and Biases*. Cambridge: Cambridge University Press.

Law, C.K., Yip, P., Chan, W., Fu, K.W., Wong, P., and Law, Y.W. (2009). Evaluating the effectiveness of barrier installation for preventing railway suicides in Hong Kong. *Journal of Affective Disorders*, 114(1–3), 254–262.

Law, C.K., and Yip, P.S.F. (2011). An economic evaluation of setting up physical barriers in railway stations for preventing railway injury: Evidence from Hong Kong. *Journal of Epidemiology and Community Health*, 65(10), 915–920.

Rådbo, H., Renck, B., and Andersson, R. (2012). Feasibility of railway suicide prevention strategies; a focus group study. In C. Bérenguer, A. Grall, and C. Soares (Eds). *Advances in Safety, Reliability and Risk Management*. London: Taylor and Francis Group, 25–32.

Reynders, A., Scheerder, G., and Van Audenhove, C. (2011). The reliability of suicide rates: An analysis of railway suicides from two sources in fifteen European countries. *Journal of Affective Disorders*, 131, 120–127.

RSSB (2012). *Annual Safety Performance Report 2011/12. Report 2012*. London: Rail Safety and Standards Board.

RSSB (2013). *Samaritan/National Rail Tackling Suicide on the Railways Programme Interim Annual Report 2012*. London: Rail Safety and Standards Board.

Ryan, B. (2013) Reducing the risk of suicide or trespass on railways: Developing better interventions through understanding behaviours of people. Proceedings of the Institution of Mechanical Engineers: Part F: Journal of Rail and Rapid Transit, 227(6), 715–723.

Ryan, B., and Kallberg, V-P. (2013). Developing methodology in RESTRAIL for the preliminary evaluation of preventative measures for railway suicide and trespass. In N. Dadashi, A. Scott, J.R. Wilson, and A. Mills (Eds) *The Fourth International Rail Human Factors Conference*, March 2013. London: CRC Press, Taylor and Francis Group, 89–98.

Ryan, B., and Stedmon A.W. (2012). What is normal behaviour? How might 'gut feelings' help us identify suspicious behaviours? *8th World Congress on Railway Security*, Bratislava, Slovakia, 25-26 October 2012 (CD-Rom proceedings).

Silla, A., and Luoma, J. (2011). Effect of three countermeasures against the illegal crossing of railway tracks. *Accident Analysis and Prevention*, 43, 1089–1094.

Chapter 11
Tackling Financial and Economic Crime through Strategic Intelligence Management

Simon Andrews, Simon Polovina, Babak Akhgar,
Andrew Staniforth and Dave Fortune
Centre of Excellence for Terrorism, Resilience, Intelligence and Organised Crime Research (CENTRIC) Sheffield Hallam University, UK

Alex Stedmon
Human Systems Integration Group, Faculty of Engineering and Computing, Coventry University, UK

Introduction

Serious Organised Economic Crime (SOEC) and the associated activity of fraud is a growing multinational business without respect to national borders. It is also directly linked to terrorist organisations seeking to fund their activities and launder funds generated through illegal sources such as drugs, human trafficking and firearms sales. As a consequence, in the European Union (EU) alone it costs its Member States billions of euros annually as a result of being targeted by illegal operations. The ability to discover and develop sophisticated new weapons to detect and fight these crimes, based on a cooperative and collaborative strategy across these states is imperative. Each European Member State police force and Financial Intelligence Unit (FIU) has its own Financial SOEC and fraud monitoring system. Although existing pan-EU systems, such as Europol's Secure Information Exchange Network Application (SIENA) and the Financial Intelligence Unit's (FIU.NET) provide information exchange and some crime-matching analysis, further steps need to be taken to provide a more coherent and coordinated approach for detecting and deterring SOEC and fraud. This chapter describes an approach for strategic intelligence management that will add value to SIENA and FIU.NET through taking a systems perspective for increasing the effectiveness of communication between Member States by developing an agreed common language (or taxonomy of understanding) for SOEC and fraud, with automated multi-lingual support. By appropriating and applying existing business tools and analysis techniques to the illegal businesses of SOEC and fraud, better intelligence is possible to help Member States target these crimes and criminals. However, to be effective at the multinational level requires these systems to be comprehensively integrated into one pan-European system. Such a system would federate the large volumes of SOEC and fraud information from existing systems and other key sources across the EU, into a single shared inventory. This facility would be capable of capturing even the low-level and low-intensity crimes, thus providing Member States with a tool for counter-terrorism on a spectrum of criminal activities. It is in fulfilment of this need that the Economic criMe PReventIon for a Strengthened European Society (EMPRISES) project was developed.

The Enterprise Architecture of SOEC and Fraud

SOEC and fraud can be described as components of a sociotechnical system or more specifically an Enterprise Architecture (EA) (Ross, Weill and Robertson, 2006). EA provides a holistic view of agents across business, economic, social and technological dimensions. As an EA, SOEC consists of 'business enterprises' in a similar fashion to any legitimate enterprise (e.g. business development, resources, staff, technologies, business infrastructure and resilience models) to which the vast body of EA knowledge can be applied. However, the distinction for SOEC is that the transactions it engages in are inherently unbalanced and directed against the interests of others (i.e. the victims) be they individuals, business organisations or wider society. Put simply, SOEC and fraud activities accept the risks associated with breaking the law as a cost to their business process and a legitimate part of their business model. It is this economic risk and its adverse effects on others that distinguishes SOEC and fraud from other forms of economic activity. It is through this that such activities can be delineated from other legitimate business activities and then criminal activities stopped.

EMPRISES thus exploits the state of the art in EA so that businesses can understand their own transactions better (e.g. to safeguard against cyber-terrorism and insider threats) and so that security agencies can identify anomalies and patterns of illegal activity. This basis enables us to extend the state of the art by applying the same processes to SOEC and fraud, adding meaning to the data that illegal activities generate and thereby discovering their fundamental enterprise anatomy. Once explicated, their supply and consumer chains can be identified and trapped, or the potential victim(s) alerted.

To illustrate this, Figure 11.1 takes the best practices from The Open Group Enterprise Architecture Framework (TOGAF), reflecting the organisational structures of SOEC including its protagonists (e.g. criminal enterprises) and involuntary agents (e.g. the victims) they transact with, the local enforcement agencies (LEAs), and EU-wide and national enforcement bodies (TOGAF, 2011).

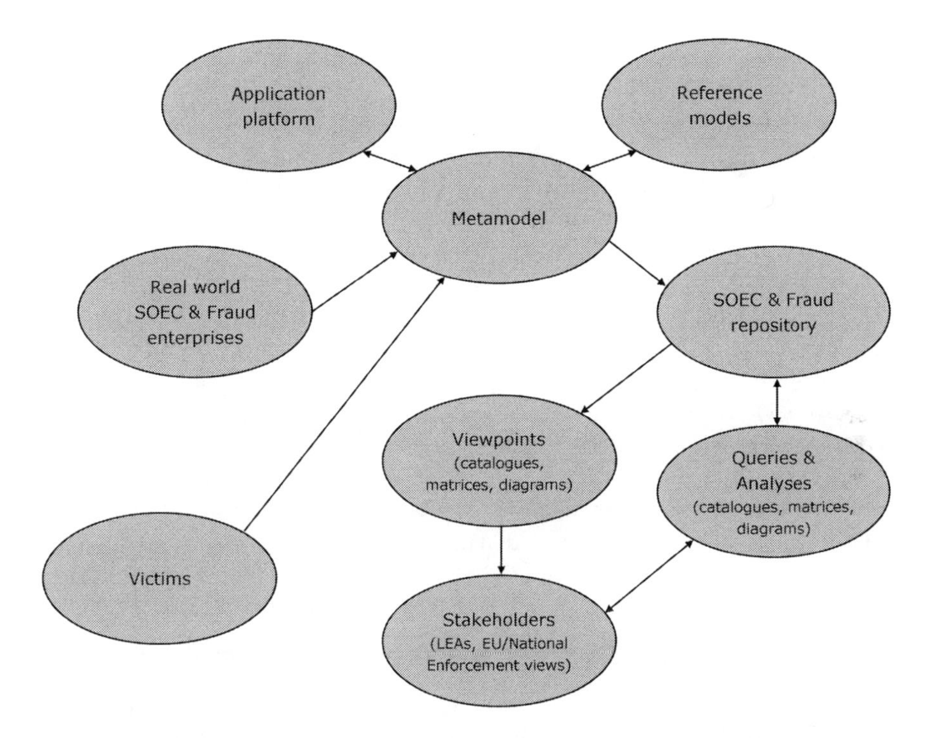

Figure 11.1 The SOEC and fraud architecture framework (adapted from TOGAF, 2011)

The SOEC and Fraud Transaction Concept

To deepen our understanding of SOEC and fraud transactions, the EMPRISES Architecture incorporates a Transaction Concept (TC) (Launders, 2012; Polovina, 2012; Polovina and Andrews, 2011) that is based on Resource-Events-Agents (REA) principles and identifies the 'real-world' components in enterprise transactions, how they transact (e.g. events) and their medium of transaction (e.g. specific resources) (Vymětal and Scheller, 2012). Economic events and economic resources embody economic scarcity that underpins the value and the costs of each transaction, and their effect on the wider system. Each economic resource in a SOEC or fraud transaction would thus capture the adverse impact on the EU economies, loss of state revenues, and the overall social and political impact in the wider system. Economic events relate to the effects of illegal activities on exchange of the resources, identifying the victim (individual, corporate or jurisdiction) of lost resources.

Agents in the model are delineated according to their being characterised as either internal or external agents. The internal agent is the illicit propagator of the SOEC or fraudulent activity while the external agent is the victim. Adding these semantics to the consequences of the transaction distinguishes the legitimate agents from the illegal or rogue agents in the system. The TC also captures the pragmatics as well as semantics of SOEC, thus minimising the impact of a 'cat-and-mouse' scenario as SOEC enterprises or fraudsters might attempt to beat the detection. They may achieve this by capturing apparently contradictory evidence on the adverse impact of their enterprises as the impact is captured at a generic level of economic event or resource, rather than a specific event or resource (Stamper, 1996). To capture such expressivity, Conceptual Graphs (CGs) that are conceptual structures that marry the creativity of humans with the productivity of computers can be used (Polovina, 2007) and can be parsed from and to natural language (International Standards Organisation, 2012). Additional rigour for the logic of CGs is provided at a mathematical level by Formal Concept Analysis (FCA) as demonstrated in other applications (Andrews and Polovina, 2011).

In conjunction with the TC, Computable General Equilibrium (CGE) analysis provides a basis for understanding the direct and indirect economic impact of SOEC and fraud. It is well respected in many other fields and particularly in fiscal studies (Sandmo, 2005; Sennogoa, 2006) and this economic model fits in closely with the behaviour of SOEC enterprises, their enterprise architecture, fraudulent transactions, and the derived data monitoring architecture that underpins the basis of a federated analysis model. CGE has demonstrated a useful capacity to capture the intrinsic mechanisms of the economy and to translate these inefficiencies through productive structures and institutional sectors. Introducing the previously notified economic distortions and computing its chained effects in the whole economic system provide a means of estimating a more accurate total effect in terms of welfare state losses (Turner, 2010).

Existing Systems: SIENA and FIU.NET

Europol is at the already at the centre of an approach to coordinate and assist all Member States to maximise the benefits of collective data sharing and analysis to allow a fuller information and intelligence picture to be acquired and understood. Europol's SIENA system is the existing platform for the exchange of operational information (in the form of structured data) between Europol and its partners.

Indeed, Goal 2 of Europol's 2013 work programme: Becoming an EU Criminal Information Hub, states that the priority is to strengthen Europol's financial intelligence capabilities by linking money flows to criminal activities and following up on the identified links. Analysis and information management will be further enhanced through the evolution of SIENA and the delivery of a new generation Europol analysis system. In order to achieve this goal a system utilising a data structure called Universal Messaging Format II (UMFII) will be used as part of the EU information management strategy. UMFII systems will be the common framework for the structural cross-border information exchange between LEAs and thus promote interoperability. It will also shorten response times, improve the data quality and improve resource efficiency.

In addition FIU.NET for Europe and the Egmont Group internationally work in coordinating and facilitating information exchange between Financial Investigation Units (FIU) on a national and international level. Representatives from Member States on FIU.NET range from the Serious Organised Crime Agency (SOCA) which is a law enforcement agency and not a police agency in the UK; the Executive Service of the Commission for the Prevention of Money Laundering and Monetary offences (SEPBLAC) which is coordinated by Bank of Spain; and the National Intelligence unit of the Finish Police Service. In fact most, but not all Member States are members of FIU.NET while all are members of Europol. FIU.NET allows members to exchange information on economic crime using their bespoke MA3tch system. This system allows FIUs to share data as it is converted into uniform anonymised information using filters so that no personal data is available. The information can then be shared with other specific members and not automatically shared with the whole network of FIU members. Europol, on the other hand, utilises a system of Analytical Work Files (AWF) to process and analyse the data and intelligence it receives from its members. Using this data it supports and helps coordinate Member States by looking at a high strategic level to help tackle all types of serious cross-border criminality. While both systems work independently they also acknowledge the need for closer cooperation given the high level of finances associated with serious organised crime.

Building on these valuable approaches, the FIU.NET and SIENA systems would be accessible via EMPRISES end-user monitoring systems such as the Finish National Police Results Data System (PolStat) and the Police Information System (Patja), the West Yorkshire Police intelligence analysis system in the UK, and the Central Intelligence Analysis Unit in Spain.

E-PEUMS

In order to deliver this combined solution EMPRISES will implement a bespoke Pan-EU Monitoring System (PEUMS). More specifically the EMPRISES PEUMS (E-PEUMS) implementation architecture would take advantage of existing solutions across SIENA and FIU.NET. Figure 11.2 illustrates the E-PEUMS pilot system involving the integration of existing LEA systems for five Member States.

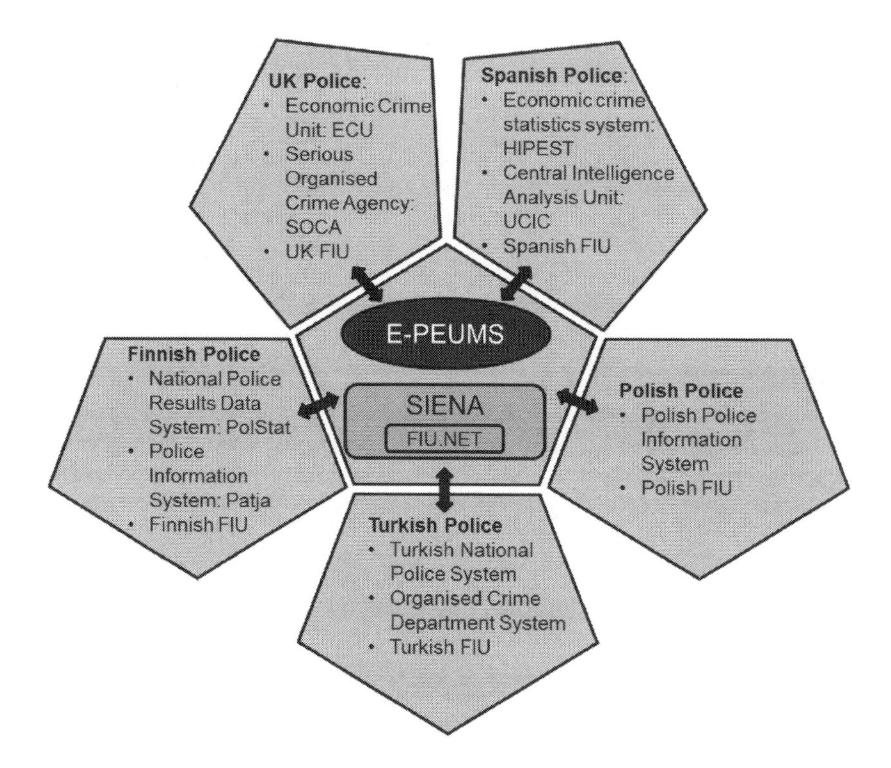

Figure 11.2 E-PUEMS pilot system integrating five Member States

The current use of specialised tools and technologies for the monitoring, detection, evaluation and deterrence of SOEC and fraud is limited mostly to local and (some) national systems. And although in the EU there is basic data exchange as described earlier, there is currently no central repository of SOEC and fraud that can be shared by Member States. Using the SOEC and Fraud Architecture Framework described in Figure 11.1, E-PEUM will comprise of a suite of new tools, technologies and techniques to provide new methods of monitoring, detection, evaluation and deterrence of SOEC and fraud, based on a shared inventory and taxonomy. Functionalities will include the investigation of effective interventions in SOEC and fraud (to inform new guidelines and methods of combating and deterring such crimes); reporting of trends in SEOC and fraud; the identification of differences in EU/Member State-based legislation and tax law; identifying common modus operandi, situational assessments, economic evaluations of market damaged; alerts to new organised investment fraud schemes, new threats of a particular SOEC and new trends, predictions of new types of crime by extrapolation of trends and new crime methods; visualising the management structure of known groups and gangs; early warning of new SOEC and fraud by matching SOEC components across several Member States.

The tools might also include such components as fuzzy logic and probabilistic analysis (for predictive analysis, trends and situation assessment) (Kim and Bishu, 2006; PR-OWL:, 2012); CGs for (criminal) transaction modelling such as the identification of key agents, resources and facilitators for SOEC and fraud (Du, Song and Munro, 2006; Jedrzejek, Falkowski and Bak, 2009; Mifflin et al., 2004); FCA for pattern finding for modus operandi and indicator analysis, threat detection, taxonomy visualisation, predictive analysis (Kirda, 2010; Snášel, Horák and Abraham, 2008; Thonnard, 2011); Social Network Analysis (SNA) for the detection and analysis of organised crime (OC) groups and OC activity (Fox, 2012; McNally and Alston, 2006; SAS, 2009); extending SNA with FCA: Formal Conceptual Network Analysis to provide SNA with enhanced capabilities for analysis of OC group–group interaction and OC hierarchies, extending CGs with FCA for CG-FCA to identify incomplete transactions, supply chains and transactional hierarchies, identification of missing agents in transitions (Andrews and Polovina, 2011); linked data analysis for detecting financial pathways and supply and economic food chains (Larreina, 2007); Fuzzy Cognitive Mapping (FCM) for determining weighted cause-and-effect relations and actions (Carvalho and Tomè, 1999); and machine learning for diagnostic analysis of suspected SOEC activity and economic impact (Schrodt, 1995).

The pan-European inventory of SOEC and fraud would be accessible and queried via a dedicated user interface employing a Create/Retrieve/Update/Delete system. By using a simple ontology-based visualisation of the SOEC and fraud repository, end-users will have a clear view of the underlying data structure and relationships therein. For inventory queries, new FCA-based visual analytics will make the most of the underlying SOEC and fraud ontology, allowing semantic, relational, hierarchical, recursive and propagating queries possible beyond the current state of the art in traditional database systems. An E-PEUMS dashboard can be created that provides a common, unified, interface to the tools in the toolkit. In order to support pan-European response, its service-oriented architecture (SOA) framework would be extended to incorporate the dashboard as a front-end web-service. This will create a pan-European service, greatly extending the opportunities for collaboration and cooperation in the monitoring, detection, evaluation and deterrence of SOEC and fraud. A Transaction Analytics of Criminal Enterprises (TrACE) component will focus on the TC as described earlier, and co-exist with the tools described above to provide a complementary analytical view that is directly based on transaction concepts modelled in CGs and supported by FCA.

Multilingual support for the system will be provided in a 'many language to many language' form. The agreed taxonomy means that Internationalisation (I18N) is suitable as a web-based (i.e. globally interoperable) standard for the display of data and results of analysis. I18N allows any language to be plugged-in to an LEA's interface with no need to re-write any code. It is simple to switch between languages at run-time as web services are delivered to different Member States with keywords dynamically configured based on the agreed taxonomy. The provision of multilingual capabilities for data extraction is more challenging but again is aided by the agreed taxonomy. Multilingual support will be developed based on text and data parsing searching for taxonomy matches. As keywords are identified, inferences can automatically be made between text words, for example, to identify data values using natural language semantics for each supported language and rules based on the EA ontology.

FCA for the detecting SOEC

A major challenge for detecting SOEC is the efficient and effective scanning of the environment for strategic early warnings of irregular activities (Andrews et al., 2013). In part, the challenge is due to vast and increasing amounts of potentially relevant data that is accessible (Criminal Intelligence Service Canada, 2007; Europol 2011). In addition, typical questions posed and analyses conducted are not always of a straightforward numerical/statistical nature, but rather necessitate a more conceptual or semantic approach. This poses major issues for the development of a system such as E-PUEMS in order to support the human user. New developments in computational intelligence and analytics have opened up new solutions for meeting this challenge with FCA and its faculty for knowledge discovery and ability to intuitively visualise hidden meaning in data (Andrews and Orphanides, 2010a; Valtchev, Missaoui and Godin, 2004). This is particularly important in environmental scanning where signals are typically weak and occluded by background noise and where the data are often incomplete, imprecise or unclear (Andrews et al., 2013). However, the potential for FCA to reveal semantic information in large amounts of data is beginning to be realised by developments in efficient algorithms and their implementations (Andrews, 2011; Outrata and Vychodil, 2012) and by the better appropriation of diverse data for FCA (Andrews and Orphanides, 2010b; Becker and Correia, 2005).

Connections between formal concepts in a formal context can be visualised using a concept lattice (Figure 11.3). Concept lattices are developed through an understanding of sub-concept and super-concept constructs and the underlying relationships between them. This is an intuitive way of discovering hitherto undiscovered information in data and portraying the natural hierarchy of concepts that exist in a formal context.

A concept lattice can provide valuable information when the user understands how to read it (and this, in itself, can present major human factors challenges). Each of the nodes in Figure 11.3 represent formal concepts and it is the convention that formal objects are labelled slightly below the node and formal attributes slightly above the nodes. As an example, the node that is labelled with the formal attribute 'Asia Pacific' shall be referred to as Concept A. To retrieve the extension of Concept A (the objects that feature with the attribute 'Asia Pacific') the user starts at the node where the attribute is labelled and traces all paths that lead down from the node. Any objects along the pathway are the objects that have

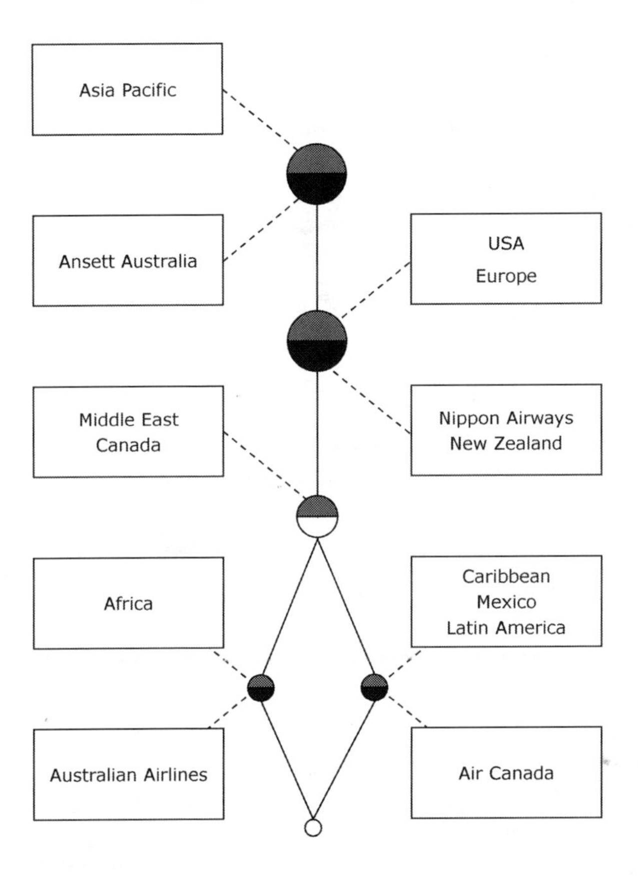

Figure 11.3 A lattice of airline routes and destinations (from Andrews et al., 2013)

that particular attribute. Thus Concept A can be interpreted as 'All airlines fly to Asia Pacific'. Similarly, the node that is labeled with the formal object 'Air New Zealand' shall be referred to as Concept B. To retrieve the attributes of Concept B the user will collect the attributes 'USA', 'Europe', and 'Asia Pacific'. This can be interpreted as 'The Air New Zealand airline flies to USA, Europe and Asia Pacific'. Looking at the node in the centre of the lattice labelled 'Middle East' and 'Canada', it becomes apparent that although Air New Zealand and Nippon Airways both fly to Europe, USA and Asia Pacific, only Air Canada and Austrian Airlines fly to Canada and the Middle East.

Although the airline context is a small example of FCA, visualising the formal context shows that concept lattices provide richer information than perhaps interrogating a cross-tabulated matrix alone. This type of hierarchical intelligence that is gleaned from FCA is not so readily available from other forms of data analysis.

To represent Organised Crime (OC) with FCA it is necessary to consider what are suitable as the objects of study and what are attributes of those objects. For example, the objects could be instances of crime or types of crime and the attributes could be properties of these crimes. A formal context can be created from recorded instances of crime or from subject matter expert or domain knowledge regarding the types of OC. Alternatively, for horizon scanning or situation assessment purposes, objects could be represented by activities or events that may be associated with OC based on predicted future technologies systems and processes.

When it comes to detecting OC activities, often seemingly inconsequential activities might underline a specific form of hostile intent: the purchase of fluorescent lighting tubes for the cultivation of cannabis, hydrogen peroxide or fertiliser for improvised incendiary devices, and even the alleged hiring of a retail unit in the run up to the Nairobi attack in 2013. Typically, those conducting such activities do not want to be suspected before their plan is completed and will go to various lengths to conceal their identities and their intent. With the purchase of specific items, groups may distribute their purchasing behaviours over time and space (or perhaps go to as far as setting up an agricultural business so that the purchase of large amounts of fertiliser does not look so suspicious – as with the case of Anders Breivik's attack on Oslo). However, it may still be possible to detect these activities using Frequent Itemset Mining (FIMI) (Goethals and Zaki, 2004) where the notion of frequency of occurrence of a group of items (the so-called item-set) is akin to FCA with objects being represented by instances. If we monitor the purchasing of fertiliser, FIMI can be used to automatically highlight possible clusters, thus alerting the possibility of hostile intent. The item-set (attributes) need to be carefully considered and may be a combination of quantity of fertiliser purchased, location of purchase (stores or towns/areas for delivery from websites) and time frames. Thus we may be automatically alerted of a number of purchases occurring in a particular time frame and/or being delivered to a particular geographical area that could be serving as a depot.

Although the computation required to carry out the analysis is intensive, recent developments in high-performance concept mining tools (Andrews, 2011; Outrata and Vychodil, 2012) mean that this type of monitoring could be carried out in real-time situation assessment. The outputs of the analysis would be suitable for visualising on a map (Figure 11.4) and end-users can be provided with the ability to alter parameters such as geographical area size and time frame, as well as being able to select different OC activities to analyse.

By creating a formal context for OC (and, by extension, hostile intent) using information such as that from the EU survey (United Nations, 2002) it may be possible to reveal hidden dependencies between different types of OC activities. Such dependencies are often called association rules and are inherent in FCA, being the ratio of the number of objects in concepts that have connections. In a similar way, the social network analysis in Chapter 12 considers similar dependencies amongst known suspects. Using this approach, we could investigate the association between drugs trafficking and the use of violence by OC gangs (Figure 11.5).

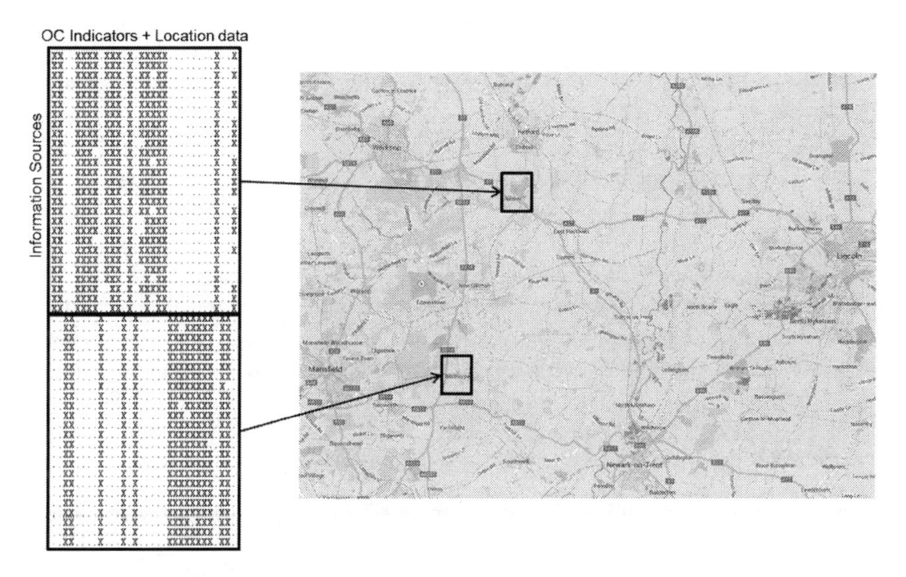

Figure 11.4 Visualising OC activities from frequent item-sets (from Andrews et al., 2013)

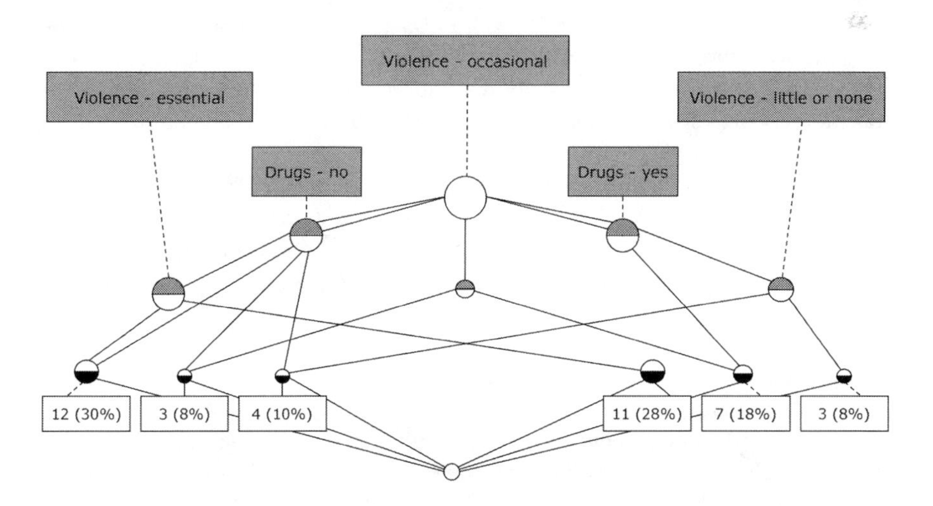

Figure 11.5 A concept lattice showing the association between drugs trafficking and the use of violence (from Andrews et al., 2013)

The numbers used here represent the number of OC gangs. Using FCA tools such as 'ConExp' (Yevtushenko, 2000) that was used to produce the lattices in this chapter, it is possible to investigate all associations between attributes in a formal context by calculating and listing association rules (for more information on these see Andrews et al., 2013). Typically, if we are carrying out an exploratory analysis of attributes, we will be interested in rules that show a strong association and that involve a (statistically) significant number of related objects. The notions of dependency and association can be taken a step further by analysing the links between situations, events and activities and the occurrence or emergence of OC. There are many known indicators of OC (Europol, 2011) but FCA may provide a means of discovering new, less obvious ones. The problem may be considered akin to a classification problem, either by classifying instances as OC or not OC, or by classifying instances as particular types within the larger OC descriptor. While there exist several well-known techniques of classification such as those in the field of machine learning (Frank and Asuncion, 2010), FCA has shown potential in this area (Ganter and Kuzntesov, 2000) and, with the evolution of high-performance algorithms and software (Andrews, 2011; Outrata and Vychodil, 2012) FCA may provide an approach that can be applied to large volumes of data in real-time situation assessment. With appropriate data (that may even be used as training data) deeper analyses should be possible for OC, to reveal possible sets of indicators for OC that can be used as part of a 'horizon scanning' system to detect or predict the emergence of OC in the future.

Although the work presented here is mainly of a propositional nature, it shows potential for FCA to be applied in the domain of detecting and monitoring SOEC. The culmination of the FCA for SOEC may be in the creation and implementation of an SOEC 'threat score card'. Using known and newly discovered indicators, association rules can be used to provide a weighting of those indicators. The resulting 'score' can be implemented as part of a horizon scanning system for the detection of SOEC types and situation assessment of the possible emergence of future SOEC threats if certain environmental conditions (indicators) pertain. Indeed, this is the proposed role of FCA in the new European ePOO-LICE project (Pastor, 2013) where it will play a part as one of several data analysis tools in a prototype pan-European OC monitoring and detection system.

Conclusion

By agreeing a common language of SOEC and fraud, sharing data and collaborating in the development of pan-European tools and techniques, existing monitoring systems across the EU can be integrated to provide a more effective force to detect and deter such crimes. EMPRISES focuses on adding value to the existing efforts of SIENA and FIU.NET by providing an improved commination and new forms of cooperative analysis. Existing global and EU businesses, governments and markets use sophisticated models, tools and techniques to detect trends and predict opportunities. These tried and tested approaches can be applied to the pan-EU 'businesses' of SOEC and fraud, thus providing the Member State LEAs with a better insight and understanding of the crimes and criminal groups that they are investigating and new means of detecting and deterring such crimes.

References

Andrews, S. (2011). In-close2: A high performance formal concept miner. In S. Andrews, S. Polovina, R. Hill, and B. Akhgar (Eds), *Conceptual Structures for Discovering Knowledge – Proceedings of the 19th International Conference on Conceptual Structures (ICCS)*. Berlin and Heidelberg, Springer, 50–62.

Andrews, S., Akhgar, B., Yates, S., Stedmon, A., and Hirsch, L. (2013). Using Formal Concept Analysis to detect and monitor organised crime. Flexible query answering systems. *Lecture Notes in Computer Science*, 8132(2013), 124–133.

Andrews, S., and Polovina, S. (2011). A mapping from conceptual graphs to Formal Concept Analysis. In S. Andrews, S. Polovina, R. Hill, and B. Akhgar (Eds), *Conceptual Structures for Discovering Knowledge – Proceedings of the 19th International Conference on Conceptual Structures (ICCS)*. Berlin and Heidelberg, Springer, 63–76.

Andrews, S., and Orphanides, C. (2010a). Knowledge discovery through creating formal contexts. *IEEE Computer Society*, 455–460.

Andrews, S., and Orphanides, C. (2010b). Fcabedrock, a formal context creator. In M. Croitoru, S. Ferre, and D. Lukose (Eds), *18th International Conference on Conceptual Structures, ICCS 2010*, 6208/2010 LNCS. Berlin Heidelberg: Springer, 181–184.

Becker, P., and Correia, J.H (2005). The Toscana J suite for implementing conceptual information systems. *LNCS*, 3626, Berlin Heidelberg: Springer, 324–348.

Carvalho, J.P., and Tomè, J.A. (1999). Rule based fuzzy cognitive maps: Fuzzy causal relations. In M. Mohammadian (Ed.), *Computational Intelligence for Modelling, Control and Automation,* Amsterdam, The Netherlands: IOS Press, 276–281.

Criminal Intelligence Service Canada (CISC) (2007). *Strategic Criminal Analytical Services – Strategic Early Warning for Criminal Intelligence.* Technical report. Ottawa: Central Bureau.

Du, X., Song, W., and Munro, M. (2006). Semantics recognition in service composition using conceptual Graph. *Proceedings of the 2006 IEEE/WIC/ACM International Conference on Web Intelligence and Intelligent Agent Technology (WI-IATW '06)*, Washington, DC., IEEE Computer Society, 295–298.

Europol. (2011). *EU Organised Crime Threat Assessment: Octa 2011*. Technical report, file no. 2530–274. The Hague: Europol.

Frank, A., and Asuncion, A. (2010). *UCI Machine Learning Repository* http://archive.ics.uci.edu/ml.

Fox, B. (2012). Defeating organized crime with social network analytics. Presentation at *Himss12 Annual conference and exhibition* 17–19 September 2012, Singapore.

Ganter, B., and Kuzntesov, S.O. (2000). Formalizing hypotheses with concepts. In B. Ganter and G.W. Mineau (Eds) *ICCS, LNAI*. Berlin and Heidelberg: Springer-Verlag, 342–356.

Goethals, B., and Zaki, M. (2004). Advances in frequent itemset mining implementations: Report on fimi '03. *SIGKDD Explorations Newsletter*, 6(1), 109–117.

International Standards Organisation (ISO) (2012). *ISO/IEC 24707 – Common Logic*. Available: http://metadata-standards.org/24707/.

Jedrzejek, C., Falkowski, M., and Bak, J. (2009). *Graph Mining for Detection of a Large Class of Financial Crimes*. Available at: http://ceur-ws.org/Vol-483/paper4.pdf.

Kim, B.J., and Bishu, R.R. (2006). Uncertainty of human error and fuzzy approach to human reliability analysis. *International Journal of Uncertainty, Fuzziness and Knowledge-Based Systems*, 14(2006), 111–129.

Kirda, E. (2010). *Root Causes Analysis*. Available at: http://www.wombat-project.eu/WP5/FP7-ICT-216026-Wombat_WP5_D12_V01_RCA-Technical-survey.pdf.

Larreina, M. (2007). *Detecting a Cluster in a Region Without Complete Statistical Data, using Input-Output Analysis: The Case of the Rioja Wine Cluster*. Available at: http://www.st-andrews.ac.uk/~www_crieff/papers/dp0706.pdf.

Launders, I. (2012). *The Transaction Graph: Requirements Capture in Semantic Enterprise Architectures*. Saarbrücken, Germany: Lambert Academic Publishing.

McNally, D., and Alston, J. (2006). *Use of Social Network Analysis (SNA) in the Examination of an Outlaw Motorcycle Gang*. Available at: https://www.ncjrs.gov/App/Publications/abstract.aspx?ID=236225.

Mifflin, T., Boner, C., Godfrey, G., and Skokan, J. (2004). A random graph model for terrorist transactions. *Aerospace Conference, 2004, IEEE,* 3258-3264.

Outrata, J., and Vychodil, V. (2012). Fast algorithm for computing fixpoints of galois connections induced by object-attribute relational data. *Information Sciences*, 185(1), 114–127.

Pastor, R. (2013). *epoolice: Early Pursuit Against Organised Crime Using Environmental Scanning, the Law and Intelligence Systems*. Available at: https://www.epoolice.eu/.

Polovina, S. (2007). An introduction to conceptual graphs. In U. Priss, S. Polovina and R.Hill (Eds) *Conceptual Structures: Knowledge Architectures for Smart Applications*. Berlin–Heidelberg:, Springer Lecture Notes in Artificial Intelligence, 1–15.

Polovina, S. (2012). The transaction concept in enterprise systems. In S. Andrews and F Dau (Eds) *Proceedings of the 2nd CUBIST Workshop. The 10th International Conference on Formal Concept Analysis (ICFCA 2012)*. Leuven, Belgium: 43–52.

Polovina, S., and Andrews, S. (2011). A transaction-oriented architecture for structuring unstructured information in enterprise applications. In V Sugumaran (Ed.), *Intelligent, Adaptive and Reasoning Technologies: New Developments and Applications*, Hershey, PA: IGI-Global, 285–299.

PR-OWL (2012). *A Bayesian Framework for Probabilistic Ontologies*. Available at: http://www.pr-owl.org/.

Ross, J.W., Weill. P., and Robertson, D. (2006). *Enterprise Architecture as Strategy: Creating a Foundation for Business Execution*. Boston, MA: Harvard Business School Press.

Sandmo, A. (2005). The theory of tax evasion: A retrospective view. *National Tax Journal*, 58(4), 643–663.

SAS (2009). *Social Network Analysis*. Available at: http://www.sas.com/resources/product-brief/social-network-analysis-insurance-brief.pdf.

Schrodt, P.A. (1995). *Patterns, Rules and Learning: Computational Models of International Behavior*. Available at : http://polmeth.calpoly.edu.

Sennoga, E.B. (2006). *Essays on Tax Evasion*. Atlanta, GA: Georgia State University.

Snášel, V., Horák, Z., and Abraham, A. (2008). Understanding social networks using formal concept analysis. *Proceedings of the 2008 IEEE/WIC/ACM International Conference on Web Intelligence and Intelligent Agent Technology – Volume 03 (WI-IAT '08)*, Washington, DC.

Stamper, R. (1996). Signs, norms, and information systems. In *Signs at Work*, Berlin: Walter de Gruyter, 349–397.

Thonnard, O. (2011). *Root Causes Analysis: Experimental Report*. Available at: http://www.wombat-project.eu/WP5/FP7-ICT-216026-Wombat_WP5_D22_V01_Root-Cause-Analysis-Experimental-report.pdf. Accessed 20 August 2014.

TOGAF (2011). *Content Metamodel*. Available at: http://pubs.opengroup.org/architecture/togaf9-doc/arch/chap34.html. Accessed 20 August 2014.

Turner, S.C. (2010) *Essays on Crime and Tax Evasion, Paper 64*. Available at: http://digitalarchive.gsu.edu/econ_diss/64. Accessed 20 August 2014.

United Nations (2002). *Global Programme Against Transnational Organized Crime. Results of a Pilot Survey of Forty Selected Organized Criminal Groups in Sixteen Countries*. Technical report. United Nations Office on Drugs and Crime.

Valtchev, P., Missaoui, R., and Godin, R. (2004). Formal concept analysis for knowledge discovery and data mining: The new challenges. In P. Eklund (Ed.), *Second International Conference on Formal Concept Analysis: Concept Lattices, volume 2961 of Lecture Notes in Computer Science*. Berlin Heidelberg: Springer, 352–371.

Vymětal, D., and Scheller, C.V. (2012). MAREA: Multi-Agent REA-Based Business Process Simulation Framework. *ICT for Competitiveness 2012*, Karviná, Czech Republic.

Yevtushenko, S.A. (2000). System of data analysis concept explorer (in Russian). In *Proceedings of the 7th National Conference on Artificial Intelligence KII-2000*. 127–134.

Chapter 12

Competitive Adaptation in Militant Networks: Preliminary Findings from an Islamist Case Study

Michael Kenney
Graduate School of Public and International Affairs, University of Pittsburgh, USA

John Horgan
International Center for the Study of Terrorism, Pennsylvania State University, USA

Cale Horne
Covenant College, Lookout Mountain, USA

Peter Vining
International Center for the Study of Terrorism, Pennsylvania State University, USA

Kathleen M. Carley
Center for Computational Analysis of Social and Organizational Systems (CASOS),Carnegie Mellon University, USA

Mia Bloom and Kurt Braddock
International Center for the Study of Terrorism, Pennsylvania State University, USA

Introduction

There is widespread agreement among scholars and practitioners that terrorism scholarship suffers from a lack of primary-source field research (Horgan, 2012). Moreover, terrorism studies have largely failed to integrate ethnographic research into computational modelling efforts that seek to represent and predict terrorist behavior. A growing number of scholars and practitioners recognize the value of mixed methods and interdisciplinary approaches to studying actors that engage in political violence themselves or support the use of political violence by like-minded groups. The vast majority of terrorism scholarship is based on secondary sources, which restricts the database from which scholars can develop theories and test hypotheses. The project outlined in this chapter addresses these shortcomings by combining the strengths of ethnographic field research with sophisticated computational models of individual and group behavior. This chapter provides an interdisciplinary framework from which to study the behavior of militant groups that either carry out acts of political violence themselves or support the use of violence by others. Specifically, we analyze data from news reports and interviews concerning the militant activist group Al-Muhajiroun (AM). Using competitive adaptation (Kenney, 2007) as a comparative organizational framework, this project focuses on the process by which adversaries learn from each other in complex adaptive systems and tailor their activities to achieve their organizational goals in light of their opponents' action. Our approach combines the analytical richness of ethnographic research with computational modelling to provide a meso-level model of militant networks that function in complex adaptive systems. This chapter presents preliminary results of AM, a former Islamist group in the United Kingdom that was banned by British authorities in 2010.

Organisational Learning and Competitive Adaptation in Militant Groups

As conceived in this study, terrorist groups engage in acts of political violence against civilian non-combatants in order to terrorize a wider audience, generally in pursuit of some political aim (Hoffman, 2006). Militant groups like AM do not engage in terrorism themselves but their rhetoric highlights the efficacy of violence for achieving certain political objectives, such as repelling perceived foreign invaders of Muslim lands. What separates militant networks like AM from more general political activists is the focus on violence in their discourse and their stated aim of overthrowing Western (and non-Western) governments to create a global Islamic caliphate based on Shariah (Islamic law) (Raymond, 2010; and author interview (Kenney) with AM leader, London, November 4, 2010).

Definitions aside, terrorist and militant networks alike are frequently characterized by opacity, decentralization, fluidity and illegality. It is precisely these qualities that give merit to network approaches for detecting changes in groups that engage in or support the use of political violence. Network analysis can provide predictive insights about how structural and relationship variables influence emergent group behavior. Network analysis can also assist practitioners in the prediction of violent behavior, including terrorism. Using network analysis, recent work by Magouirk and Atran (2008) demonstrates that the assumption of top-down indoctrination fails to account for important horizontal sources of radicalization in Jemaah Islamiyah (a Southeast Asian militant Islamist terrorist organization). Those in leadership roles do indeed indoctrinate junior members, but small group dynamics play an equally important role as members facilitate and reinforce the indoctrination of each other, thus contributing to emergent violence.

Jordan et al. (2008) and Vidino (2007) similarly find that the perpetrators of major terrorist attacks, as with the Madrid bombings and the Hofstad Group, may not be formally affiliated with global jihadist movements. In sum, proscribed groups tend not to behave like formal bureaucracies, making the flexibility of network approaches to terrorist detection all the more critical. Related, and contrary to conventional wisdom, scholars are finding that militant social networks do not always form for the purpose of carrying out violent acts. Recent works by Perliger and Pedahzur (2009, 2011), Sageman (2004, 2008) and Rodriquez (2005) all suggest that social processes within existing nonviolent networks are responsible for causing groups to drift towards the use of violence. Thus in order to predict an established group's likelihood of exhibiting violent behaviors, scholars must focus on the internal structure, relationships and decision-making of the group and how external variables (such as the joining of a new member, the removal of an existing member, or changes in environmental constraints) affect these dynamics.

Cumulatively these studies suggest that terrorist and militant groups are amorphous and in a state of flux, subject to change as a function of internal group dynamics or in response to external stimuli. Few studies, however, specifically examine how unconventional groups learn in their unique and conflict-based environments. Nevertheless, some recent work offers useful insights into how militant and terrorist organizations learn and adapt within the adversarial environments in which they operate. Jackson et al. (2005a, 2005b) examined organizational learning in several terrorist groups, including the Provisional Irish Republican Army, Aum Shinrikyo, Jemaah Islamiyah, Hezbollah, and the radical environmentalist Animal Liberation Front and Earth Liberation Front. Separate studies by Hamm (2005, 2007) draw on court documents contained in the 'American Terrorism Study' database and the criminological literature on social learning to explore how some violent political extremists acquire the skills to perform their tradecraft. While these studies offer insights into how numerous militant groups train their members and develop certain technological innovations, they do not systematically examine the internal processes of group learning and interpretation, as experienced by militants themselves. Moreover, these studies do not take into account the broader competitive environments in which militant groups operate.

Drawing on organizational and complexity theory, Kenney (2007) describes how organizational knowledge is leveraged by competing networks that interact in complex adaptive environments. Kenney dubs this process 'competitive adaptation', which explains how organizational learning occurs within an environment that is typically (though not always) characterized by hostility and multiple actors pursuing opposing goals. A network-based theoretical approach to the study of militant groups allows for modelling of both the internal organizational dynamics of militant groups and broader strategic interactions between militant groups and governments. Competitive adaptation is thus the framework from which we approach our study of militant groups, and we employ ethnographically based network analysis as our primary tool when modelling this framework.

Research on various militant networks offer qualitative descriptions of group structures and how those structures may account for group behavior (Horgan and Taylor, 1997; Kenney, 2007; Wiktorowicz, 2005). We seek to expand on these findings by using a mixed-methods approach to analyze one particular militant network. Quantitative metrics such as degrees of separation allow us to measure the connectedness between network leaders and rank-and-file members. As we discuss below, connectedness has important implications for network hierarchy, specifically the leader's ability to exert influence over his followers. 'Betweenness centrality' measures the extent to which each network member (or agent) links disconnected groups in the network. Agents scoring high in betweenness centrality serve as gatekeepers, connecting otherwise disparate nodes to the broader network. This facilitates information-sharing throughout the network, which, in turn, contributes to organizational learning. 'Eigenvector centrality' measures each agent's connection to other, well-connected nodes in the network. Agents scoring high in this measure have the ability to disperse information and mobilize resources rapidly in response to problematic situations and changing conditions (Hanneman and Riddle, 2005). Together with qualitative analysis of ethnographic data, these quantitative metrics allow us to study the evolution of the militant network over time. In the competitive adaptation framework, we expect that social network properties, such as connectedness, betweenness centrality, and eigenvector centrality for a political movement vary across time in response to changes in their environments.

When studying organizational learning and competitive adaptation in militant and terrorist networks, we seek to examine these networks in their entirety, beyond the social context. Specifically, we recognize that in order to understand group dynamics, learning, evolution, decision-making and emergent behavior, it is necessary not only to examine the roles and relationships of individual agents and groups within organizations, but also how those agents relate to locations in space, as well as the knowledge and resources leveraged by agents within organizations, in order to fulfill group tasks. Carley (1999) writes that an organization can be described as an 'ecology of networks' that continually evolves as agents within the organization learn, move and interact. A network of social roles within an organization might appear very different from a network of knowledge and expertise, which in turn might be very different from the network of resources or geographic proximity. Kenney's (2007) work on competitive adaptation similarly emphasizes the importance of organizational properties beyond those associated with individual human agents, arguing that the flow of knowledge, routines and artifacts within organizations is as important as the flow of personnel. We conceptualize militant networks consistently with these arguments and expect that they 'learn' when their participants receive information about their activities, process this information through knowledge-based artifacts, and apply the information to their practices and activities.

The case study presented in this chapter provides a detailed, yet preliminary, analysis of these two hypotheses by focusing on one militant network that, under a variety of names and organizational platforms, has been remarkably active over the past fifteen years in spite of being targeted for disruption by British authorities (Raymond, 2010; Wiktorowicz, 2005).

Case Study: The Evolution of Al-Muhajiroun

Al-Muhajiroun (AM) is not a terrorist organization but a political activist group that pushes the boundaries of free speech and association with belligerent, violence-laced rhetoric and inflammatory public demonstrations that sometimes result in criminal charges being filed against its members and associates. In recent years numerous AM-affiliated individuals have been convicted of inciting racial hatred, solicitation to murder, and terrorist fundraising by overstepping the bounds of legally permissible speech at their provocative rallies (Simcox et al., 2010). Indeed, the call to violence, under certain conditions, is a cornerstone of AM's rhetoric. The group has consistently supported the use of political violence overseas (outside of Britain) in what it maintains is a defensive, Islamically correct response to the aggressive foreign policies of Western states, including Britain and the United States. Beyond certain activists' incendiary protest speeches, other AM-affiliated individuals have been convicted of more direct involvement in violence over the years, including arson attacks and attempted petrol bombings in Britain (Simcox et al., 2010). Moreover, Al-Muhajiroun's goal has always been to replace Western governments with a global Islamic caliphate, an objective that, to be successful, would require political violence on a grand scale.

Founded in 1996 by Omar Bakri Mohammed and officially disbanded for several years in 2004, former AM members continue to engage in their provocative brand of political activism in the United Kingdom under the banner of several successor groups included in our study. Work by Wiktorowicz (2005) indicates that AM not only exhibits adaptive behavior as it interacts with British authorities, but that the relatively liberal political and social environments of the United Kingdom often condition these interactions, providing both advantages and disadvantages to each side. Wiktorowicz shows how freedom of the press in the United Kingdom has been a double-edged sword for AM, allowing the group to publicize its ideas to potential recruits, but also resulting in AM's widespread condemnation in British society, damaging its local operations (Wiktorowicz, 2005). AM's ostracism in Britain led to it being banned from using public venues under the AM name, loss of the group's charitable organization status in the United Kingdom, and increased police scrutiny of group activities, resulting in arrests of its members and associates.

In response to the negative ramifications of publicity, AM has adopted a strategy of organizational proliferation, diversification and obfuscation in order to spread the group's ideology and connect with potential recruits while avoiding the costs associated with the AM label, and without risking organizational death in the event of a police crackdown (Wiktorowicz, 2005). Kenney (2009) discusses this adaptive dynamic between the British government and AM in depth. Following the disbanding of AM in 2004, the group's leadership established two new groups called Al Ghurabaa (The Strangers) and the Saved (or Saviour) Sect, both of which attracted many AM members. When these successor groups faced the threat of legal proscription by the British government, former AM leaders created the 'Ahlus Sunnah wal Jamaah', an invitation-only Internet discussion forum (Kenney, 2009). More recently, interviews and ethnographic data in this research suggest that former AM members and associates have created several new platforms or groups to facilitate their ongoing activism, including Muslims against Crusades, Supporters of Sunnah, and Salafi Media (source: author interviews (Kenney) with AM members, November–December 2010 and June 2011, field notes November–December 2010, June 2011).

Whereas AM's name changes are a clear example of adaptive behavior within the competitive environment in which it operates, it is equally interesting that groups such as AM often fail to adapt, or learn the wrong lessons, despite their experiences. Kenney (2010) explains that militant groups might fail to adapt within their environments not only due to simple mistakes and human error, but potentially due to the underlying structures, ideologies or rules guiding an organization's behavior. The religious underpinnings of AM clearly condition the incentive structure of individual AM members and leaders, influencing how they adapt or fail to do so (Kenney, 2007). Kenney (2010) notes that AM's religiosity has led its leadership to conclude that their need to respond to external pressure, including pressure that may result in the imprisonment of individual activists, is limited. As one leader puts it, 'we believe Allah's will is there to protect us' and that their fate is already predetermined (Kenney, 2010, p. 924). From these examples of both adaptive and non-adaptive behaviours, we can see that AM is an interesting and relevant case study of how a militant group evolves and adapts (or fails to evolve and adapt) in a Western democracy.

Primary and Secondary Source Data for Network Analysis

The data used in this study have been collected from a variety of primary and secondary sources, including newspaper articles and original interviews with 69 respondents. Kenney conducted the interviews during several months of fieldwork in Britain. For this research, he met with and formally interviewed 41 individuals that were actively affiliated with AM or one or more of its successor groups. Kenney interviewed an additional 28 respondents that included government officials, researchers, and former AM members that have since left the group. In addition to these interviews, Kenney conducted ethnographic fieldwork in London, attending political demonstrations and public dawah (preaching) stalls organized by AM's current successor groups and socializing with activists at several locations.

From the newspaper articles we constructed a thesaurus of known AM members and associated Islamists, representing a total of 364 individuals (agents). In order to protect the anonymity of our respondents, we did not use any interviews for this purpose. We also created additional thesauri of AM front and successor groups (n = 353), events (n = 27), locations (n = 940), resources (n = 139), tasks (n = 560) and knowledge attributes (n = 3,240). These data were then semantically processed using the text analysis program 'AutoMap' (Carley et al., 2011a). AutoMap has been used by other researchers to extract networks representing mental models (Carley, 1997), semantic networks (Kim, 2011), and social networks (Frantz and Carley, 2008) from text extracts. It has been applied to a variety of domains ranging from email assessment (Frantz and Carley, 2008) to command post operations (Chapman et al., 2005) to media framing for stem cell research (Kim, 2011). By way of illustrating AutoMap's functionality in this research, we used the software program's word proximity command to identify a link between Mohammed Babar and Omar Bakri from the following sentence in the newspaper articles dataset: '[Mohammed] Babar, 31, told the court that he met Omar Bakri Muhammad, the exiled leader of Al-Muhajiroun, a radical Islamist group, during a visit to Britain' (Woolcock, 2006, p. 16).

Social network analyses were conducted using ORA, a network analytics program with integrated network statistics, graph analytics and visual analytics for performing traditional social network analysis as well as dynamic network analysis (Carley et al., 2007, 2011b). ORA is used to assess both the social network data (e.g. who is connected to whom) and the meta-network data (e.g. connections among who, what, why, how, where, and when). It has been widely used by researchers in many countries where illustrative applications include covert networks (Carley et al., 2009), public health organizations (Merrill et al., 2010), emergency care units (Effken et al., 2011), citation networks (Meyer et al., 2011), and stem cell research (Kim, 2011). Using various statistical procedures, graph-based metrics, and visual-based assessments the analyst can assess the data for a single network or a set of networks, determining which actors, groups or locations are critical; changes in this criticality over time; spatial characteristics of behavior; and emergent leaders and group capabilities. The analyst first visualizes the network, runs analyses, and then interprets the results based on the meaning of the various measures such as betweenness and eigenvector centrality. Many statistical procedures and their interpretations are defined in the ORA help documentation (Borgatti and Lopez-Kidwell, 2011; Carley et al., 2011b; Wasserman and Faust, 1994).

The preliminary findings presented below are based primarily on an ORA analysis of the network extracted by AutoMap from 1,079 AM-related newspaper articles published between 1996 (the year AM was formed in Britain) and 2009. At several points in the discussion below we compliment these social network measures with qualitative data drawn from the first author's (Kenney's) interviews and field notes. These interview data have not yet been processed by AutoMap, nor analysed by ORA. This will be done in the next phase of the project, the findings for which will be reported in subsequent publications. In this chapter, we draw on the interview data and field notes to add qualitative depth to the quantitative network measures drawn out by ORA. The newspaper articles used in this analysis were collected from Lexis Nexis Academic, an electronic database that contains full-text articles from over 2,000 newspapers throughout the world published from 1980 onwards. Duplicate articles and those not primarily concerned with AM were excluded from the dataset. In the following discussion, these data are divided into three time periods corresponding to major events in the group's history. Network A runs from AM's founding in 1996 through October 4, 2004, when the group voluntary disbanded under government pressure. Network B begins with the 7/7 bombings on the London Underground in 2005, an event that sparked Omar Bakri's flight to Lebanon, through to July 16, 2006, the day before the British government officially banned AM's successor organizations, Al Ghurabaa and the Saved Sect. Network C runs from this ban through the end of 2009, the cut off point for our Lexis Nexis data collection.

Preliminary Findings for the Al-Muhajiroun Network

In each network presented below (Table 12.1) nodes represent specific AM members and associated Islamists (a total of 364 individuals) that we extracted from the newspaper dataset using AutoMap. Segmenting the AM networks into three distinct time periods corresponding to major events in the group's history allows us to track changes in the social network measures over time. This adds an essential dynamic component to our understanding of AM's evolution, particularly when combined with our qualitative analysis of the ethnographic data.

Table 12.1 summarizes the changing relationship between Omar Bakri, AM's founder and spiritual leader, and other agents in the wider AM network. The numbers in each column provide the cumulative totals (i.e., two degrees of separation gives the cumulative total of rows 1 and 2, and so forth). The totals given in the bottom row indicate the total number of agents detected in the network, including isolates that were not connected to Bakri in this analysis.

Table 12.1 Omar Bakri's sphere of influence

Degree of separation	Network A 1 Jan 1996–4 Oct 2004	Network B 7 Jul 2005–16 Jul 2006	Network C 17 Jul 2006–31 Dec 2009
1	19	13	7
2	65	32	14
3	83	41	18
4	84	42	23
Isolates	102	38	68
Total	**186**	**80**	**91**

Note: The figures presented in rows 1 to 4 are cumulative (the row provides the total number of agent nodes connected to Bakri during each time period).

Several characteristics of Bakri's connectedness to the network merit discussion. First, during any time period, if an agent is not connected to Bakri, that agent is not connected to anyone in the network (i.e. that node is an 'isolate' and the Lexis Nexis newspaper data do not support a relationship between these individuals and Omar Bakri. While it is possible that some isolates are not involved in AM, these individuals are still included in the dataset of potential Islamist associates because they were identified as such through a thorough hand coding and verification process. Second, at any point in time, every agent in the network who is connected to Bakri is connected to him by no more than four degrees of separation. Third, Network A is a superset of Networks B and C. In other words, no new agents connect to Bakri subsequent to the first time period. These findings underscore the significant, yet evolving, impact Bakri's leadership has had on the AM network and Bakri's connectedness to other agents in the AM network changes over time, particularly after he left Britain for Lebanon.

Figure 12.1 depicts this changing relationship. As might be expected, Bakri's direct connections to other agents in the network declines significantly following his move to Lebanon, shortly after the 7/7 attacks on the London Underground and the growing pressure he faced from British authorities. In Network A, Bakri is connected to 83 of 84 agents by no more than three degrees of separation. From Network A to Network B, where the July 7 2005 cutoff approximates with Bakri's move to Lebanon, his total connections within the network drops by exactly half, from 84 to 42.

While everyone remains connected to Bakri within four degrees, the proximity of these connections declines following Bakri's move to Lebanon and the ban on AM's successor groups, Al Ghurabaa and the Saved Sect. In Network A, 65 individuals are connected to Bakri by no more than two degrees of separation;

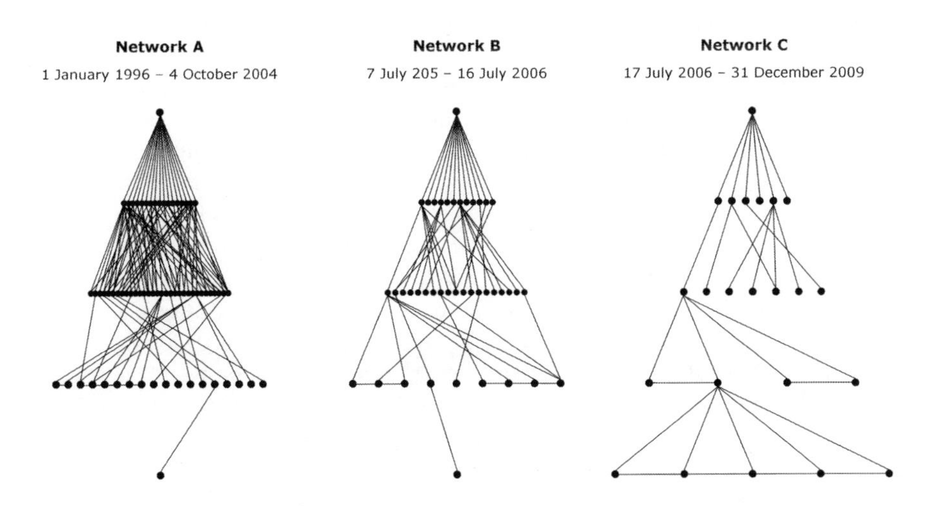

Figure 12.1 Omar Bakri's changing sphere of influence

in Network C this number drops to 14 individuals. The ability to influence other agents does not normally extend beyond the second degree, suggesting that Bakri's influence over AM members in Network C is less than a quarter of his original influence in Network A. This shift suggests a gradual distancing between Bakri and the rest of the AM network, which has continued to evolve since he left Britain. Such a shift is consistent with the ethnographic data uncovered by the first author (Kenney) during his fieldwork in London. Numerous interviews with AM leaders and members, and ethnographic observation of these and other network figures at several protests, public dawah stalls, educational lessons, and Internet chat rooms strongly suggest that Bakri's leadership has changed from direct oversight to a position of (geographically removed) symbolic leadership (source: author (Kenney) interviews with AM members, November–December 2010 and June 2011, field notes November–December 2010, June 2011).

Several of Bakri's long-standing students and AM veterans still based in Britain, including Anjem Choudary and Abu Izzadeen, have essentially replaced their mentor as day-to-day leaders of the evolving AM network. More recently, a new, third generation of leaders has emerged that have had little or no direct contact with Omar Bakri. These individuals became involved with AM through one of the successor groups, such as Islam4UK and Street Dawah, that were created after Bakri left Britain. These young leaders also formed their own groups, with Muslims against Crusades being the most prominent example. They receive guidance from senior AM veterans like Choudary but not, apparently, from Bakri himself.

Omar Bakri remains important in the AM network but his role is now limited to delivering audio and video lectures that his followers can access online and advising the senior AM veterans that occasionally seek his counsel. Interview and ethnographic data from this research also suggest that while Bakri's departure from London initially represented a major blow to his British students, they eventually adapted to this setback by learning new ways of communicating with their spiritual leader via online communications technologies, and by the emergence of new operational leaders that assumed day-to-day authority for directing AM's operations in Britain (source: author (Kenney) interviews with AM members, November–December 2010 and June 2011, field notes November–December 2010, June 2011). Government efforts to disrupt Al-Muhajiroun, though initially successful, gradually weakened as the AM network adapted to the setback by learning how to function effectively in a new, more hostile counter-terrorism environment without enjoying regular access to their former operational leader.

Other measures of leadership allow a comparison of Bakri's role in the network to the roles of other network elites. Agents in the network may score high along one or more dimensions of leadership without ever holding a position of formal leadership in the organization. Because informal leaders can be critical to organization functions, various metrics of network elites can aid in identifying agents important to the network, and in detecting meaningful changes in the network across time. For example, betweenness centrality measures the extent to which a given node (an agent) constitutes the most efficient path between other nodes in the network. For all node pairs that have a shortest path to a particular node, betweenness centrality is calculated as the percentage of node pairs that pass through this node. This metric assesses which agents present the 'best paths' between other agents, suggesting that individuals ranking high in this metric are likely to serve as brokers or gatekeepers between different subgroups within the network (Table 12.2).

Table 12.2 rank orders AM agents in terms of their betweenness centrality across the three time periods. During any given period, Bakri ranks behind five to nine other agents in the network. While Osama bin Laden's and Mohammed Omran's high betweenness centrality rankings in the AM networks are most likely artifacts of the newspaper data, the high betweenness centrality of several other agents merits discussion.

Anjem Choudary, Abu Izzadeen, Abu Uzair, and Abdul Saleem were all long-standing students of Omar Bakri's, each of whom played important roles in AM and its various successor groups. For example, Abdul Saleem served as a key broker for AM, connecting different militants in Britain and Pakistan, even organizing the

Table 12.2 Betweenness centrality across networks A, B and C

Rank	Network A		Network B		Network C	
1	Abdul Saleem	0.042	Abdul Saleem	0.078	Abu Izzadeen	0.013
2	Osama bin Laden	0.037	Abu Hamza	0.060	Osama bin Laden	0.012
3	Abu Hamza	0.035	Abu Izzadeen	0.051	Anjem Choudary	0.011
4	Hassan Butt	0.020	Osama bin Laden	0.039	Abu Hamza	0.009
5	Saladhuddin Amin	0.013	Saladhuddin Amin	0.027	Mohammed Omran	0.009
6	Abu Qatada	0.011	Omar Bakri	0.020	Omar Khyam	0.008
7	Omar Sharif	0.011	Omar Sharif	0.018	Abdul Saleem	0.007
8	Waheed Mahmood	0.010	Abu Uzair	0.018	Abu Qatada	0.007
9	Mohammed Omran	0.010	Hassan Butt	0.016	Omar Bakri	0.005
10	Omar Bakri	0.009	Anjem Choudary	0.011	Mohammed Babar	0.005

Note: Figures presented are normalized ratios.

movement of British Muslims into Pakistan, and from there into the insurgency in Afghanistan. Once back in Britain, Saleem acknowledged that he received military training in Afghanistan and Pakistan, and sought to recruit other young Muslims to do the same. Interestingly, according to interviews in this research, Saleem is no longer involved in AM-related activities, suggesting the network may have lost access to this important resource (source: author interview with AM member, June 2011). Unlike Saleem, Abu Hamza al-Masri was not Bakri's student but his associate and erstwhile competitor. Abu Hamza led the militant Supporters of Shariah group, which collaborated with AM in some conferences and political protests until Hamza's arrest for terrorism-related offences in 2004. Following the dissolution of Hamza's Supporters of Shariah group, some of his students migrated to the AM network, where they remain active in AM's successor groups (source: author interview with AM member, June 2011).

Eigenvector centrality offers a very different measure of agents' elite status within a network (Table 12.3). This measure calculates the degree to which a given node is considered central to the network to the extent that its neighbours are central.

Table 12.3 Eigenvector centrality across networks A, B and C

Rank	Network A		Network B		Network C	
1	Abdul Kahar Kalam	1	Abdul Kahar Kalam	1	Anjem Choudary	1
2	Omar Sharif	1	Omar Bakri	1	Abdul Saleem	1
3	Abdul Karim	1	Richard Reid	1	Willie Brigette	1
4	Younis al Hayyari	1	Abdul Karim	1	Omar Bakri	0.960
5	Abu Obeida	1	Younis al Hayyari	1	Saladhuddin Amin	0.918
6	Ramadan Shallah	1	Ezzit Raad	1	Jawad Akbar	0.787
7	Ezzit Raad	1	Fadal Sayadi	1	Omar Khyam	0.704
8	Abdul Koyair	1	Abdul Koyair	1	Waheed Mahmood	0.407
9	Abdul Qassim	1	Anjem Choudary	0.626	Abu Izzadeen	0.390
10	Mohammed Salim	1	Abu Izzadeen	0.562	Abu Hamza	0.343

Note: Figures presented are normalized ratios.

Well-connected agents connected to other well-connected agents score high on this metric, while the formula discounts nodes possessing many connections, as well as accounting for the fact that most nodes will have some connections. The eigenvector centrality measure is calculated using the largest positive eigenvalue of the adjacency matrix representation. Notably, Omar Bakri's eigenvector centrality ranking actually improves in networks B and C, after he leaves Britain for Lebanon. While Bakri's direct connections to the network rank-and-file suffer while in exile, his continued association with the AM leadership (i.e. other well-connected nodes) maintains his high scores on this measure. This is consistent with the first author's interviews and ethnographic data, which found that AM leaders in Britain maintained regular contact with Bakri through various communications technologies. Moreover, several current AM leaders even visited Lebanon, where they sought to meet with their religious leader (source: author interviews with AM members, November–December 2010 and June 2011, field notes November–December 2010, June 2011).

Others, such as Richard Reid and Omar Sharif, appear to score high in eigenvector centrality as a function of their involvement in highly publicised, terrorism-related incidents rather than playing a central role in AM. Richard Reid was the shoe-bomber that tried and failed to bring down a transatlantic flight from Paris to Miami several months after the 9/11 attacks. Omar Sharif attempted, and failed, to ignite a suicide bomb vest outside a bar in Tel Aviv in 2004 (Wiktorowicz, 2005). While both Reid and Sharif reportedly attended AM lessons or rallies, they were not formally affiliated with the movement (*Al Jazeera*, 2003; Brown, 2002). Such individuals may be part of tight cliques, meaning they are highly connected to others in their group while they do not encounter discounts for connections to many nodes in the AM networks that they do not possess. Similarly, Omar Khyam, Jawad Akbar, Saladhuddin Amin, and Waheed Mahmood were convicted of belonging to a terrorist cell that planned to bomb different 'soft targets' in London. Interviews conducted in this research suggest that while several members of this cell had previously navigated AM circles, they were not actively involved in the group at the time of their apprehension (source: author (Kenney) interviews with AM members, November–December 2010). Likewise, Ezzit Raad, arrested in 2005 for his role in a terrorist plot in Australia, and Younis al Hayyari, an Al-Qaeda affiliate shot dead in Saudi Arabia in 2005, were part of terrorist cliques in their respective countries and do not appear to have been involved with AM in Britain.

Conclusions

The findings discussed in this chapter, though preliminary, are significant not only for what they tell us about a banned Islamist group that openly espouses the use of political violence against Western governments under certain conditions, but for highlighting the value of mixed methods in studying such militant networks more generally. Unlike much of the literature in terrorism studies, the research project described here combines quantitative and qualitative analysis of primary and secondary source data, all focused on a single case study with national security implications for Britain and other countries. In blending quantitative analysis of secondary source newspaper articles with qualitative analysis of primary source interviews and field notes, we have uncovered findings that neither approach could reveal in isolation.

Using quantitative social network measures, as applied to the newspaper dataset, we analyzed AM at the macro level. This allowed us to identify changes in leadership relations over time corresponding to major events in the group's development. The results of this analysis were consistent with the notion that social network properties within movements such as AM vary across time in response to changes in their environments. By measuring the degrees of separation between Omar Bakri and other network nodes we were able to track the declining density of Bakri's AM social ties and his corresponding loss of influence over his followers after he left London for Lebanon. Geographic proximity did matter for Bakri's leadership and his departure from London decreased his connectedness with the British-based members of his network.

Quantitative social network measures were also useful at the micro level of analysis dealing with specific nodes. From the betweenness centrality measure we identified several key brokers in the AM network, such as Abdul Saleem and Abu Uzair, that have not received as much media attention as other, more prominent network figures, including Anjem Choudary, Abu Izzadeen, and Omar Bakri himself. Interestingly, this finding was generated from a newspaper dataset, underscoring the ability of quantitative network measures to generate valuable insights from secondary data.

These quantitatively generated insights were supported, and extended, by qualitative analysis of the first author's (Kenney's) field research in Britain. Primary source interviews and field notes not only confirmed that Abdul Saleem had been a key broker for AM during the time periods under analysis, supporting the quantitative analysis, but that Saleem, unlike other AM leaders, is no longer actively involved in the group, suggesting the militant network has lost an important, centrally connected node. The implications of Saleem's departure from AM remain to be explored in future research. However, the qualitative analysis presented in this chapter regarding Omar Bakri's move to Lebanon suggest that the implications of Saleem's exit may not be far-reaching. Drawing on interviews and field notes, we found that Bakri's British-based students successfully adapted to their leader's move to Lebanon by learning how to function in a more hostile counter-terrorism environment. They did so by maintaining ties with their spiritual leader through online communications technologies, developing new day-to-day operational leaders in Britain, and creating numerous successor groups to Al-Muhajiroun that even today continue the network's public dawah and political activism.

As this analysis suggests, what matters is not what AM's leaders and members choose to call themselves, but how they adapt their activities in response to government pressure. To understand this competitive, adaptive dynamic, a mixed-methods approach that combines quantitative and qualitative analysis of primary and secondary data to analyze changes in network relations and activities over time is indispensable.

Acknowledgement

This research is supported by a grant from the Office of Naval Research (ONR), U.S. Department of the Navy, No. N00014-09-1-0667. Any opinions, findings, or recommendations expressed in this material are those of the authors and do not necessarily reflect the views of the Office of Naval Research. We thank Devan Zeger for extensive help in the preparation of this manuscript.

References

Al Jazeera (2003). Interview with Omar Bakri Mohammed. Mecca, 1 May 2003).

Borgatti, S.P., and Lopez-Kidwell, V. (2011). Network theory. In P. Carrington and J. Scott (Eds), *The Sage Handbook of Social Network Analysis*. Sage Publications: London, 40–54.

Brown, J.A. (2002). Muslim radicals plan meeting about 9/11. *The Boston Globe*, A7 (10 September 2002).

Carley, K.M. (1997). Extracting team mental models through textual analysis. *Journal of Organizational Behavior*, 18, 533–538.

Carley, K.M. (1999). On the evolution of social and organizational networks. In S.B. Andrews and D. Knoke (Eds), *Research in the Sociology of Organizations on Networks in and Around Organizations*. JAI Press, Inc. Stamford, CT, 3–30.

Carley, K.M., Diesner, J., Reminga, J., and Tsvetovat, M. (2007). Toward an interoperable dynamic network analysis toolkit. *Decision Support Systems*, 43, 1324–1337.

Carley, K.M., Martin, M.K., and Hancock, J.P. (2009). *Dynamic Network Analysis Applied to Experiments from the Decision Architectures Research Environment,* In P. McDermott and L Allender (Eds) *Advanced Decision Architectures for the Warfighter: Foundation and Technology*. Partners of the Army Research Laboratory Advanced Decision Architectures Collaborative Technology Alliance, HPB-Ohio, Columbus, OH, 15–34.

Carley, K.M., Columbus, D., Bigrigg, M., and Kunkel, F. (2011a). *AutoMap User's Guide 2011*. Carnegie Mellon University, School of Computer Science, Institute for Software Research, Pitsburgh, PA Technical Report, CMU-ISR-11-108.

Carley, K.M., Reminga, J., Storrick, J., and Columbus, D. (2011b). *ORA User's Guide 2011*. Carnegie Mellon University, School of Computer Science, Institute for Software Research, Pitsburgh, PA, Technical Report, CMU-ISR-11-107.

Chapman, R.J., Graham, J.M., Carley, K.M. Rosoff, A., and Paterson, R. (2005). Providing insight into command post operations through sharable contextualized net-centric visualizations and analysis. In *Proceedings of the 2005 Human Interaction with Complex Systems Symposium*, Greenbelt, VA, 17–18 November 2005, http://www.cws-i.com/papers/chapman-graham-carley-rosoff-paterson-hicss2005.pdf, accessed 11 August 2014.

Effken, J.A., Carley, K.M., Gephart, S., Verran, Joyce A., Bianchi, D., Reminga, J., and Brewer, B.B. (2011). Using ORA to explore the relationship of nursing unit communication to patient safety and quality outcomes. *International Journal of Medical Informatics*, 80(7), 507–517.

Frantz, T.L., and Carley, K.M. (2008). Transforming raw-email data into social-network information. In C.C. Yang, H. Chen, M. Chau, K. Chang, S.D. Lang, P.S. Chen, R. Hsieh, D. Zeng, F.Y. Wang, K. Carley, W. Mao, and J. Zhan (Eds) *Proceedings of the 2008 LNCS 5075 Workshop*, Berlin Heidelberg, Springer, 413–420.

Hamm, M. (2005). *Crimes Committed by Terrorist Groups: Theory, Research and Prevention.* U.S. Department of Justice, Washington, DC.

Hamm, M. (2007). *Terrorism as Crime: From Oklahoma City to Al-Qaeda and Beyond.* New York University Press, New York.

Hanneman, R.A., and Riddle, M. (2005). *Introduction to Social Network Methods.* University of California Riverside, Riverside, CA.

Hoffman, B.M. (2006). *Inside Terrorism*, revised and expanded edn. Columbia University Press, New York.

Horgan, J. (2012). Interviewing the terrorists: Reflections on fieldwork and implications for psychological research. *Behavioral Sciences of Terrorism and Political Aggression*, 4(3), 195–211.

Horgan, J., and Taylor, M. (1997). The Provisional Irish Republican Army: Command and functional structure. *Terrorism and Political Violence*, 9, 1–32.

Jackson, B.A., Baker, J.C., Chalk, P., Cragin, K., Parachini, J.V., and Trujillo, H.R. (2005a). *Aptitude for Destruction Volume 1: Organizational Learning in Terrorist Groups and its Implication for Combating Terrorism*. RAND, Santa Monica, CA.

Jackson, B.A., Baker, J.C., Chalk, P., Cragin, K., Parachini, J.V., and Trujillo, H.R. (2005b). *Aptitude for Destruction Volume 2: Case Studies of Organizational Learning in Five Terrorist Groups*. RAND, Santa Monica, CA.

Kenney, M. (2007). *From Pablo to Osama: Trafficking and Terrorist Networks, Government Bureaucracies, and Competitive Adaptation.* Penn State Press, University Park, PA.

Kenney, M. (2009). *Organizational Learning and Islamist Militancy.* National Institute of Justice, U.S. Department of Justice, Washington, DC.

Kenney, M. (2010). Dumb yet deadly: Local knowledge and poor tradecraft among Islamist militants in Britain and Spain. *Studies in Conflict and Terrorism*, 33, 1–22.

Kim, L. (2011). Media framing for stem cell research: a cross-national analysis of political representation of science between the UK and South Korea. *Journal of Science Communication*, 10(3), 1–17.

Magouirk, J., and Atran, S. (2008). Jemaah Islamiyah's radical madrassah networks. *Dynamics of Asymmetric Conflict*, 1, 1–16.

Jordan, J., Mañas, F.M., and Horsburgh, N. (2008). Strengths and weaknesses of grassroot Jihadist networks: The Madrid bombings. *Studies in Conflict and Terrorism*, 31, 17–39.

Merrill, J., Keeling, J.W., and Carley, K.M. (2010). A comparative study of 11 local health department organizational networks. *Journal of Public Health Management and Practice*, 16(6), 564–576.

Meyer, M., Zaggl, M.A., and Carley, K.M. (2011). Measuring CMOT's intellectual structure and its development. *Computational and Mathematical Organization Theory*, 17(1), 1–34.

Perliger, A., and Pedahzur A. (2009). *Jewish Terrorism in Israel*. Columbia University Press, New York.

Perliger, A., and Pedahzur A. (2011). Social network analysis in the study of terrorism and political violence. *Political Science and Politics*, 44, 45–50.

Raymond, C.Z. (2010). *Al Muhajiroun and Islam4UK: The Group Behind the Ban. Working Paper*. International Centre for the Study of Radicalisation and Political Violence, King's College London, UK.

Rodriquez, A.J. (2005). The March 11th Terrorist Network: In its Weakness Lies its Strength. *Proceedings of 25th International Sunbelt Social Network Conference*, 16-20 February 2005, Redondo Beach, CA.

Sageman, M. (2004). *Understanding Terror Networks*. University of Pennsylvania Press, Philadelphia, PA.

Sageman, M. (2008). *Leaderless Jihad: Terror Networks in the Twenty-First Century*. University of Pennsylvania Press, Philadelphia, PA.

Simcox, R., Stuart, H., and Ahmed, H. (2010). *Islamist Terrorism: The British Connections*. The Centre for Social Cohesion, London.

Vidino, L. (2007). The Hofstad Group: The new face of terrorist networks in Europe. *Studies in Conflict and Terrorism*, 30, 579–592.

Wasserman, S., and Faust, K. (1994). *Social Network Analysis: Methods and Applications*. Cambridge University Press, Cambridge, MA.

Wiktorowicz, Q. (2005). *Radical Islam Rising: Muslim Extremism in the West*. Rowman and Littlefield, Lanham, MD.

Woolcock, N. (2006). The Al-Qaeda supergrass who wanted to wage war in Britain. *The Times*, London, 24 March 2006, 16.

Chapter 13
Evaluating Emergency Preparedness: Using Responsibility Models to Identify Vulnerabilities

Gordon Baxter and Ian Sommerville
School of Computer Science, University of St Andrews, UK

Introduction

Preventing acts of terrorism requires being able to accurately and reliably predict when, where and how the terrorists will strike. When prevention is not possible the aim is to mitigate the severity of the consequences of the terrorist acts. Terrorists normally attempt to cause maximum disruption and achieve significant publicity by attacking systems that they perceive as being of great importance (physically and psychologically). Here we are talking about systems in the most general sense of the term, rather than just technological systems. More particularly, we are talking about sociotechnical systems, and encompass both organisations and infrastructure within our broad definition.

The responses to emergency situations, such as terrorist attacks, can be characterised by three common traits (Goughnour and Durbin, 2008):

- The information sources and the people who use them are geographically widely dispersed.
- The information which rapidly changes as the situation develops needs to be made widely distributed as soon as it becomes available (and has been validated).
- The types of information that have to be managed (processed, analysed and reviewed) are quite disparate and may keep changing over time as the situation develops.

We know a lot about the different ways in which terrorists can strike, but often know less about the when and where until it may be too late to prevent at least some damage from occurring. To protect the public effectively against a terrorist attack therefore requires a high degree of emergency preparedness. This is based on co-ordinated functioning, co-operation and communication at global, regional, national and local levels, and multi-agency working (Veness, 2012). The details for how to respond to particular emergency situations are normally laid out in contingency plans. These plans can only be accurate if the relevant people with the appropriate skills work together to develop them. This is because the plans have to cover many different aspects of the situation, such as how to achieve shared awareness and understanding across the various agencies involved (Convertino et al., 2011; Wu et al., 2013).

One way of making sure that a system has an appropriate level of emergency preparedness is to learn from other, similar, emergency situations that have happened in the past. The London Resilience Team, for example, developed new plans and established new facilities based on lessons learned in the aftermath of 2005's London bombings (Lewis, 2012). It is also possible to observe and learn from others as they practice emergency management. The Beijing Olympic Games organising committee used this approach. They sent staff to monitor the emergency management operations at the Sydney and Athens summer Olympics, and the Turin winter Olympics so that they could apply what they had learned to the Beijing games (Jiwei et al., 2010).

The notion of responsibility is particularly important when handling emergency situations. When analysing South Carolina's census data on preparedness to deal with terrorism, however, Pelfrey (2007) found that while states were assigned significant responsibilities, most of the relevant federal documents focused instead on goals and objectives. The problem is exacerbated because multiple agencies are normally involved in responding to terrorist attacks so responsibilities are distributed across and within the different agencies.

A better understanding of the way that responsibilities are managed in multi-agency situations is therefore important. There may be some responsibilities that are shared across agencies and there may be some responsibilities that fit naturally into the remit of more than one agency. Understanding how the different agencies deal with these situations can generate lessons that can be incorporated into contingency plans. This should improve the shared awareness and understanding among the agencies of how responsibilities are managed. In this way it should be easier for agencies to work together in emergency situations, and step in to help each other where appropriate.

In the next section we introduce our conceptualisation of the notion of responsibility. We then go on to describe a graphical technique for modelling responsibilities that can be used in analysing sociotechnical systems. We discuss how it can be used, and illustrate its use with some simple examples. We then identify the classes of vulnerabilities that are associated with responsibilities, as well as those that are associated with the use of resources. We show how responsibility modelling can be used to identify vulnerabilities in contingency plans, before discussing its role in informing the design of table-top exercise scenarios, and shaping the nature of live table-top exercises as they evolve. We conclude by summarising how performing vulnerability analyses (using responsibility modelling) can help improve emergency preparedness.

Responsibility

The tools and techniques that are available to help analyse failures (see Leveson, 1995) are often based on concepts such as goals, and describe how these are achieved (or not) using abstractions such as tasks and functions. These tools and techniques are very good at dealing with technical failures, but are less well suited to dealing with the many failures where the attributed causes lie within the sociotechnical system per se. There is some evidence, for example, that the London bombings in 2005 were at least partly attributable to sociotechnical failures (Intelligence and Security Committee, 2006). We have therefore been using the notion of responsibilities, which is more abstract than goals and objectives, to express the softer aspects of system performance. It is also a concept that system stakeholders are generally more comfortable with, and find easier to comprehend and discuss.

Our definition of responsibility is a pragmatic one. A responsibility is a 'duty, held by some agent, to achieve, maintain or avoid some given state, subject to conformance with organisational, social and cultural norms' (Sommerville, Storer and Lock, 2009, p. 181). In the majority of cases the agents that are assigned (or allocated) responsibilities will be people or organisations. The agents could, however, also be a technological device or a software application. If the responsibility is to raise an alarm when smoke is detected, for example, this could be assigned to an automatic smoke detection device.

In order to discharge a responsibility an agent will normally need to utilise some resources. These resources can be concrete (such as people or physical objects) or abstract (time, or a piece of information). Responsibilities are normally discharged in a specific context in which there are several norms that define and constrain how the agents are supposed to operate. These may include contingency plans, standard operating procedures, and statutory regulations which govern how the components of the sociotechnical system should work.

Responsibility Modelling

We have developed a graphical representation technique called Responsibility Modelling (RM) for analysing systems and organisations. A responsibility model shows how responsibilities are assigned to agents and what resources are needed to discharge those responsibilities. The basic concepts are illustrated in Figure 13.1 which shows that the agent '*Security guard*' is assigned the responsibility '*Check for unauthorised access*' and uses the resource '*CCTV*' to discharge that responsibility.

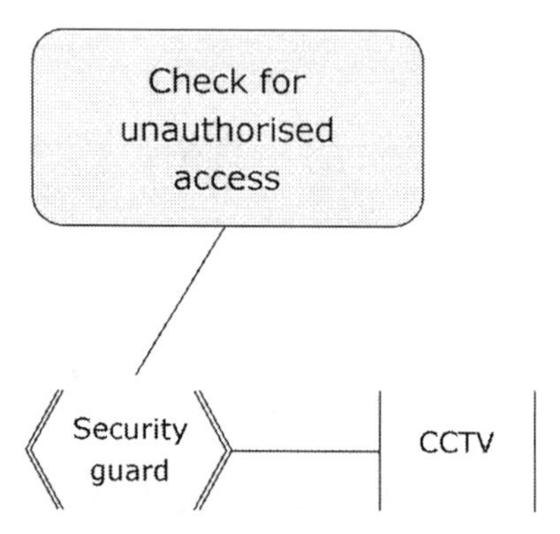

Figure 13.1 Basic elements of a responsibility model
The agent (*Security guard*) discharges the allocated responsibility (*Check for unauthorised access*) using a resource (*CCTV*).

Responsibility models can be developed for prospective analysis, as well as retrospective analysis. Depending on the purpose of the analysis, access to resources to create the models in the first place, and the amount of time and effort available to do the analysis, three methods can be used to collect data for the models:

- Stakeholder interviews and user requirements elicitation. Stakeholders are asked about their responsibilities and the resources they use to discharge them (see also Chapter 19).
- Document analysis. Documents such as organisational structure charts, local operating procedures and contingency plans are analysed to identify responsibilities, agents and resources. Some care should be exercised here when interpreting the results: there are usually differences between the way that the procedures describe how people are supposed to work, and what people really do in practice (Dekker, 2006).
- Archive data. Post hoc accident reports and incident debriefings can sometimes be used to identify responsibilities where systems have failed.

To create the model any drawing package can be used, or they can be drawn by hand. The information that has been collected is graphically represented to show how responsibilities are allocated, and how resources are used. For large sociotechnical systems, such as those that provide and support Critical National Infrastructure (CNI), it is often easier to present the system as a set of related responsibility models, rather than try to model the whole system in one diagram. Each level of responsibility can be represented by its own responsibility model diagram(s). These models, which reflect the different levels of responsibilities, are usually organised hierarchically. We have found it most useful to adopt a risk analysis style of approach to help identify where to focus our attention. We analyse in detail those responsibilities where the failure to discharge that responsibility has a high likelihood and can lead to serious adverse consequences that may affect the overall functioning of the system. It is often impossible to perform a highly detailed analysis all of the responsibilities associated with a large system to create a comprehensive model in a timely manner.

Identifying Vulnerabilities

All systems and organisations contain vulnerabilities. If these vulnerabilities are not identified, they can persist as latent failures in the sociotechnical system (Reason, 1990). In other words, they may be unknown or not considered to be a problem until they are triggered by real events. These latent failures can reduce the resilience of the sociotechnical system to events such as a terrorist attack, because when they become active they adversely affect the system's ability to deal with the consequences of that event. One of the benefits of responsibility models is that they can be used to identify potential vulnerabilities in responsibilities, agents and resources.

Responsibility Vulnerabilities

There is a set of system vulnerability types associated with responsibilities (and agents), each of which can contribute to system failures. We have extended Sommerville's (2007) original list of six main types of responsibility and agent-related vulnerabilities, based on our subsequent observations.

Unassigned responsibilities
Unassigned responsibilities are probably the easiest type of vulnerability to identify from a responsibility model. An unassigned responsibility is simply one that has not been assigned to any agent. If a responsibility has no links associated with it in the responsibility model, this suggests that there is a vulnerability and that that particular responsibility will not be discharged.

Duplicated responsibilities
Duplicated responsibilities arise when a particular responsibility is assigned to more than one agent. These can arise when systems are decomposed in a way that leads to exactly the same responsibility appearing in separate parts of the system. They can also occur when multiple agencies have to work together, particularly if they are dealing with an acute problem, where the discharging of responsibilities may not always be obvious, leading to one agent attempting to discharge a responsibility that somebody else is either currently discharging, or has already discharged.

Diffusion of responsibility
The notion of diffusion of responsibility comes from social psychology (Latané and Nida, 1981). Diffusion of responsibility can occur in multi-agency environments when each agency may believe that one of the other agencies has been assigned a particular responsibility. The net result is that nobody discharges the responsibility because they expect another agency to do it. This vulnerability includes elements of unassigned responsibilities and duplicated responsibilities and is specific to multi-agency requirements.

Uncommunicated responsibilities
Often responsibilities are associated with roles, and an agent may have several roles. Uncommunicated responsibilities arise when an agent who has been assigned a particular role is not told about the responsibilities that are associated with that role.

Misassigned responsibilities
Sometimes an agent may not have the capabilities (knowledge, skills and attributes) or resources to discharge the assigned responsibility. In this case, it is described as a misassigned responsibility. These can occur when a situation evolves in such a way that responsibilities are (re-)allocated to those agents that are available at that point in time so that they can be discharged, rather than waiting for the originally allocated agent to become available.

Responsibility overload
When an agent is assigned several responsibilities they may not have the physical or mental capacity to be able to discharge all of those responsibilities within the required time frame. This situation is described as responsibility overload.

Responsibility fragility
Also described as responsibility brittleness, responsibility fragility occurs when a particular responsibility is assigned to an agent, but if that agent is not available, there is no back-up agent who can discharge that responsibility. This sort of situation arises when a responsibility is assigned to a dedicated individual (e.g. the Health and Safety Officer) because they have the particular capabilities to discharge it, and that individual happens to be absent or unavailable for some reason.

Responsibility conflict
Sometimes agents have to fulfil multiple roles within an organisation and the responsibilities associated with those roles may conflict. If, for example, a person has the responsibility for managing a project (making sure it is delivered on time and within budget), but also has the responsibility for technical assurance of that project (making sure that the application or system has been rigorously developed, tested and so on), these two responsibilities are likely to come into conflict.

Resource Vulnerabilities

The vulnerabilities associated with resources relate to their production and consumption, and their availability and accessibility. If an agent consumes a particular resource to discharge some responsibility, for example, then this resource has to be provided somehow. If the resource is produced by another agent, this indicates that that agent should be assigned the responsibility to produce that resource.

If an agent needs to use information resources to help them discharge their responsibility as part of the response to an event such as a terrorist attack, consideration must be given to how that agent accesses these resources. If the information resource is only available online, for example, and the network connection to that resource has been wiped out, the agent will be unable to discharge that particular responsibility unless they can access the required information by other means.

If an agent has to use physical resources, such as transport for relocating or evacuating people after a terrorist attack, consideration has to be given to how to access that transport. If the road network becomes blocked or even destroyed, for example, this may make it impossible to get access to the transport vehicles.

Managing Vulnerabilities

In an ideal world, there would be a perfect match between responsibilities, agents and resources. Responsibilities would always be discharged in a timely manner by agents with the appropriate combination of knowledge, skills and attributes. The way that responsibilities get discharged on the ground, however, will often diverge from written procedures and plans. Part of the reason for this is that procedures and plans cannot cover the contingencies of all possible situations. What usually happens is that people will adapt the plans and procedures to deal with the particular situations at hand.

Managing the different types of responsibility vulnerabilities may not always be straightforward. Duplicated responsibilities, for example, are a vulnerability, but a system can be made more resilient if back-up agents are assigned to critical responsibilities. The secondary agents must understand, however, when and how they should step in to discharge those responsibilities.

Similarly, a misassigned responsibility will not necessarily be discharged ineffectively (if at all). As an event unfolds, some agents may become unavailable for various reasons. This may mean that their assigned responsibilities get dynamically re-assigned to one of the available agents (rather than waiting for the original agent). In these situations the inherent flexibility and adaptability of people often means that some way will be found to discharge that responsibility, at least partially, using workarounds as appropriate.

Agents will usually be assigned several responsibilities. The way they often deal with this is to assign a priority to each of their responsibilities and focus on those with the highest responsibility first. These priorities may be changed dynamically as events unfold, however. In the aftermath of an event such as a terrorist attack, the system has to operate using the available means. Some agents may therefore end up being assigned responsibilities that they do not have the required competence to be able to fully discharge. The decision about prioritising and possibly postponing (or abandoning) a particular responsibility has to therefore consider whether the available (scarce) agents and resources might be better deployed elsewhere to discharge other, possibly more important responsibilities.

Some responsibilities have deadlines. It is therefore important to consider how agents should deal with these. If a responsibility is discharged too early, for example, there may not be any adverse consequences. If that responsibility is not discharged before its deadline, however, consideration should be given to whether that responsibility should be abandoned. If abandoning it would bring the system to a halt, there may be some merit in carrying on after the deadline, even if this means that the system provides a degraded level of service.

Problems can also arise when synchronisation of the discharging of responsibilities is necessary. An agent may be allocated the responsibility for keeping an information resource that is used by a second agent up to date. The second agent may therefore need to be kept informed of any changes to the information resource, so that they do not end up using old information to make responsibility-related decisions that are suboptimal. This could mean that that responsibility is discharged ineffectively, because any actions based on those decisions may turn out to be incorrect and have to be revisited or corrected at a later stage.

Identifying Vulnerabilities in Contingency Plans

Contingency plans are normally designed with a view to ensuring the resilience of a system. In other words, they are designed to help ensure that the system keeps running (possibly at a reduced level of service) when a major event (accident, incident, failure, terrorist attack etc.) occurs. It is now widely accepted at national levels that CNI, for example, needs to be resilient, and quickly restorable to normal use in the event of failures (Scottish Government, 2011; UK Government Cabinet Office, 2011).

The way to achieve resilience involves making sure that responsibilities in contingency plans are allocated and discharged appropriately and in a timely manner. Unfortunately, however, contingency plans are often inconsistent and incomplete and contain vulnerabilities. These vulnerabilities can be identified using responsibility modelling (Lock, Sommerville and Storer, 2009).

Figure 13.2, for example, shows part of the Responsibility Model that was created by analysing the flood contingency plan for Carlisle in the UK. The model focuses on the responsibilities associated with the evacuation of people from properties. Responsibilities are shown as shaded rounded rectangles (e.g. *Co-ordination*), and the agents are shown in angle brackets (e.g. *Police*).

If we examine Figure 13.2 more closely, we can see that all the lower level responsibilities have been allocated to an agent, except for the responsibility *Collect Evacuee Information*. This unallocated responsibility was not obvious in the textual contingency plan, but drawing the responsibility model highlighted this potential problem. In this case, we might expect that one of the agencies (*Police, District Council*, etc.) would have been assigned the responsibility to collect the evacuee information. A failure to collect this information could have led to a situation in which the emergency services kept revisiting high-risk properties (e.g. schools, nursing homes and so on) that had already been evacuated which would affect the resources available for visiting un-evacuated properties.

Figure 13.2 shows high-level responsibilities, which is why resources are omitted. In Figure 13.3, the responsibility *Initiate Evacuation* (previously shown in Figure 13.2) has been elaborated to show the resources that *Silver Command* would need in order to discharge it. The decision to initiate an evacuation is made using information that comes from a risk assessment and flood warnings that are both issued by the Environment Agency.

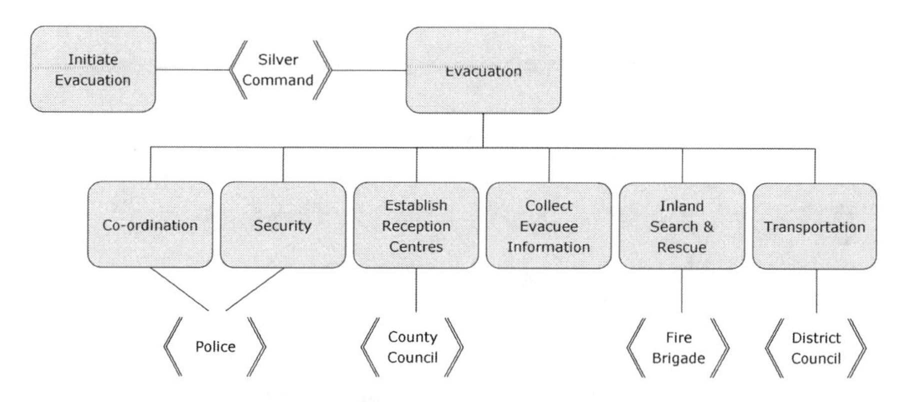

Figure 13.2 High-level responsibility model for flood evacuation plans in Carlisle, England

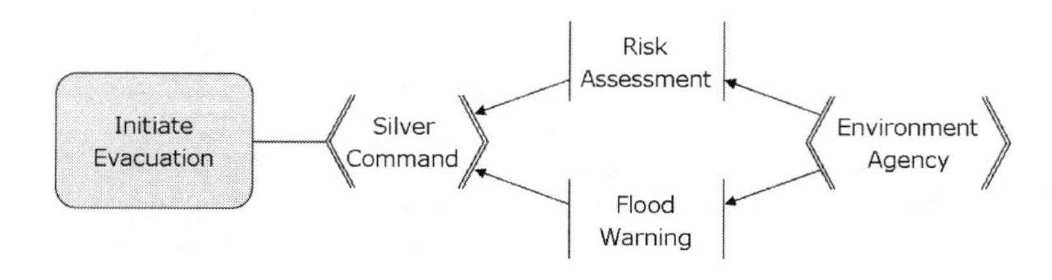

Figure 13.3 Responsibility model showing resource associations

The directional arrows between agents and resources indicate how those resources are used. In Figure 13.3, the arrows point into the resources *Risk Assessment* and *Flood Warning* from the *Environment Agency*. This shows that the Environment Agency produces those resources. The fact that the arrows from these resources point into *Silver Command* shows that *Silver Command* reads these resources (and subsequently uses the information in them to initiate an evacuation). Note that the responsibility for the *Environment Agency* to produce the *Risk Assessment* and *Flood Warning* information resources has been omitted for simplicity.

Even though the example in Figure 13.3 has been somewhat simplified, it illustrates how responsibility modelling can be used to identify vulnerabilities, such as unassigned responsibilities. The models provide a common representation that can be used by stakeholders as a basis for discussing various aspects of how the system works.

Testing Contingency Plans with Table-top Exercises

We have already noted that contingency plans contain vulnerabilities. Once contingency plans are written they therefore need to be exercised, tested and re-tested to make sure they are effective in maintaining business continuity (Duncan et al., 2011). In practice, the level of testing of these plans varies somewhat. The 2002 survey by *Contingency Planning & Management* magazine and Strohl Systems (cited in Cerullo and Cerullo, 2004), for example, showed that, of the organisations surveyed:

- 15 per cent performed only IT specific tests on their business continuity plans.
- 8 per cent performed table-top walk-throughs.
- 8 per cent performed call list tests, business unit tests or enterprise, full-scale tests.
- 58 per cent used a combination of these three methods.
- 10 per cent did not test their plans at all.

Exercises such as table-top simulations, and tactical decision games (Crichton and Flin, 2001) normally involve evaluating the use of contingency plans in a simulated emergency. The exercises are used to develop and refresh the knowledge and skills needed to deal with emergency situations: judgement and decision-making; developing shared understanding of problems; and building up patterns of known problems (so they can be both identified and responded to more quickly in the future).

The utility of table-top exercises has been illustrated by Jarrett's (2003) assessment of the 'Pale Horse' bioterrorism response exercise and Hallett's (2010) discussion of managing waterfront security in Australia. Although Jarrett focused mostly on issues related to dealing with casualties, both he and Hallett noted several general lessons about the problems that arise when people with different backgrounds and objectives have to co-operate and collaborate in a high-pressure environment. In particular they highlighted problems of multidisciplinary communication; who has authority over what; and the control and dissemination of information. Hallett also noted that the exercises have helped to establish working optimal communication channels between participants and increased the understanding of how other organisations are likely to respond to a particular situation.

Table-top exercises are often conducted manually, but can be supported or even completely implemented using IT solutions. Mooney et al. (2012), for example, describe a relatively lightweight, low-cost table-top exercise software application, eTableTop, that was developed to help train UK police in how to handle major incidents. At the other end of the scale, virtual world simulation technologies are being used to train security staff and emergency responders to deal with terrorist activity (see Stedmon et al., 2012 for a brief review)

Designing and planning scenarios for table-top exercises is a major undertaking. The scenarios need to be credible and sufficiently simple that they can be handled using the available resources, while being difficult enough to present a challenge to the participants (Lacoursiere, 2006). There have been some efforts to support the automatic development of table-top exercise scenarios. The CyberSMART tool (Marshall, 2009), for example, is used to collate data from multiple sources, and then supports the use of that data in developing exercises. Similarly, Payne and Koch (2011) focus on the collection and analysis of data in an organisation's security plan. The organised data is then used to inform a table-top exercise.

We believe there is a role for responsibility modelling here. The responsibility vulnerabilities and risks of a particular system configuration can be analysed by combining responsibility modelling with keyword-based approaches, such as Hazard and Operability Studies (HAZOPS) (Kletz, 1999). A systematic approach is used, adapting standard HAZOPS keyword phrases such as 'Too early', and 'Too late'. The analyst takes a checklist of the keyword phrases and applies them to each of the different elements of the responsibility model in turn (Lock et al., 2009) to examine the possible consequences (and the probability of those consequences occurring). Table 13.1 illustrates how the keywords are interpreted for responsibilities. This approach allows the analyst to ask 'what-if?' questions, such as 'What if this responsibility is discharged too early?' and 'What if this resource is not available?' The answers to these questions can then be used to help define the scenarios for table-top exercises.

Table 13.1 Keyword interpretations for categories of hazards associated with responsibilities that should be considered during vulnerability analysis

Keyword	Interpretation
Early	What would happen if the responsibility was discharged before the required time?
Late	What would happen if the responsibility was discharged after the required time?
Never	What would happen if the responsibility was never discharged?
Incapable	What would happen if the agent did not have the capability to discharge the responsibility?
Insufficient	What would happen if the responsibility was not fully discharged?
Impaired	What would happen if the responsibility was discharged incorrectly?

Source: adapted from Lock et al., 2009.

The vulnerability analysis can be performed on the contingency plan for a terrorist attack. This will help to highlight those parts of the plan that need to be tested, and how they should be tested. If there is an unassigned responsibility identified, for example, then a scenario should be designed to create the situation where that responsibility has to be discharged to see how the system copes.

The responsibility models can also be used to influence how the scenario evolves during the table-top exercise. The models show how agents use particular resources to discharge their responsibilities. One way this could be used is to look at the models, and identify a responsibility that is deemed to be critical to a successful response to a terrorist attack. Then, during the table-top exercise, access to the resources used to discharge that responsibility could be removed or limited in some way (e.g. to simulate a network communication failure).

Summary

Emergency preparedness is recognised as the best defence to deal with the threat of terrorist attacks. Emergency situations are characterised by disparate types of dynamically changing information that has to be managed and distributed to widely dispersed users. The way that emergency situations should be handled is normally documented in contingency plans. These plans invariably contain several vulnerabilities. Identifying these vulnerabilities through responsibility modelling is just one way of improving emergency preparedness.

We consider the creation of responsibility models to be the start of the process, rather than the end. The models provide a common representation that can be used as a basis for discussions between stakeholders. This has the potential benefit of bringing together multiple agencies without the pressures that are intrinsic to any emergency situation. This should help to develop and facilitate communication and shared understanding of problems and responsibilities between agencies.

The process of creating responsibility models is relatively straightforward, and does not require highly specialised skills. Anyone who has a basic understanding of risk management, and can use a drawing package (or even draw freehand) should be able to create a responsibility model. We believe that appropriate tools are needed, however, and have developed a stand-alone web-based tool that can be used to create responsibility models.

Understanding the vulnerabilities associated with responsibilities, and being able to identify them, is useful in three ways. First, it makes it easier to identify problems within contingency plans so that they can be appropriately managed. Second, it provides a basis for creating scenarios for exercising the contingency plan (e.g. in a table-top exercise) to examine how responders manage in the face of the identified vulnerabilities. Third, it provides suggestions about how particular vulnerabilities can be introduced during live table-top exercises (e.g. by removing an agent, or removing access to an information resource).

Systems and organisations evolve over time. As they adapt to counter changes in the threat posed by terrorists, contingency plans will be changed accordingly. This means that responsibilities will evolve, and responsibility models (and vulnerability analyses) will therefore need to be updated. We believe this is a small price to pay, however, for helping maintain the maximum level of emergency preparedness to deal with terrorist threats.

References

Cerullo, V., and Cerullo, M.J. (2004). Business continuity planning: A comprehensive approach. *Information Systems Management*, 21, 70–78.

Convertino, G., Mentis, H.M., Slavkovic, A., Rosson, M.B. and Carroll, J. (2011). Supporting common ground and awareness in emergency planning management: A design research project. *Transactions on Computer–Human Interaction*, 18, 1–34.

Crichton, M. and Flin, R. (2001). Training for emergency management: Tactical decision games. *Journal of Hazardous Materials*, 88, 255–266.

Dekker, S. (2006). Resilience engineering: Chronicling the emergence of confused consensus. In E. Hollnagel, D. Woods and N. Leveson (Eds), *Resilience Engineering: Concept and Precepts*. Ashgate: Aldershot, 77–92.

Duncan, W.D., Yeager, V.A., Rucks, A.C. and Ginter, P.M. (2011). Surviving organizational disasters. *Business Horizons*, 54, 135–142.

Goughnour, D.A. and Durbin, R.T. (2008). Device independent information sharing during incident response. *Proceedings of IEEE Conference on Technologies for Homeland Security*. Waltham, MA: IEEE, 486–491.

Hallett, M. (2010). Australian perspectives on waterside security. *Proceedings of the International Waterside Security Conference (WSS)*. IEEE, 1-6.

Intelligence and Security Committee (2006). *Report into the London Terrorist Attacks on 7 July 2005*. Norwich: HMSO.

Jarrett, D. (2003). Lessons learned: The "Pale Horse" bioterrorism response exercise. *Disaster Management & Response*, 1, 114–118.

Jiwel, Z., Ligong, T., Xuqing, X., Hong, Z. and Zhiguo, Z. (2010). The emergency management experience of Beijing 2008 Olympic games. *Proceedings of 2010 IEEE International Conference on Emergency Management and Management Sciences (ICEMMS)*. Beijing, China: IEEE Press, 535–537.

Kletz, T. (1999). *HAZOP and HAZAN: Identifying and Assessing Process Industry Standards*.Rugby: Institution of Chemical Engineers.

Lacoursiere, J.P. (2006). A risk management initiative implemented in Canada. *Journal of Hazardous Materials*, 130, 311–320.

Latané, B. and Nida, S. (1981). Ten years of research on group size and helping. *Psychological Bulletin*, 89, 308–324.

Leveson, N. (1995). *Safeware: System Safety and Computers*.Wokingham, UK: Addison-Wesley.

Lewis, S. (2012). Emergency preparedness – working in partnership. *Journal of Terrorism Research*, 3, 13–16.

Lock, R., Sommerville, I. and Storer, T. (2009). Responsibility modelling for civil emergency planning. *Risk Management*, 11, 179–207.

Lock, R., Storer, T., Sommerville, I. and Baxter, G. (2009). Responsibility modelling for risk analysis. *Proceedings of European Safety and Reliability (ESREL) 2009*. Prague, Czech Republic. 1103–1109.

Marshall, J. (2009). The cyber scenario modelling and reporting tool (CyberSMART). *CATCH '09 Proceedings of the 2009 Cybersecurity Applications & Technology Conference for Homeland Security*. Washington, DC: IEEE Computer Society.

Mooney, J.S., Griffiths, L., Patera, M., Roby, J., Ogden, P. and Driscoll, P. (2012). An electronic tabletop 'eTableTop' exercise for UK police major incident education. *Proceedings of the 12th IEEE International Conference on Advanced Learning Technologies.* Washington, DC: IEEE Computer Society, 40–42.

Payne, P.W. and Koch, D.B. (2011). A counter-IED preparedness methodology for large event planning. *Proceedings of the IEEE International Conference on Technologies for Homeland Security*. Washington, DC: IEEE, 185–189.

Pelfrey, W.V. (2007). Local law enforcement terrorism prevention efforts: A state level case study. *Journal of Criminal Justice*, 35, 313–321.

Reason, J. (1990). *Human Error*. Cambridge: Cambridge University Press.

Scottish Government, The (2011). *Secure and Resilient: A Strategic Framework for Critical National Infrastructure in Scotland.* Edinburgh: The Scottish Government.

Sommerville, I. (2007). Models for responsibility assignment. In G. Dewsbury and J. Dobson (Eds), *Responsibility and Dependable Systems*. London: Springer,165–186.

Sommerville, I., Storer, T. and Lock, R. (2009). Responsibility modelling for civil emergency planning. *Risk Management*, 11, 179–209.

Stedmon, A.W., Lawson, G., Saikayasit, R., White, C. and Howard, C. (2012). Human factors in counter-terrorism. In P. Vink (Ed.) *Advances in Social and Organisational Factors*. CRC press: Boca Raton, FL, 237–246.

UK Government Cabinet Office, The (2011). *Keeping the Country Running: Natural Hazards and Infrastructure*. London: The UK Government Cabinet Office.

Veness, D. (2012). Introduction: emergency preparedness. *Journal of Terrorism Research*, 3, 3–5.

Wu, A., Convertino, G., Ganoe, C., Carroll, J.M. and Zhang, X. (2013). Supporting collaborative sense-making in emergency management through geo-visualization. *International Journal of Human–Computer Studies*, 71, 4–23.

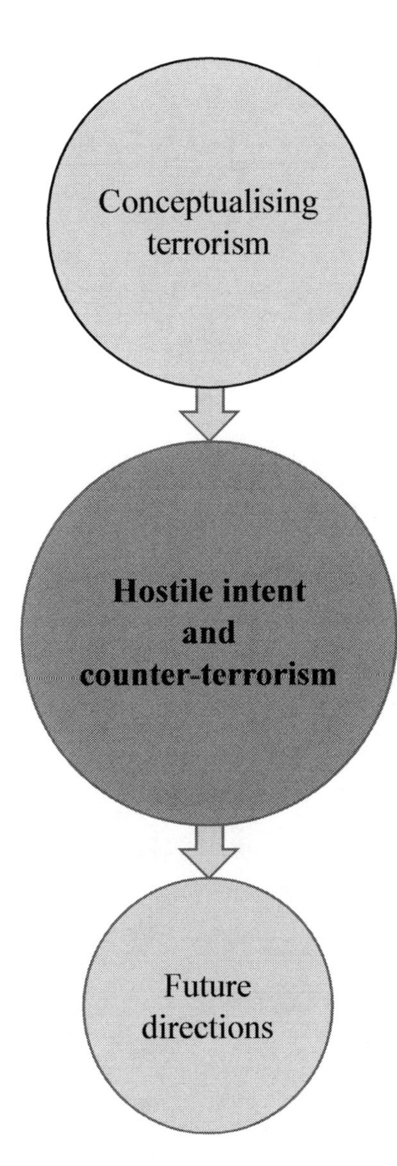

Sociocultural factors

Part 4, Sociocultural factors, discusses how many sociocultural factors can influence the occurrence and nature of terrorism. It addresses aspects such as home-grown terrorists and conflict engaged citizens; terrorism targeting schools and female suicide terrorism, and also contains an innovative chapter which challenges views of existing models of terrorism by drawing analogies to parasitic infection.

Chapter 14

Unintended Consequences of the 'War on Terror': Home-grown Terrorism and Conflict-engaged Citizens Returning to Civil Society

John Parkinson OBE

Centre of Excellence in Terrorism, Resilience, Intelligence and Organised Crime Research (CENTRIC), UK

Andrew Staniforth

Detective Inspector, North East Counter Terrorism Unit

Introduction

On Monday 17 September 2001, within a week of the catastrophic 9/11 terrorist attacks on the United States, President George W Bush convened a meeting of his senior advisors. The meeting followed a review of diplomatic and military plans by his War Council at Camp David over the previous weekend. The President stated that: 'The purpose of this meeting is to assign tasks for the first wave of the war against terrorism. It starts today' (9/11 Commission, 2004, p. 367).

The 'War on Terror'

By 7 October 2001, the military response phase of the so-called 'War on Terror' began with the bombing of Afghanistan by US and UK air forces. The Taliban had refused to cooperate and meet the demands of the US in handing over Osama Bin Laden for trial and closing down the Al Qa'ida bases and training facilities. In London, MI5 intelligence agency chiefs were not entirely happy with President Bush's call for a 'War on Terror' (Andrew, 2009). Security Service Director General, Sir Stephen Lander, sought to reassure staff on 27 September 2001, stating that:

> This is a war on terrorism in the same sense as we talk about a war on drugs. A military response is obviously under consideration, but you should be reassured that political, humanitarian and intelligence and law enforcement responses are also high on the UK agenda. (Andrew, 2009, p. 623)

During November 2001 the United States armed forces had troops on the ground in Afghanistan. On 9 November, Mazir-e-Sharif in northern Afghanistan fell to a coalition assault by Afghan and US forces (9/11 Commission Report, 2004). Four days later, the Taliban had fled from Kabul and by early December all major cities had fallen to the coalition forces. On 22 December, Hamid Karzai, a Pashtun leader from Kandahar, was installed as the chairman of Afghanistan's interim administration. Within just ten weeks, Afghanistan had been liberated from the rule of the Taliban and the core of Al Qa'ida had been disrupted and dispersed, though Bin Laden and his followers had escaped from their mountainous hideout in the Tora Bora caves. The primary objectives of the Afghanistan invasion had been achieved.

Shortly after Christmas 2001, the British Prime Minister Tony Blair wrote to Director General Stephen Lander:

> The Government and the British people are fortunate to be served by security and intelligence organisations whose professionalism is admired and, by our enemies, feared throughout the world. My thanks on behalf of the Government for all you are doing. We all have cause to be grateful for your efforts. (Andrew, 2009, p. 624)

Those in charge of operating the national intelligence machinery across the British government welcomed the much-needed additional finances that were made available to tackle the threat from Al Qa'ida-inspired terrorism. With the invasion and continued presence of troops on the ground in Afghanistan and later in Iraq, this provided terrorists with the propaganda they required to recruit new and willing volunteers to their extremist cause. Unbeknown to UK security forces at the time, a series of deadly and determined terrorist plots by British citizens were already in process and the Al Qa'ida-inspired genre of international terrorism would threaten the security of the British citizens throughout what became known as the first decade of New Era global terrorism. These were plots that brought new terrorist tactics to Britain, fuelled by resentment and revenge for the so-called 'War on Terror' overseas. These were also threats of such magnitude and ambition that they required a unique collaborative approach by UK domestic law enforcement and intelligence agencies.

The first evidence of British born citizens being inspired to become suicide bombers was on 30 April 2003 when 22 year-old Asif Muhammad Hanif from London and 27-year-old Omar Khan Sharif from Derby, attacked a bar called 'Mike's Place' in downtown Tel Aviv near the American Embassy. The Palestinian militant group Hamas claimed responsibility, but little is known about the roots of when and how Hanif and Sharif were inspired to act for the cause or how much the War on Terror conflicts overseas contributed to the motivation of their action. This is in stark contrast to the wealth of information now known about those who went on to murder 52 innocent people in the London bombings on 7 July 2005. The ringleader of the 7/7 terrorist cell bombers, Mohammed Siddique

Khan, had been covertly filmed by anti-terror officers in 2004 talking to men who were plotting to plant and detonate a fertilizer-based bomb (Hayman, 2009). Other evidence given at the Inquest into the bombings during 2011 in London showed that Khan had been in attendance at a suspected terrorist training camp in 2001, referred to by Special Branch as Operation WARLOCK (Hayman, 2009) that indicated Khan's potential early links with extremism. The starkest evidence of the influences on Khan came from his prepared suicide video that was released on the Arab television station Al Jazeera on 1 September 2005. He said:

> Your democratically elected governments continuously perpetuate atrocities against my people all over the world. And your support of them makes you directly responsible, just as I am directly responsible for protecting and avenging my Muslim brothers and sisters. Until you stop the bombing, gassing, imprisonment and torture of my people we will not stop this fight. We are at war and I am a soldier. (Hayman, 2009, p. 145)

Although an abhorrent message to most citizens, it clearly articulated his sentiments and those of his colleagues who were inspired by Al Qa'ida. Almost exactly a year after the bombings, a second suicide video, this time from Shazzad Tanweer, was released on Al Jazeera:

> What you have witnessed now is only the beginning of a string of attacks that will continue to become stronger until you pull your forces out of Afghanistan and Iraq. And until you stop your financial and military support to America and Israel. (Hayman, 2009, p. 165)

Perhaps this message from Tanweer was prophetic or simply a sign that many young men had become radicalised to the point of direct engagement in the planned acts of terror that were yet to come. One issue was however becoming clear to UK security forces – the conflict in Afghanistan and Iraq was contributing to the recruitment and radicalisation of British citizens.

Transatlantic terror

Overnight on 9 August 2006, large numbers of police officers were deployed across High Wycombe, London and Birmingham as part of an operation to disrupt a major terrorist plot. A total of 24 terrorist suspects were arrested and questioned in relation to the commission, instigation and preparation of an alleged transatlantic terrorist attack of 9/11 proportions. The police and security service had been monitoring the activities of British-based Al Qa'ida-inspired terrorists for several years and the executive action taken overnight was the result of the largest operation ever conduced by the Metropolitan Police Service and MI5.

A potentially catastrophic terrorist attack had been disrupted and intelligence and evidence had been painstakingly gathered under Operation OVERT. Despite the number of arrests that morning the security forces could not be confident that they had managed to capture all of the terrorist conspirators in their raids. In order to warn the public of, and protect people from, an imminent attack, the terrorist threat level issued by the Joint Terrorism Analysis Centre (JTAC) and the Security Service (MI5) was raised to 'CRITICAL – an attack is expected imminently', following a decision made at an emergency meeting of the Cabinet Office Briefing Room (COBR) chaired by then Home Secretary, Dr John Reid. Speaking outside New Scotland Yard to a gathered press, then Deputy Commissioner of the Metropolitan Police Service, Paul Stephenson, stated 'We cannot stress too highly the severity that this plot represented. Put simply this was intended to be mass murder on an unimaginable scale' (BBC, 2006). The Deputy Commissioner went on to reveal more about the terrorist plote:

> We are confident that we have disrupted a plan by terrorists to cause untold death and destruction and to commit, quite frankly, mass murder. We believe that the terrorists' aim was to smuggle explosives on to aeroplanes in hand luggage and to detonate these in flight. We also believe that the intended targets were flights from the United Kingdom to the United States of America. (BBC, 2006)

In the weeks and months that followed, the British public would come to learn more about the realities of the terrorist plot as the police carried out extensive searches of buildings and vehicles, focusing on a 354-acre wood near High Wycombe. The huge scale of the searches and resources required were a significant logistical challenge, with more than 200 police officers on site every day. A variety of evidence was gathered including potential bomb-making equipment and chemicals, computers, telephones and portable storage media such as memory sticks, CDs, and DVDs. Operation OVERT was gathering pace and investigators and prosecutors building their case against the conspirators were soon to identify the primary drive and motivation behind the terrorist plot.

Western Foreign Policy

The covert surveillance activity under Operation OVERT had served to identify three primary terrorist cell members: Abdulla Ahmed Ali (the leader of the terrorist cell), Tanvir Hussain, and Assad Sarwar from High Wycombe. All three men were well acquainted. Ali was an Engineering graduate who choose not to go into industry after leaving City University in London in 2002 but instead had decided to pursue business opportunities in Pakistan (BBC, 2009a). In contrast, Sarwar turned down a place at university in Chichester, Sussex because he became homesick, and his second attempt to progress undergraduate study, first with a sports science course, then earth sciences, also failed because he found the work too challenging. Ali and Sarwar were to meet in 2003 when they both went to deliver aid to refugee camps on the Pakistan and Afghanistan border. As a result of the US and UK armed forces presence in Afghanistan, many Afghans decided to flee from areas of conflict, compounding a decades-old refugee crisis. In support of the worsening situation, the Islamic Medical Association, a charity shop located in Clapton, East London, raised money and collected equipment to send to the refugee camps (BBC, 2009b). Ali and Sarwar were volunteers, distributing such aid in what seems to have been a sincere humanitarian endeavour, but their shared experiences at the refugee camps radically altered their worldview. Ali, married with a son, was shocked by the appalling conditions in the camps, where he witnessed many people dying. Sarwar would also be dismayed at what he had seen at the camps, later believing that the aid work in which he was engaged was an ineffective way of helping. Both Ali and Sarwar were angered about the situation and decided to tackle what they believed was the root cause of the death and misery they witnessed at the camps – Western foreign policy.

The anger felt by Ali and Sarwar turned them against the UK and the US and both men began to relate to the anti-Western rhetoric of radical Islamism. They started to move in Islamist circles that were increasingly calling for attacks on America and Britain. In particular, by March 2005, Ali was suspected of communicating with Rashid Rauf, a primary influential Al Qa'ida figure who had fled to Pakistan from Birmingham in 2002 following the issue of a warrant for his arrest in connection with the stabbing of his uncle (Staniforth, 2012). Security forces suspected that Rauf had come to be engaged in numerous Al Qai'da plots including the London 7/7 suicide bombings and the failed 21/7 bombings in London during July 2005. As the security service continued to monitor those individuals preaching violent extremism and calling for violent jihad against the UK, Ali became of significant interest and was part of a broader growing network of radical Islamists threatening UK national security.

The presence of Al Qa'ida-inspired home-grown British terrorists was a shock to the communities of High Wycombe, the area which had been the centre of police and media activity for Operation OVERT. The Chief Constable of Thames Valley Police, Sara Thornton, later explained:

> Operation Overt was a significant success for the police and security services in disrupting a plot to commit mass murder and preventing any lives being lost. The handling of the investigation and media interest was achieved to an extremely high professional standard and rather than dividing the police from Muslim communities, opportunities have been taken to improve relationships and build new lines of communication. The many positive outcomes that have been achieved are due in no small part to the mature and reasonable response of the community living in High Wycombe. (Thornton and Mason, 2007, p. 36)

The former Director General of the Security Service, Dame Eliza Manningham-Buller, commented that it was almost inevitable that the search for terrorists within the Muslim community would lead to social tensions, but stated that: 'it must not be allowed to deflect the police, nor the Security Services, from continuing the intelligence work which is necessary and proportionate to match the terrorist threat' (Rayment, 2007).

Operation OVERT served to cement in the minds of those in authority that UK foreign policy had become a key driver for the radicalisation and recruitment of British citizens and was an unintended consequence of the War on Terror. Manningham-Buller said the radicalization of teenage Muslims: 'from first exposure, to extremism, to active participation in terrorist plotting' (Staniforth, 2012, p. 295) was now worryingly rapid and it was vital that the Government rose to the challenge of trying to change the attitudes that: 'lead some of our young people to become terrorists' (Staniforth, 2012, p. 295). Yet beyond the comments of the former Director General of MI5, members of local communities were not just being recruited and radicalised in the UK, they were already receiving instruction as part of their terrorist training and this needed to be tackled with some urgency. The Director General of MI5, Jonathan Evans remarked that the growing use of the Internet by Al Qa'ida to inspire and exploit teenage Muslims in the UK was almost like a form of 'child abuse' (Hayman, 2009). The youngest ever convicted terrorist in the UK, Hammaad Munshi, who was 16 at the time, also originated from Yorkshire. Aabid Khan, an older man working from Pakistan, groomed Munshi, encouraging him to go to camps and learn techniques about avoiding security and building improvised explosive devices. Both were subsequently convicted of terrorist offences after protracted investigations, which demonstrated the power and reach of this extremist ideology.

Exporting Terror

During 2006 the International Institute for Strategic Studies in London published a report called *The Military Balance* (Langton, 2006). It estimated that as many as 10 per cent of the 20,000 insurgents fighting the conflict in Iraq were foreign-born. It was also revealed that up to 150 radicals from Britain had travelled to Iraq to join up with a 'British Brigade' that had been established by Al Qa'ida leaders to fight coalition forces. It appeared that the flow of young men from Western Europe to Iraq was increasing. The supply chain of volunteers to join the ranks of the insurgency was not just restricted to the UK. In France, Pierre de Bousquet de Florian, the head of the French domestic security service, revealed that 15 young French men remained in and around Iraq that they were aware of and of these at least 9 had been killed. This raised two key issues for the UK government: first, how were these men travelling to Iraq and receiving their training? The second was the realisation that, while many of the volunteers may die in the theatre of conflict in Iraq, some may well survive their experiences and return to the UK with military training and hardened combat experience. Such conflict-engaged citizens could put their skills and experiences to unlawful use when they returned to their local communities in the UK, continuing their violent jihadist crusade.

To address this new security challenge, the British government introduced new powers under the Terrorism Act 2006 to stop British citizens swelling the ranks of insurgencies overseas. While the preventive intentions of the Prime Minister were clear, violent jihadist training camps were already in operation in the UK. Assistant Commissioner of Specialist Operations (ACSO) at New Scotland Yard stated that 'we latched on to one group of terrorists running outdoor training camps and put them under surveillance' going on to say that 'what we discovered was mind-boggling' (Hayman, 2009). The training camp investigation was developed under Operation OVERAMP, a two-year joint MI5 and police operation focusing upon Attila Ahmet, 43 and from Bromley, Kent and Mohammed Hamid, 50 and from Hackney, East London. ACSO Hayman revealed that Hamid was a veteran of Pakistan military camps who called himself 'Osama bin London', going on to describe Ahmet as 'an associate of the jailed preacher Abu Hamza'. ACSO Hayman also revealed that Hamid had key connections with other UK terrorist plots:

> Hamid had taken two of the 21/7 bombers on a paintballing trip only three weeks earlier. All the bombers had spent camping weekends in the Lake District in Cumbria and Hamid had called or texted them at least 173 times in the months before the attacks. (Hayman, 2009, p. 181)

It was clear from the information gathered at an early stage of Operation OVERAMP that the training camps being arranged were not 'soft-touch' camps but a cover for terrorism (Hayman, 2009).

To progress the covert investigation, the police and MI5 mounted intrusive surveillance operations and an undercover officer was able to infiltrate the group. This provided excellent coverage of the activities of Hamid and Ahmet, which later led to their arrest. Operation OVERAMP provided unique challenges for the investigators and prosecutors. Deborah Walsh, Deputy Head of the Crown Prosecution Service Counter Terrorism Division, revealed that

> Operation OVERAMP was the first prosecution for offences contrary to sections 6 and 8 of the Terrorism Act 2006. It was only possible to charge these offences because of the evidence of the undercover officer who attended the training sessions. (Staniforth, 2012, p. 289)

At the Central Criminal Court in London and at Woolwich Crown Court during 2007 and 2008, seven men investigated under Operation OVERAMP, including Hamid and Ahmet, were convicted of terrorism-related offences. Hamid was found guilty of three counts of soliciting murder and three counts of providing terrorist training and was jailed on 7 March 2008 under special Imprisonment for Public Protection powers. The judge said Hamid should serve seven-and-a-half years but that he cannot be released until he has reformed. Ahmet, pleaded guilty at the Old Bailey in September 2007 to three counts of soliciting to murder and was sentenced to 6 years and 11 months imprisonment. Kibley da Costa, aged 24, of south-west London was found guilty of providing terrorist training, two counts of attending terrorist training and one count of possessing a record containing information likely to be useful to a terrorist and was sentenced to 4 years and 11 months' imprisonment. Mohammed Al-Figari, aged 44, of East London, was found guilty of two counts of attending terrorist training and two counts of possessing a record containing information likely to be useful to a terrorist and was sentenced to four years and two months' imprisonment. Kader Ahmed, aged 20, was found guilty at Woolwich Crown Court on 20 February 2008 of two counts of attending terrorist training and was sentenced to three years and eight months' imprisonment. Yassin Mutegombwa, aged 22, of Norwood, south-east London, pleaded guilty to attending a terrorist training camp and was sentenced to three years and five months' imprisonment, and Mohammed Kyriacou, aged just 18, of south-east London pleaded guilty to attending a terrorist training camp and was sentenced to three years and five months' imprisonment. In an earlier trial, Hassan Mutegombwa, aged 20, the brother of Yassin, was also found guilty of seeking funding for terrorism overseas while attending one of the camps and was jailed for 10 years.

The landmark case brought the first conviction under new provisions provided by the Terrorism Act 2006 which created the offence of providing terrorist training or attending a place for terrorist training. With this case a new dimension to UK counter-terrorism was introduced. It was known soon after the events of 9/11 that around 120 terrorist training camps were located in Afghanistan alone (Staniforth and Police National Legal Database, 2009). These camps were sited in secure locations; they were lightweight, very mobile and often moved around to avoid identification and capture. Given the terrain they were difficult to track by military units and recruits would typically receive training between one and six months (Staniforth and Police National Legal Database, 2009).

Operation OVERAMP showed that the UK camps were not clandestine in nature and were being carried out in what was effectively the complete open. People attending these camps did so in full view of everyone. Further investigations led to the police and security service identifying legitimate venues, such as outward-bound centres and paint-balling facilities, that were being used by extremists to train and indoctrinate their followers. While these venues were not terrorist training camps per se, it was the purpose of the individuals attending that gave them a new and sinister cachet and meant that those people could be prosecuted for the new offence. The 'training camps' were also used to assess whether potential recruits to the Islamist cause could adapt to trekking over hills and sleeping in tents in the UK, providing an indication of whether they then could operate in harsher conditions to be found in training camps on the mountainous border regions of Pakistan and Afghanistan. The camps also provided a cover for attack plans and logistics to be progressed as well as the raising of funds for terrorist purposes. They provided new challenges to those charged with the responsibility of delivering homeland security.

Returning Home

While Operation OVERAMP uncovered the terrorist training tactics of UK-based extremists supporting the Islamist cause, the struggle against terrorism overseas in Afghanistan and Iraq continued to fuel a particular hatred for Britain amongst Islamic sympathisers. On one level the notion that men and women recruited to local regiments of the armed forces were engaged thousands of miles away in direct combat with insurgents drawn from the very same communities back home was bizarre. Meanwhile the counter-terrorism machinery of government faced a widening mission arising from the same emerging reality. Not only did they have to stop terrorist attacks from succeeding in the UK from British-based extremists but they also had to prevent would-be insurgents from leaving the UK to fight in conflicts overseas again British soldiers. This broader mission would take another challenging turn as anti-terror police officers raided a series of premises in Birmingham to disrupt a terrorist plot to kidnap and behead a British Muslim soldier. Not long back from his six-month tour in Iraq, the young Muslim soldier was puzzled when police called at his family home in Birmingham (Staniforth, 2012). Officers described how a gang from his home town was allegedly plotting to abduct him and then force him on video to 'apologise' for what he had done in Iraq. After this propaganda coup the gang intended to video themselves executing their hostage. His murder would be seen worldwide on the Internet as a warning to other British Muslims regarded by the kidnappers as 'traitors' (Staniforth, 2012).

On the 31 January 2007 the police and MI5 conducted Operation GAMBLE and arrested Pervaiz Khan, the leader of the kidnapping terrorist plot. Khan, aged 36, was born in Derby, England and had turned from a young man whose main passions were Sunday league football and cricket into a radical jihadist after making several trips to Pakistan. Raised in Birmingham and holding both British and Pakistani passports, Khan was married with three young children. His involvement with Al Qa'ida changed him into a homicidal obsessive (he was codenamed 'Motorway Madness' by MI5) who spoke of the July 7 bombers as 'brothers' (Staniforth, 2012). MI5 spent a total of 8,500 hours collating and translating conversations recorded by a device placed inside his house (Staniforth, 2012) During February 2008, Khan, together with four other individuals, was found guilty of committing terrorism offences. Khan admitted his guilt to charges of conspiracy to murder and kidnap and was sentenced to life imprisonment. According to the testimony of his co-defendant, Zahoor Iqbal, Khan was a man who in his 20s liked to drink, smoke and go clubbing. Iqbal said his near-lifelong friend then transformed in his early 30s from a non-practicing Muslim into an Islamist fanatic after visiting Pakistan, revealing that he became more anti-West, blaming Britain for the Israel-Palestine issue, Kashmir, and civilians dying in Iraq and Afghanistan (Staniforth, 2012).

The Muslim soldier plot represented yet another change in the tactics being used by UK-based Islamists, proving nothing was predictable in the current struggle against terrorism at home or overseas. During May 2013, such tactics were demonstrated in front of the world of social media as Drummer Lee Rigby was killed by Islamic fundamentalists near his army barracks in Woolwich, London. The violent jihadists used the gathered public with their smartphones as a platform to communicate their motivation and justification for the brutal murder of the British soldier.

While the foreign policy of the British government served to fuel the fires of extremist propaganda, drawing individuals towards adopting extreme Islamist perspectives in support of Al Qa'ida, another potent threat to the UK's national security began to emerge from an unlikely and disturbing source. At 4 p.m. on the afternoon of 23 September 2012, Fraser Rae, a 28-year-old male from Johnstone in Renfrewshire, Scotland, walked towards the Central Mosque in the south side of Glasgow shouting 'I will blow this place up' (*Scotland Herald*, 2012, p. 1). Wearing a backpack, Rae passed through the Mosque car park to the main entrance where he terrified a 15-year-old boy, saying 'Run, run, run, I've got a bomb' (*Scotland Herald*, 2012, p. 1). Rae made his way into the Mosque where the caretaker heard him shouting and swearing as he went down corridors towards the main prayer room. As Rae entered the prayer room, in which 30 adults and children were present, he yelled 'Christians can do it too, boom' (*Scotland Herald*, 2012, p. 1). When the adults in the prayer room approached Rae he was standing holding a bottle in his left hand with his right hand tucked behind his back and into the waistband of his trousers under his jacket. He said 'Stay back, I've got a gun … I'll shoot you' (*Scotland Herald*, 2012, p. 1). Rae then launched into a tirade of racial abuse aimed at those present in the prayer room, implying he had a bomb in his backpack. Despite threats to their personal safety, members of the Mosque restrained Rae and physically escorted him off the premises. Police officers attending the incident arrested Rae and searched his backpack seizing a number of items including a 'Scream' mask, gloves and scissors (*Scotland Herald*, 2012, p. 2). Rae was escorted to Cathcart police station where he was detained and later interviewed. The police investigation did not uncover a firearm and after a search by a specialist dog to identify any traces of explosive materials, none were found.

On 20 December 2012, Rae appeared at Glasgow Sheriff Court. The court heard that Rae, born 8 August 1984, was a former soldier and Iraq veteran, serving with the Argyll and Sutherland Highlanders. Defence lawyer Mark Chambers said his client was remorseful and that his actions were out of character although it was revealed that Rae had suffered psychologically after leaving the Army and that he had become abusive towards Pakistani people in general being in Britain, suggesting they were responsible for the bombings in Iraq (*Scotland Herald*, 2012). The court also heard that Rae had said to the police: 'I was in Iraq and all they did was bomb. My brothers in the army got blown up' (*Scotland Herald*, 2012, p.1).

Rae pleaded guilty to threatening to blow up the mosque and implying he had a bomb by concealing his arm behind his back. He also admitted to carrying scissors, shouting and swearing and making racist remarks to police. Rae was sentenced to 28 months imprisonment for contravening section 38(1) of the Criminal Justice and Licensing (Scotland) Act 2010 (threatening and/or abusive behaviour) and section 49(1) Criminal Law (Consolidation) (Scotland) Act 1995 (carrying a knife) (). Speaking following sentencing, John Dunn, Procurator Fiscal for West Scotland said:

> Everyone has the right to live free from violence, threats, intimidation or the fear of harassment or abuse stemming from the prejudice of another person. Hopefully the conviction of and sentencing of Fraser Rae will encourage the public to report all hate crimes to the police. They can have the confidence that all such hate crimes will be investigated carefully and prosecuted robustly. (*Scotland Herald*, 2012, p. 2)

Reinforcing the public message of the Procurator Fiscal, Strathclyde Police Detective Inspector Joe Mckerns said 'We will not tolerate any acts targeted against minority ethnic communities and will take robust and prompt action' (*Scotland Herald*, 2012, p. 2).

The actions of Rae were abhorrent, causing unnecessary alarm and distress to innocent members of the public. Surprisingly little media coverage or comment was made by authorities concerning the key issues behind the motivations and experiences of Rae that either pushed or pulled him towards adopting such extreme and prejudicial behaviour. Whatever the particular circumstances in this case, it is evident that engagement in conflict overseas had resulted in psychological issues from which a danger to the public had arisen. Whether failure of authorities to accurately assess the potential threat to the public from Rae, or a lack of support from family, friends or wider society, Rae may also consider himself to be a victim. Rae is not the first British citizen to return home following engagement in conflict overseas to suffer psychological issues from his experiences, nor will he be the last. Many armed forces personnel who have returned home to their local communities are deeply affected by their exposure to conflict. A complex malaise of social, political and economic factors confronts troops return to their communities.

Conclusion

While those in authority have triumphantly announced the return of armed forces personnel from Afghanistan during 2014, marking what they view as the close of the War on Terror, we should perhaps not be so optimistic nor hasty as to comment on whatever political or humanitarian success looks like in such complex conflicts. With new violent jihadist fronts opening in Mali and Algeria during January

2013, shortly followed by the Boston Marathon bombing in the United States by home-grown extremists in April 2013, we would be well advised to regard the last decade of New Era global terrorism as merely being the first. While the unintended consequences of a decade of War on Terror have taught all in authority a great deal, many lessons from this experience have yet to be learned and passed on to the next generation of government policy makers and security practitioners. The secure future of our increasingly interconnected and interdependent world remains unpredictable, yet it is a global community in which the violent struggle against extreme jihadist terrorism continues to raise critical concerns for our national security. The return of conflict engaged citizens from theatres of war now raises new concerns and challenges for the future.

References

9/11 Commission Report (2004). *Final Report of the National Commission on Terrorist Attacks Upon the United States (Authorized Edition)*. W.W. Norton & Company: New York City, US.

Andrew, C. (2009). *The Defence of the Realm: The Authorised History of MI5*. Penguin Books: Harmondsworth.

BBC (2006). *BBC News 24: Live Coverage of the Airlines Terror Plot* (10 August 2006).

BBC (2009a). *Profiles: Airline Plot Accused*. Available at: http://news.bbc.co.uk/1/hi/uk/7604808.stm, accessed 15 January 2013.

BBC (2009b). *Liquid Bomb Plot: What Happened?* Available at: http://news.bbc.co.uk/1/hi/uk/8242479.stm, accessed 6 February 2013.

Hayman, A. (2009). *The Terrorist Hunters*. Bantam Press: London.

Langton, C. (2006). *The Military Balance*. 106(1). Oxford: Routledge

Rayment, S. (2007). *Nuclear Alert by ex-head of MI5.* Available at: http://www.telegraph.co.uk/news/uknews/1556812/Nuclear-alert-by-ex-head-of-MI5.html, accessed 14 August 2014.

Scotland Herald (2012). *Ex-Soldier Jailed after Threat to Bomb Mosque*. Available at: http://www.heraldscotland.com/news/home-news/ex-soldier-jailed-after-threat-to-bomb-mosque.19741622, accessed 8 February 2013.

Staniforth, A. (2012). *Routledge Companion to UK Counter-Terrorism*. Routledge: Abingdon, UK.

Staniforth, A. and Police National Legal Database (2009). *Blackstone's Counter Terrorism Handbook*, 2nd edn. Oxford University Press: Oxford.

Thornton, S., and Mason, L. (2007). Community cohesion in High Wycombe: A case study of Operation Overt. *Policing: A Journal of Policy and Practice*, 1(1), 57–60.

Chapter 15

Parasites, Energy and Complex Systems: Generating Novel Intervention Options to Counter Recruitment to Suicide Terrorism

Mils Hills and Ashwin Mehta
Northampton Business School, The University of Northampton, UK

Introduction

Having been developed through presentation at a range of defence research fora in the United Kingdom, in 2012 Hills published a thought-piece titled 'A new perspective on the achievement of psychological effects from cyber warfare payloads: the analogy of parasitic manipulation of host behavior' (Hills, 2012). Informed by personal experience working within the UK counter-terrorism and wider defence and security community – and constantly irritated by superficial media coverage that describes terrorists and other criminals as 'brainwashed' – Hills sought to both challenge conventional explanations by means of Analogical Research (AR) as well as to encourage scholarly collaboration. This chapter has arisen from interaction stimulated by the 2012 publication and the authors introduce further elaboration around the analogy of parasitic infection shaping or driving undesirable behaviour. It is intended to once again catalyse reactions as well as suggest the potential for practical interventions in policy.

AR seeks to shed new light on intractable problems by exploiting analogous contexts. In this case, viewing suicide terrorism as being the product of rational (to the perpetrator) actions driven by powerful bundles of memes that influence cognitive processes. Parasites change behaviour in animal models. Subtle mechanisms (changes at the endocrinal level) produce profound effects (from timid to aggressive traits). At the same time, the human brain can achieve extraordinary effects when driven by extreme belief: in other words, when infected by memes. The anthropologist Claude Levi-Strauss has written eloquently of the incredible impact that the human mind can inflict on the body of those who believe they are victims of magic: ultimately psychological thoughts can lead to believers dying from physiological effects.

This chapter continues and expands the line of AR first advanced in print in 2012. That paper focused on exploring the concept that bundles of beliefs (memes) can usefully be thought of as influencing behaviour in the manner that biological parasites interrupt the functioning of animal models' brains. In alliance with the known power of belief to affect the body, this chapter argues that there is a strong argument that there is the potential for memes to drive transformational and problematic behaviour: such as that of the migrating of a classroom assistant in a school to self-recruitment and sacrifice as a willing suicide bomber. In further advancing a challenging and unusual approach to understanding deviant actions, the authors seek to stimulate engagement with readers in order to advance further novel ideas for development in policy-making and other applied communities.

The Problem that Requires Explanation

The background to even considering alternative approaches to terrorism recruitment and action is driven by (a) the awareness that 'popular psychology' explanations of brainwashing are far from adequate, and (b) that even exhaustive official inquiries – drawing on security service and other reporting – were unable to shed any easy light on suicide terrorism recruitment:

- 'there is little in their backgrounds which mark them out as particularly vulnerable to radicalisation [. … they were] apparently well integrated into British society' (Home Office, 2006).
- 'the vast majority of suicide bombers, the 7/7 four included, show no sign of mental illness and have no criminal history. They are often better educated than their peers and hold respectable jobs' (Bond, 2007).
- 'The backgrounds of the 4 men appear largely unexceptional [. …]. We do not know how Khan developed his extreme views or precisely when' (Home Office, 2006).

And as this vignette could have included, we also don't know why these extreme views developed. Given the lack of knowledge about the mechanisms (other than descriptive accounts of the watching of extreme videos, influence of charismatic preachers and so on) – there is obviously analytical space available to (and a need for) more satisfying explanations. The transformation of 'normal' individuals into 'abnormal' states where there is no clear pathology to explain the substantive change surely demands explanation. As it so happens, the authors are as interested in the emergence of problematic workplace behaviour (e.g. in terms of risk-taking in financial investments or in the management of inherently hazardous operations) as in the emergence of suicide terrorists – yet this latter social phenomenon sets a high bar in terms of ability to make sense of a challenging issue and so is the focus of our current thinking.

Let's open the exploration of the explanatory power of analogical reasoning of parasites by presenting a short insight into how one notable example alters the actions of grasshoppers:

> The parasitic Nematomorph hairworm, *Spinochordodes tellinii* (Camerano) develops inside the terrestrial grasshopper, *Meconema thalassinum* (De Geer) (*Orthoptera: Tettigoniidae*), changing the insect's responses to water. The resulting aberrant behaviour makes infected insects more likely to jump into an aquatic environment where the adult parasite reproduces. […]
>
> The findings suggest that the adult worm alters the normal functions of the grasshopper's central nervous system (CNS) by producing certain 'effective' molecules. In addition, in the brain of manipulated insects, there was found to be a differential expression of proteins specifically linked to neurotransmitter activities. The evidence obtained also suggested that the parasite produces molecules from the family Wnt [proteins] acting directly on the development of the CNS (Biron et al., 2005: 2117).

The extraordinary alterations in animal behaviour, driven by parasitic manipulation of the complex functioning of brain activity, are quite something to the uninitiated. There are a wide range of changes – from the strange to the rather gruesome. While extremes of behavioural change in a grasshopper might be easily dismissed, humans, as animals, are far from immune from influence by parasites. Indeed, an emerging clinical literature suggests that there is very strong evidence that some human ailments, disorders and behaviours (e.g. schizophrenia or a tendency towards schizophrenia) may be caused by parasitic infection (Mortensen et al., 2007).

However, this chapter will restrict its scope to the analogy of parasitic infection. But we endorse the idea that bundles of ideas (beliefs) have transforming effects on a range of mental processes (emotional state; trusted sources of information; risk versus reward and other activities) – and that these in turn produce potentially extraordinary effects on behaviour. Given that Levi-Strauss explored (originally in the 1960s) how magic can kill, the link between entrenched belief and the causing of death is not new.

The magic that Levi-Strauss describes does not need to be 'real' in an objective sense – if everyone in a community believes in it, then it is real in the effects on those, for example, accused of possession:

- Cannon showed that fear, like rage, is associated with a particularly intense activity of the sympathetic nervous system. This activity is ordinarily useful, involving organic modifications that enable the individual to adapt to a new situation. But if the individual cannot avail any instinctive or acquired response to an extraordinary situation (or to one which he conceives of as such), the activity of the sympathetic nervous system becomes intensified and disorganised; it may, sometimes within a few hours, lead to a decrease in the volume of blood and a concomitant drop in blood pressure, which result in irreparable damage to the circulatory organs. The rejection of food and drink, frequent among patients in the throes of intense anxiety,

precipitates this process; dehydration acts as a stimulus to the sympathetic nervous system, and the decrease in blood volume is accentuated by the growing permeability of the capillary vessels. These hypotheses were confirmed by the study of several cases of trauma resulting from bombings, battle shock, and even surgical operations; death results, yet the autopsy reveals no lesions (Levi-Strauss, 1963).
- 'There is, therefore, no reason to doubt the efficacy of certain magical practices. But at the same time we see that the efficacy of magic implies a belief in magic. The latter has three complementary aspects: first, the sorceror's belief in the effectiveness of his techniques; second, the patient's or victim's belief in the sorceror's power; and, finally, the faith and expectations of the group' (Levi-Strauss, 1963: 168)

The power of belief then – often invoked in situations as varied as positive thinking for lifestyle enhancement through to longevity amongst patients with terminal diagnoses and the ability of the religious to accommodate stress – has the potential to condemn to death or exculpate others.

Parasites achieve their effects and conceal their existence through a myriad of subtle and sophisticated operations (endocrinal, neurochemical). If, for example, a parasite can transform the risk-taking behaviour, preferential food choice and emotional reward of complex animal models, perhaps bundles of memes (beliefs working as though they were biological parasites) could be changing the risk profile, preferential information sources and emotional rewards that, for example, suicide terrorists (and their supporters) access and gain respectively.

One example from the fascinating activities of parasites will buttress the analogical cause. The normally risk-averse mouse or rat, when subject to parasitic infection by *Toxoplasma gondii*, actively seeks out opportunities to be predated by cats. This enables the infecting organism to continue its life cycle in the cat. The mouse or rat will, for example, pursue and attack cats when they detect their odour as well as adopting changed, less covert behaviours: including the speed at which they move and reduced situational awareness and alertness (lower levels of anxiety). All of these behaviours make it more likely that they will be predated by cats, achieving the outcome that drives the parasite's actions.

The authors' contention is that a package of ideas (memes) could exercise profound changes if they could produce changes in brain chemistry and therefore concrete behaviour. We think that the possibility that this could reinvigorate the study of terrorism/criminal/deviant behaviour is worth exploring. The following section muses over potential further insight that could lead to modelling, developing of new options for intelligence direction, collection and analysis and other practical measures.

The Energetic Triggering of Extremism

In an extension to the parasitic analogy, 'infection' is presented as achieved through the movement of energy. Individuals susceptible to energised ideas are excited by what could be seen as an ideational version of free radical particles. To conceptualise the impact of an extremist event on human systems, the formation, fermentation and eventual release of an extremist ideology – culminating in a mass casualty event – might helpfully be viewed in terms of energy equilibria along a free-radical pathway, i.e. initiation, propagation and termination.

In quantum mechanics, an excited state of an electron in orbit around a nucleus is reached by stimulation using external energy. Upon relaxation, a quantum of energy is released. Presuming global equilibrium, the increasing release of energy from recent mass impact events (and 'spectacular' graphic terrorist activities) have led to a release of discrete quanta of negative energy – a consequence of relaxation in the system. In free-radical chemistry such energy can be used to break a two-electron covalent bond, resulting in highly reactive species that possess a single electron, ready to react with other bonded species around it.

The free radical, then, bound with the receptive molecule from the host, is the nascent parasitic infection – the equivalent of the nematode worm egg. For the unsusceptible, with no molecular opportunities for free radicals to bond with, the infection does not take place: these individuals are immune. No doubt this immunity may be compromised or change in future, but this would explain why some individuals are inspired by, say, graphic imagery of the suffering of oppressed peoples, battlefields and the aftermath of mass casualty attacks and find that these events are a call to action – while others are unaffected. Perhaps there are different forms of (analogical) free radical particles: some work to produce a high magnitude of action in a very short period of time, others combine over an extended period to gradually develop their impact (the initial infection).

The combination of quantum mechanics with parasitic infection then enables us to conceptualise the connection between infected extremists and the host system as being one where energy is moved around the system through propagation, which can involve complex interactions, before that energy is released. The analogy with parasites would be, for example, when the infected grasshopper (normally with an inherent fear of water) hops into a pool enabling the parasitic worm to advance to its next stage of life – but leaving the grasshopper dead: the energy leaves it. Perhaps the analogy with this and the suicide bomber seems a little too obvious, but it is powerful.

Friedrickson (1983) usefully classified the range of microbial relationships as including the following:

- Mutual
- Commensalistic
- Competitive (e.g. parasitic versus predator/prey)
- Amensalistic

- Independent

Here, then, the infected/free-radicalised individual may (depending on variables of infection which we have yet to fully consider and/or conditions in the environment) interact with their surrounding systems and their fellow humans in differing forms. For example, then, the relationship may be one of mutual benefit (symbiosis): both benefit in some manner. This model might be more suited to the relationship between an organisation and society, where there may be a consideration of risk, cost and benefit to produce moderate or extreme behaviours (Wintrobe, 2006). Here, then, individuals under a degree of parasitic influence may challenge and enhance the resilience of an organisation – in the same way that immune systems benefit from exposure to pathogens, building resistance to viruses and bacteria and thereby improving the fitness of the body.

Commensalistic or catalytic relationships are those where one party benefits (or changes), while the other is left unchanged. This would be a situation where the infected ('parasitised') individuals extract benefit from the wider system while they themselves are unaffected, or where the parasitic infection achieves the results it needs without damaging or changing the behaviour of the host. The analogy here could be where a covert grouping evades control mechanisms – where that system is unaware that within it malign forces are present.

The parasitic relationship is the initiating idea behind this chapter. Consideration needs to be given as to whether the infection is active (e.g. the result of targeted delivery) or more passive (the free radical concept introduced above), where infected individuals receive infection as a result of a broadcast delivery (a scatter-gun versus sniper strategy).

Amensalism relationships are those where one party benefits, while another is destroyed. The biological parallel here is with the killer phenomenon in microorganisms. Yeasts (e.g. *Saccaromyces cerevisiae*) secrete a metabolite which has antimicrobial properties (Bajaj et al., 2013; Polonelli and Morace, 1986). Here, one organism is protecting itself at the expense of another or others. In order to maintain their fitness, to achieve their ambitions, they seek to cause harm or loss to others, investing energy into the generation of toxins. Perhaps this is the condition where covert, malign groups seek to destroy others in a system.

Independent – here, a relationship could be coincidental: individuals may find themselves caught up in supporting, for example, the ambitions of an individual infected by a parasite whose behaviour exerts demands on others. They may be entirely unaware that they are enabling the achievement of the gains of another.

Conclusion

This chapter has advanced extensions to the analogy of parasitic infection as a means of explaining the recruitment of individuals to extremist causes. Our mission has been to further promote innovative thinking for pertinent communities and others, yet guided by the need to ultimately generate practical outputs. The authors plan to explore the headline directions detailed above in more detail and to engage both with potential end-user groups and the wider literature on parasitology.

References

Bajaj, B.K., Raina, S., and Singh, S. (2013). Killer toxin from a novel killer yeast *Pichia kudriavzevii RY55* with idiosyncratic antibacterial activity. *Journal of Basic Microbiology*, 53, 645–656.

Biron, D.G., Marché, L., Ponton, F., Loxdale, H.D., Galéotti, N., Renault, L., Joly, C., and Thomas, F. (2005). Behavioural manipulation in a grasshopper harbouring hairworm: a proteomics approach. *Proeedings of the Royal Society B: Biological Sciences*, October 22, 272(1577): 2117–2126.

Bond, M. (2007). The psychology of bombers. *Prospect Magazine*, online, http://www.prospectmagazine.co.uk/magazine/thepsychologyofbombers/#.UoVhk_nIZqM.

Fredrickson, A.G. (1983). Interactions of microbial populations in mixed culture situations. In H.W. Blanch, T. Papoutsakis, and G. Stephanopoulos (Eds) Foundations of Biochemical Engineering, American Chemical Society, Washington, 201–227.

Hills, M. (2012). A new perspective on the achievement of psychological effects from cyber warfare payloads: The analogy of parasitic manipulation of host behavior. *Journal of Law & Cyber Warfare*, 1(1), 208–217.

Home Office (2006) *Report of the Official Account of the Bombings in London on 7th July 2005*. London: The Stationery Office, available at: http://www.official-documents.gov.uk/document/hc0506/hc10/1087/1087.pdf.

Levi-Strauss, C. (1963). *Structural Anthropology 1*. Penguin: Harmondsworth, UK.

Mortensen, P.B., Nørgaard-Pedersen, B., Waltoft, B.L., Sørensen, T.L., Hougaard, D., and Yolken, R.H. (2007). Early infections of *toxoplasma gondii* and the later development of schizophrenia. *Schizophrenia Bulletin*, 33(3), 741–744.

Polonelli, L., and Morace, G. (1986). Re-evaluation of the yeast killer phenomenon. *Journal of Clinical Microbiology*, November, 866–869.

Wintrobe, R. (2006). Extremism, suicide terror and authoritarianism, *Public Choice*, 128, 169–195.

Chapter 16
Terrorist Targeting of Schools and Educational Establishments

Emma Bradford
Department of Psychological Sciences, University of Liverpool, UK

Margaret A. Wilson
Institute for Security Science and Technology, Imperial College London, UK

Introduction

While terrorist attacks on schools are rare, the targeting of children causes widespread outrage and these incidents receive extensive media coverage. For example, on September 1, 2004, Chechneyan rebels stormed a primary school in Beslan, Russia. Over 1,200 Russian citizens were held hostage for a period of three days and, at the end of this school-targeted terrorist siege, 331 Russian citizens had been killed and a further 700 wounded. The dead included nearly 200 children and the event itself highlighted how vulnerable educational institutions can be to terrorist attacks. While being one of the most notorious examples of school-targeted terrorism, Beslan was by no means the first of its kind: rather the incident represents a recent example of a long, though intermittent, series of attacks against school-related targets. For example, on May 8, 1970, Palestinian terrorists launched an armed attack against an Israeli school bus. Nine children and three adults were killed, with nineteen more being seriously wounded.

Reports suggest that the frequency of attacks on educational institutions has been increasing steadily since 2000, with the end of the decade seeing approximately six-times more attacks and quadruple the casualties worldwide (O'Malley, 2007). In 2010, the United Nations Educational, Scientific and Cultural Organisation (UNESCO) reported a further significant increase in the number of terrorist attacks on educational institutions since 2007 (O'Malley, 2010). These attacks have taken many forms including armed assaults, assassinations, bombings, hostage takings, chemical attacks, mass poisoning, sexual attacks and arson. Attacks are not always fatal, and not always directed at the children themselves, but sometimes against teachers or property. However, the mass abduction of schoolchildren for recruitment as suicide bombers has also been reported in Pakistan (O'Malley, 2007, 2010).

Despite terrorist attacks on educational institutions being a more widespread phenomenon than previously acknowledged, the deliberate targeting of students, staff and schools is still relatively infrequent. This chapter explores the potential factors associated with the choice of educational institutions as terrorist targets.

Terrorism, Fear and Rational Choice

The basic premise of terrorist action is to bring about political, social or religious change by instilling fear. Once a person has taken the decision to use violence to instigate change, the terrorist actor must decide on an appropriate target which will fulfil their aims. In terms of Rational Choice Theory (RCT), it is believed that the decision to commit an offence is based on the assessment of the relative costs and benefits of committing that crime (Cornish and Clarke, 1986). This type of cost–benefit analysis is consistent with rational choice models of non-terrorist crime (e.g. Clarke and Felson, 1993; Cornish and Clarke, 1986), and has also proved pivotal in understanding terrorism (e.g. Anderton and Carter, 2005; Caplan, 2006; Enders and Sandler, 2006). Rational choice theory suggests that

> [A] person commits an offense if the expected utility to him exceeds the utility he could get by using his time and other resources at other activities. Some persons become 'criminals', therefore, not because their basic motivation differs from that of other persons, but because their benefits and costs differ. (Becker, 1968: 176)

When considering attacks against schools, the first question that needs to be answered is why a terrorist organisation might choose to deliberately target educational institutions over other civilian non-combatants. One could argue that specifically targeting children may reinforce the perception that terrorist violence can strike any place, at any time and against anybody (Pine, Costello and Masten, 2005). This seemingly 'random' element of terrorist violence has the ability to disseminate fear throughout a target population due to its unpredictable and indiscriminate nature. On the other hand, there are serious issues around the legitimacy of child targets which can result in severe damage to the terrorists' cause. Clearly, given rational choice, a terrorist actor can choose to target any section of the population, and with that choice comes a range of costs and benefits.

Schools as Direct versus Symbolic Targets

The first and most straightforward motivation for attacking a school may be in order to communicate a very specific grievance with that institution and its practices. In these instances, the institute of education itself or the academic curriculum might represent the incitement for an attack. For example, in March 2006, a spokesperson for the Taliban stated that schools would be attacked because of the content of their curriculum,

> The government has given teachers in primary and middle schools the task to openly deliver political lectures against the resistance put up by those who seek independence ... The use of the curriculum as a mouthpiece of the state will provoke the people against it. (Mohammed Hanif[1]).

In addition to opposition against the set curriculum, some countries and cultures are opposed to the education of women and this has led to so-called 'enemies of education' conducting attacks against female students and those that teach them. An example of such an attack occurred in southern Afghanistan in November 2008, in which assailants used a toy gun to spray battery acid into the faces of female students outside the Mirwais Nika Girls High School. More recently, attention has been drawn to the case of Malala Yousafzai, who was shot in the head by the Taliban for promoting education for girls in Pakistan.

By attacking specific institutions, this kind of action might also serve a calculated function to induce behavioural change. By creating widespread fear surrounding attendance at certain schools, terrorists can influence parental decisions regarding which school to send their children to. Therefore, by not attacking educational institutions that their organisation approves of, terrorist groups can manipulate which schools are deemed 'safe' and thereby control both access to schools and educational content.

However, the motivation for attack might be entirely unrelated to education per se. Crenshaw (2000: 386) argues that 'terrorism is also often designed to disrupt and discredit the processes of government, by weakening it administratively and impairing normal operations'. Therefore, if the aim is to demonstrate 'governmental incompetence' through its inability to protect its citizens, terrorist groups might choose to target a subsector of society that arguably requires the most protecting: children.

Therefore, it is important to remember that the ultimate 'target' of a terrorist attack is not necessarily the individual(s) who suffer physical harm. Attacks are designed to bring about political or social change and as such the intended recipients of the message of the attack are usually those capable of instigating such changes, for example, governments (Wilson and Lemanski, 2010). While some terrorist organisations do carry out assassinations of prominent political figures, the direct targeting of these individuals is often obstructed by their social position and the level of protection they are afforded as a result (e.g. Prime Minister, President). The relative inaccessibility of specific individuals means that much terrorism is aimed against civilians and the threat of further action creates leverage because the general population wants to see an end to the violence.

This is not to say that the selection of which 'uninvolved others' to target is indiscriminate; the chosen targets tend to act as representations of 'a broader ideological context' (Taylor, 1988: 88). As such, while the specific individuals

1 'Taliban statement warns Afghan government to stop politicization of education.' *Afghan Islamic Press Agency*, March 25, 2006.

targeted are often incidental, they are chosen as victims because they are symbolic representatives of the intended target (Dorn and Dorn, 2005; Horgan, 2005). State educational institutions and those who attend or work for them represent an extension of the government and thus may become a target by proxy. Put simply, attacks on schools can be used by terrorists as an indirect way of attacking the state, as has been demonstrated by Maoist insurgents in India.

This symbolic targeting can explain how being a member of a particular social, religious or cultural group can make individuals a potential target for attack without them having done anything to provoke it. So, when indirect targeting tactics are involved, it is important to consider why terrorist organisations might choose to target educational institutions over other civilian populations.

Mass Casualty Soft Targets

As a target choice, in terms of cost–benefit, schools have two immediate 'advantages'. First, by their very nature, educational institutions are places where numerous people gather on a daily basis. In fact, in many communities there are more people gathered in educational institutions than in any other location on any given day (Graham et al., 2006). Therefore the potential for violent action to yield mass casualties is high (e.g. Allanson, 1967; Wass, Williams and Gibson, 1994). Although scholars are divided on whether there has been an increase in mass casualty incidents (e.g. Crenshaw, 2006) this form of terrorism is certain to attract widespread media attention, and therefore publicity for the cause (discussed later in this chapter). In addition to guaranteeing a highly populated location for attack, schools are also relatively easy to access, providing two of the key features that Clarke and Newman (2006) define as essential in terrorist targeting.

Specific features of potential targets (e.g. accessibility, security measures) can influence a terrorist group's ability to launch a successful attack and thus can influence the likelihood of its commission (Sandler and Lapan, 1988). In terms of target selection in terrorism, the physical accessibility of a potential target (e.g. how well it is guarded) can affect the level of risk associated with an operation (Cohen, Kluegel and Land, 1981). For example, the greater the level of protection afforded to a potential target, the greater the risk, the lower the potential for success and thus the lower the likelihood of attack. Targets can therefore be categorised as 'soft' or 'hard' targets.

Hard targets tend to consist of non-civilian personnel who represent the most 'legitimate' targets for politically motivated attacks, for example, military or governmental bodies responsible for the political climate to which a terrorist group is opposed. These locations are deliberately 'hardened' to 'protect potential targets either by making attacks more costly for terrorists or by reducing their likelihood of success' (Enders and Sandler, 2006: 120). Therefore, while these locations tend to represent or house desirable targets for terrorist attacks, they involve a high risk of interception, often with lethal force, and a low probability of success and as such often deter terrorist action.

However, instead of deterring terrorism outright, target hardening can lead to displacement (Trasler, 1986) or transference (Enders and Sandler, 2006) by which terrorists shift their attention to other, more accessible 'soft' targets. Soft targets are those which have little or no security and as such present an easier alternative for a terrorist attack (Singh, 2004). It is this lack of protection which may make many schools and other educational institutions vulnerable to attack.

Legitimacy

Although terrorism targeting soft targets like schools and school-related locations increases the probability of success while simultaneously facilitating the possibility for mass casualties, terrorist attacks specifically targeting children are rare. This is because in spite of the apparent 'pros' associated with targeting schools, there are also a number of factors which very seriously limit the utility of such attacks. Numerous authors have highlighted the difference between targets of terrorist attack, both as perceived by the terrorists themselves and as perceived by the public. Specifically, victims may be seen as possessing different levels of legitimacy; from more 'legitimate' targets such as military personnel through to civilian non-combatants. Clearly, the least legitimate targets would have to be children.

Since attacks on children lack legitimacy, they are likely to cause shock and outrage in the terrorists' audience. While advances in technology have provided terrorist organisations with a medium through which to advance their message globally, the increased diversity of the audience and their gradual acclimatisation to terrorism has meant that 'terrorists must go to extreme lengths to shock' (Crenshaw, 2000: 386). Therefore, a further reason why terrorists might choose to target educational institutions is the level of shock that such an attack would generate and the subsequent intensive media coverage that it would attract, albeit balanced against the potential backlash (see below).

Media Coverage

Achieving widespread recognition has long been argued to be one of the fundamental goals of terrorist action (e.g. Thornton, 1964). While in some instances media attention may simply be desirable (Doran, 2002; Perl, 2002; Richardson, 2006), in others it is essential for a group's survival (Perl, 2002). Furthermore, recent research has argued that terrorists deliberately use the media as a method of 'armed propaganda' (e.g. Louw, 2003) and that victims are selected on the basis of their 'propaganda value' (Pereda, 2013).

While targeting children might result in greater publicity, some analysts suggest that when it comes to attacks against educational institutions, it is not the case that all publicity is good publicity. Many authors have drawn attention to the detrimental effects that attacks on non-legitimate targets can have on a terrorist cause (e.g. Silke, 2005). There are numerous published examples of incidents where the public, even those who passively 'support' a campaign, have responded with open protest to terrorist action which is perceived to lack legitimacy. For example in 1993, the Spanish public initiated 'blue ribbon' campaigns as a show of solidarity with families of individuals kidnapped by ETA. Numerous demonstrations were held to demand the liberation of the hostages and blue ribbons used throughout Spain as a symbol of protest against terrorist activities. White hands became another national anti-terrorism symbol in 1997 following the murder of a university professor by ETA. Both expressions of support for anti-terrorism efforts are illustrative of the Spanish population uniting against a common enemy, and are widely credited with playing a role in the dissolution of the organisation. Similarly, Silke (2005) places the death of two British children in Warrington as an important precursor to the cessation of IRA activities.

Interestingly, in a study of terrorist armed attacks against schools, Bradford and Wilson (2013) found that when an armed assault yielded a high number of fatalities, the terrorist group responsible for the attack was less likely to claim responsibility for their actions. It is possible that the aftermath of perhaps unintended child casualties may cause a terrorist group to reconsider the value of claiming the attack.

Loss of Support

Since the manipulation of a target audience is central to terrorist action, 'the political effectiveness of terrorism is importantly determined by the psychological effects of violence on audiences' (Crenshaw, 1986: 400). The mechanisms of terrorism require fear to act as a catalyst to encourage political or socal change. However, any terrorist organisation must consider the impact of their actions in order to maintain public sympathy for their cause.

For example, the extreme nature of the 2001 attacks on the World Trade Center resulted in reduced sympathy and support for Islamic fundamentalism and a range of secular causes (Richards, 2004). Therefore, despite the widespread publicity generated by attacks on children, the lack of legitimacy of child-targeted terrorism may lead to an erosion of support for the terrorist cause (Johnston, 2009). Specifically targeting children may cause existing supporters to distance themselves from the organisation and can subsequently lead to the withdrawal of financial support, material resources and other forms of assistance (Dorn and Dorn, 2005).

In addition to eroding their own support base, violent action can also have more widespread political consequences. In their studies on the political repercussions of 9/11, Bonanno and Jost (2006) found that traumatic events can also increase the level of political conservatism within the general public through their psychological need to manage uncertainty and threat. Therefore, if an audience finds an attack on an educational institution traumatic, it is likely that their political affiliation will move further away from an extreme terrorist position. These conservative attitudes advocate simplistic solutions (e.g. 'good' versus 'evil', 'us' versus 'them') in response to perceived threat (e.g. Altemeyer and Hunsberger, 1992; Doty, Peterson and Winter, 1991; McCann, 1997) and can lead to heightened bias against out-groups.

Research has found that perceived threat increases levels of solidarity within the threatened 'in-group' while simultaneously increasing levels of prejudice and intolerance of 'out-groups' (Huddy, 2003). The most heightened emotional arousal in response to terrorist attacks is experienced by those who are physically closest to an incident (Lowenstein, Weber, Hsee and Welch, 2001). Given the degree of attachment between caregiver and child (Bowlby, 1969) and the level of general attachment felt between society as a whole and their young, it is likely that even individuals who are not directly affected by a terrorist attack on an educational institution will feel heightened levels of threat by proxy. Indeed, the use of children as victims has been found to increase the long term impact of terrorist action on adult populations (Phillips, Featherman and Liu, 2004). This being the case, targeting children may be an effective way of ensuring that fear is further transmitted far beyond the original victims. However, this increased level of fear might lead to increased levels of solidarity with their threatened in-group and increased levels of bias against the threatening out-group.

In addition, intensified hostility toward the 'enemy' paired with a desire for revenge and amplified levels of ethnocentrism and xenophobia (Huddy, Feldman, Taber and Lahav, 2005) can ultimately lead to increased support for retaliatory military action (e.g. Arian, 1989; Herrmann, Tetlock and Visser, 1999; Jentleson, 1992; Jentleson and Britton, 1998).

Governmental Response

Finally, it is necessary to consider the way in which the authorities respond to acts of terrorism in general. The public outrage following the attacks of 9/11 made it likely that the US would respond with a major counter-terrorist campaign. It is also likely that targeting children would require a government to be seen to respond in kind. Therefore, one might question the cost–benefit analysis which precedes such an attack. When targeting schools, terrorist groups are likely to run the risk of provoking a severe governmental response and the threat of such counteraction might reasonably act as a deterrent to some organisations.

As the perceived threat of terrorism increases so does the public level of support for aggressive national security measures (both domestic and international) and implementation of increased surveillance, restricted civil liberties and stricter immigration policies (Huddy et al., 2005). This type of shift in public opinion was observed following 9/11 when there was increased public support for the war in Afghanistan (Huddy et al., 2005). Studies have shown that decision-making becomes more rigid and dogmatic under conditions of threat and this in turn produces increasingly negative views of a threatening group (Cottam, 1994; Herrmann, 1988). With public opinion on their side, governments might be more at liberty to engage in retaliatory warfare.

However, there are a growing number of scholars who are stressing the potential negative effects of a heavy handed counter-terrorist response. Rather than always acting as a deterrent, in some cases it can ultimately be viewed as a fresh affront in the eyes of the terrorist group, and importantly, other individuals newly affected by retributive actions. This will act to reinforce terrorist perceptions of a harsh, oppressive government and can cause a new subset of individuals to feel themselves or their loved ones to be threatened and treated unjustly (Horgan, 2005; Silke, 2003, 2006; Wilson, 2010; Wilson, Bradford and Lemanski, 2013). This can lead to individuals craving vengeance and seeking recruitment in a terrorist organisation that is perceived to have the power to launch a successful attack against their oppressor(s).

For this reason, recent research has argued that an overly harsh governmental response may be exactly what terrorist organisations hope for in some instances (McCauley, 2006). This phenomenon has been termed 'jujitsu politics' to reflect how terrorists can use the power of the state against itself (McCauley, 2006). With jujitsu politics, terrorists aim to hit their target hard enough to provoke a counter-attack severe enough to mobilise new supporters and amass new recruits (McCauley, 2006).

In the event of an attack against soft targets like schools, governments should therefore give careful consideration to how they launch counter-terrorism movements: overly aggressive counter-terrorism initiatives might cause a government to portray itself in a light which supports the terrorist position and as such, might mobilise new supporters for the cause rather than discourage them.

Conclusion

As discussed throughout this chapter, there are both costs and benefits associated with the deliberate targeting of schools and school children. The relative rarity of this type of attack, along with the fact that high fatality attacks are less likely to be claimed, could be argued to reflect a level of rationality in terrorist behaviour and the acknowledgement of the costs of such an action outweighing the benefits. However, despite the relative rarity of attacks against schools, recent research suggests that armed attacks against educational institutions in general have increased in frequency over the last decade, and in their overall lethality, albeit with the increase seen in staff victims (Bradford and Wilson, 2013). Therefore, for an effective examination of the evolving terrorist threat, policy makers should not ignore the possibility of deliberately school-targeted attacks.

References

Allanson, J.F. (1967). School mass disaster policies. *Journal of School Health*, 37, 285–288.

Altemeyer, B., and Hunsberger, B. (1992). Authoritarianism, religious fundamentalism, quest, and prejudice. *The International Journal for the Psychology of Religion*, 2, 113–133.

Anderton, C.H., and Carter, J.R. (2005). On rational choice theory and the study of terrorism. *Defence and Peace Economics*, 16(4), 275–282.

Arian, A. (1989). A people apart: Coping with national security problems in Israel. *Journal of Conflict Resolution*, 33(4), 605–631.

Becker, G.S. (1968). Crime and punishment: An economic approach. *Journal of Political Economy*, 76, 169–217.

Bonanno, G.A., and Jost, J.T. (2006). Conservative shift among high-exposure survivors of the September 11th terrorist attacks. *Basic and Applied Social Psychology*, 28(4), 311–323.

Bowlby, J. (1969). *Attachment and Loss*. New York: Basic Books.

Bradford, E., and Wilson M.A. (2013). When terrorists target schools: An exploratory analysis of attacks on educational institutions. *Journal of Police and Criminal Psychology*, DOI 10.1007/s11896-013-9128-8.

Caplan, B. (2006). Terrorism: The relevance of the rational choice model. *Public Choice*, 128, 91–107.

Clarke, D.B., and Felson, M. (1993). *Routine Activity and Rational Choice: Advances in Criminological Theory, Vol. 5*. London: Transaction.

Clarke, R.V., and Newman, G.R. (2006). *Outsmarting the Terrorists*. New York: Praeger Publishers.

Cohen, L., Kluegel, J., and Land, K. (1981). Social inequality and predatory criminal victimization: An exposition and a test of a formal theory. *American Sociological Review*, 46, 505–524.

Cornish, D.B., and Clarke, R.V. (1986). *The Reasoning Criminal: Rational Choice Perspectives on Offending*. New York: Springer.

Cottam, M. (1994). *Images and Intervention*. Pittsburgh, PA: University of Pittsburgh Press.

Crenshaw, M. (1986). The psychology of political terrorism. In M. Hermann (Ed.), *Political Psychology: Contemporary Problems and Issues*. London: Jossey-Bass, 379–413.

Crenshaw, M. (2000). The psychology of terrorism: An agenda for the 21st century. *Political Psychology*, 21(2), 405–420, doi: 10.1111/0162-895X.00195.

Crenshaw, M. (2006). Have motivations for terrorism changed? In J. Victoroff (Ed.), *Tangled Roots: Social and Psychological Factors in the Genesis of Terrorism*. Oxford: IOS Press, 1–14. .

Doran, M.S. (2002). Somebody else's civil war. In P. Griest and S. Mahan (Eds), *Terrorism in Perspective*. London: Sage Publications, 73–85.

Dorn, M., and Dorn, C. (2005). *Innocent Targets: When Terrorism Comes to School*. Macon, GA, USA: Safe Havens International Inc.

Doty, R.M., Peterson, B.E., and Winter, D.G. (1991). Threat and authoritarianism in the United States, 1978–1987. *Journal of Personality and Social Psychology*, 61, 629–640.

Enders, W., and Sandler, T. (2006). *The Political Economy of Terrorism*. Cambridge: Cambridge University Press.

Graham, J., Shirm, S., Liggin, R., Aitken. M.E., and Dick, R. (2006). Mass-casualty events at schools: A national preparedness survey. *Pediatrics*, 117, 8–15, doi: 10.1542/peds.2005-0927.

Herrmann, R.K. (1988). The empirical challenge of the cognitive revolution. *International Studies Quarterly*, 32, 175–203.

Herrmann, R.K., Tetlock, P.E., and Visser, P.S. (1999). Mass public decisions to go to war: A cognitive-interactionist framework. *American Political Science Review*, 93(3), 553–573.

Horgan, J. (2005). *The Psychology of Terrorism*. London: Routledge.

Huddy, L. (2003). Group membership, ingroup loyalty, and political cohesion. In D.O. Sears, L. Huddy and R. Jervis (Eds), *Handbook of Political Psychology*. New York: Oxford University Press, 511–558.

Huddy, L., Feldman, S., Taber, C., and Lahav, G. (2005). Threat, anxiety, and support of anti-terrorism policies. *American Journal of Political Science*, 49, 610–625.

Jentleson, B. (1992). The pretty prudent public: Post post-Vietnam American opinion on the use of force. *International Studies Quarterly*, 36, 49–74.

Jentleson, B.W., and Britton, R.L. (1998). Still pretty prudent: Post-Cold War American public opinion on the use of military force. *Journal of Conflict Resolution*, 42, 395–417.

Johnston, R. (2009). *Terrorist and Criminal Attacks Targeting Children*. Retrieved from www.johnstonarchive.net/terrorism/wrjp39ch.html, accessed 10 February 2010.

Louw, P.E. (2003). The 'war against terrorism': A public relations challenge for the Pentagon. *The International Journal for Communication Studies*, 65(3), 211–230.

Lowenstein, G.F., Weber, E.U., Hsee, C.K. and Welch, N. (2001). Risk as feelings. *Psychology Bulletin*, 127 (2), 267–286.

McCann, S.J.H. (1997). Threatening times, 'strong' presidential popular vote winners, and the victory margin, 1824–1964. *Journal of Personality and Social Psychology*, 73, 160–170.

McCauley, C. (2006). Jujitsu politics: Terrorism and response to terrorism. In P.R. Kimmel and C.E. Stout (Eds), *Collateral Damage: The Psychological Consequences of America's War on Terrorism*. Westport, CT: Praeger, 45–65.

O'Malley, B. (2007). *Education Under Attack 2007: A Global Study of Targeted Political and Military Violence Against Education Staff, Students, Teachers, Union and Government Officials, and Institutions*. Paris, France: UNESCO.

O'Malley, B. (2010). *Education Under Attack 2010*. Paris, France: UNESCO.

Perl, R. (2002). Terrorism, the media and the government: Perspectives, trends, and options for policy makers. In P. Griest and S. Mahan (Eds), *Terrorism in Perspective*. London: Sage Publications, 143–150.

Pereda, N. (2013). Systematic review of the psychological consequences of terrorism among child victims. *International Review of Victimology*, 19(2), 181–199.

Phillips, D., Featherman, D.L., and Liu, J. (2004). Children as an evocative influence on adults' reactions to terrorism. *Applied Developmental Science*, 8(4), 195–210.

Pine, D.S., Costello, J., and Masten, A.S. (2005). Trauma, proximity, and developmental psychopathology: The effects of war and terrorism on children. *Neuropsychopharmacology*, 30, 1781–1792.

Richards, B. (2004). Terrorism and public relations. *Public Relations Review*, 30, 169–176, doi:10.1016/j.pubrev.2004.02.005.

Richardson, L. (2006). *What Terrorists Want: Understanding the Enemy, Containing the Threat*. New York: Random House.

Sandler, T., and Lapan, H.E. (1988). The calculus of dissent: An analysis of terrorists' choice of targets. *Synthese*, 76(2), 245–261, doi: 10.1007/BF00869591.

Silke, A. (2003). Becoming a terrorist. In A. Silke (Ed.), *Terrorists, Victims and Society*. Chichester: Wiley, 29–53.

Silke, A. (2005). Children, terrorism and counterterrorism: Lessons in policy and practise. *Terrorism and Political Violence*, 17, 201–213.

Silke, A. (2006). *Terrorists, Victims and Society*, 2nd edn. Chichester: Wiley.

Singh, S. (2004). Modelling threats. *Electronic & Computer Engineering*, 23(3), 18–21.

Taylor, M. (1988). *The Terrorist*. London: Brassey's.

Thornton, T.P. (1964). Terror as a weapon of political agitation. In H. Eckstein (Ed.), *Internal War*. New York: Free Press, 71–99.

Trasler, G. (1986). Situational crime control and rational choice: A critique. In K. Heal and G.K. Laycock (Eds), *Situational Crime Prevention: From Theory into Practice*. London: Her Majesty's Stationery Office.

Wass, A.R., Williams, M.J., and Gibson, M.F. (1994). A review of the management of a major incident involving predominantly paediatric casualties. *Injury*, 25, 371–374.

Wilson, M.A. (2010). Terrorism research: Current issues and debates. In J. Brown and E. Campbell (Eds), *The Cambridge Handbook of Forensic Psychology*. Cambridge: Cambridge University Press, 571–578.

Wilson, M.A., and Lemanski, L. (2010). Forensic psychology and terrorism. In J. Adler and J. Gray (Eds), *Forensic Psychology; Concepts, Debates and Practice*, 2nd edn. Oxford: Willan Publishing, 245–263.

Wilson, M.A., Bradford, E., and Lemanski, L. (2013). The role of group psychology in terrorism. In J. Wood and T. Gannon (Eds), *Criminal Activity and Crime Reduction: the Role of Group Processes*. Abingdon: Taylor and Francis.

Chapter 17

Female Suicide Terrorism as a Function of Patriarchal Societies

Tanya Dronzina

Political Science Department, St. Kliment Ohridski, University of Sofia, Bulgaria

Introduction

Suicide terrorism is an extreme form of expression that uses the human body as a tool or weapon for killing, injuring and inflicting fear on others. The difference between this and more usual 'clinical suicide' is that in the latter instance, the individual is usually only concerned with ending their own life. Many sources refer to suicide bombing being a predominantly male activity, however this chapter argues that there is a relationship between patriarchy and female suicide terrorism. On the basis of a comparative study of Turkey (Parti Karkerani Kurdistan [PKK]), Chechnya (Chechen Rebels) and Dagestan (Dagestani insurgency) conclusions are drawn that despite different contexts, female suicide terrorism is likely to appear in societies which have at least five characteristics: a perception of acute ethnic conflict or foreign invasion (which pose dangers of national identity in part of the population); radicalized organizations that have chosen female suicide terrorism as a means of political action and mobilization; religious or cultural norms restricting the public role of women; structures of traditional society that are in a process of disintegration (sometimes as a consequence of national conflict); and disintegration caused by changes in social authority, values and priorities. By comparing social data, this chapter presents a standpoint that the motivation of individual female perpetrators goes beyond fundamental religion and ideology and is rooted in a reaffirmation of female identity as designed and tolerated by traditional patriarchal society.

Female Suicide Terrorism

The participation of women in suicide terrorism is a growing trend coupled with an increasing academic interest in understanding its causes and individual motivations. Over the last decade a number of, mainly female, social commentators have contributed to this area of inquiry (Dronzina, 2008, 2010; Victor, 2004) illustrating that while there are degrees of consensus there are still many contradictory explanations put forward for female suicide terrorism. The reason for this is possibly due to female suicide terrorism being a heterogeneous phenomenon. Examples of its mass use include:

- *Lebanon* – between 1985–1987, four parties resorted to female suicide terrorism: the Natzersit Socialist Party of Syria; Syrian Social Nationalist Party (SSNP); Lebanese Communist Party; and the Baath Party of Lebanon. The first mission was carried out by 16-year-old Sana Mehaidi (Khyadali) (different sources spell her surname in different ways) a member of the SSNP, against a convoy. Other missions were conducted against military checkpoints, hospitals and airports. Almost half of the female perpetrators were Christian.
- *Sri Lanka* – between 1989–2008, the Liberation Tigers of Tamil Eelam (LTTE, also known as Tamil Tigers) were considered champions in using female bombers with two-thirds of all suicide missions conducted by women.
- *Turkey* – between 1996–1999, the PKK conducted a number of female suicide missions, however the information is incomplete and contradictory (e.g. reliable data has only been found for 7 out of 11 attacks).
- *Chechnya* – between 2000–2004, Chechen Rebels recruited female bombers during the Second Chechen war. In the period between the first and the last female attack (6 June 2000–1 September 2004) 39 women died.
- *Palestine* – between 2002–2006, three radical organizations (the Palestinian Islamic Jihad, Al Aqsa Martyrs Brigades [the suicide wing of Al Fatah], and most recently, Hamas) resorted to female suicide bombing campaigns.
- *Iraq* – (Al Qaeda) according to the Extremis Project, a global platform for evidence-based analysis of extremism and terrorism, approximately 25 percent of all attacks perpetrated by women have occurred in Iraq (Davis, 2013)
- *Dagestan* – (Dagestan insurgency) used relatively small number of women for suicide missions.
- *Uzbekistan* – despite difficulties in obtaining reliable information, many authors consider that female bombers are used by the Islamic Movement of Uzbekistan (Bowers, Musayev and Samson, 2005)
- *China* – Uighur separatists in Xinjiang have used female suicide bombers since 2008.

From these examples it is difficult to find a common denominator in female suicide terrorism and this presents the question of whether a universal explanation might exist. In an attempt to provide an answer, this chapter concentrates on three cases within different contexts:

- *Turkey* – where a left-oriented secular nationalist party promoted the idea of autonomy, supported by part of the Kurdish community;

- *Chechnya* – when the conflict was focused on independence from Russia, but later, due to strategic reasons of the ruling elite and penetration of foreign fighters, transformed into a religious campaign;
- *Dagestan* – where most of the population does not support the Islamists' quest for secession from Russia.

While female suicide missions often provoke discussion based on whether they are terrorist or military operations, I would like to give a definition of suicide terrorism that will provide the basis for understanding throughout this chapter. Suicide terrorism is a deliberate, conscious, responsible and informed-choice action, aimed at civil or military staff or facilities, in non-war contexts with a view to modifying the behavior of a target, and where the death of the perpetrator is a precondition for the success of the operation.

Culturalist and Rationalist Explanatory Frameworks

What then are the driving forces behind female suicide terrorists? Culturalist and rationalist frameworks may offer some insights. The first concentrates mainly on culture, religion and identity. It sees culture as providing resources needed for strategic action (Emirbayer and Goodwin, 1994; Swidler, 1986) that directs researchers toward discursive and performative practices that frame suicide attacks as opportunities for unparalleled heroism, religious devotion, and personal redemption and frame acts of self-sacrifice persuasively as rational, legitimate, and necessary means to achieve desired ends (Hafez, 2006). Culturalist theories also identify the role of religion in politics and political violence. As a result, a range of expressions emerge in the literature, such as: religious terrorism, sacred terror, terror in the name of God, religiously motivated terrorism, religiously inspired terrorism, etc. For our purposes here, 'religiously legitimated terrorism' provides a basis for understanding that even when motivation is purely political, religion is widely used as a symbolic resource for legitimizing violence.

Identity and identification play a central role in culturalist theories. In this chapter, I accept the view that identity has a crucial importance for the coherence of human personality; that it is an important resource helping us to successfully situate factors within a complex world to escape uncertainty (Bauman, 2001). One of the principal questions stated by followers and adversaries of rationalist theories alike is whether martyrdom in general can be explained as a rational choice (Ferrero, 2006). Can an action be rational if it leads to the death of the perpetrator? Is there something more valuable than one's own life? Could such a sacrifice be self-interested? Are suicide terrorists fanatics, mentally unstable individuals or rational players? These discussion points across academic debate provide a variety of different responses (see Chapters 4 and 15). To a great extent they depend on how rationality is defined (Caplan, 2006). Here I accept that rationality should be understood in a broader sense and redefined as 'self-fulfillment' (Gandhi, 2005).

From this perspective, within some belief systems, there is "nothing irrational about the calculated suicide" and with regard to extremists, whilst "their goals may be different than those of most of us" they can be considered completely rational (Wintrobe, 2006: 16).

Are culturalist and rationalist theories incompatible in explaining female suicide terrorism? It can be argued that they are rather complementary. Although female choice has to do with gender identities, as well as with the restrictions and sanctions imposed on women by traditional societies, it is completely rational in so far as it is directed by the achievement of a specific goal: keeping and reaffirming female identities tolerated by patriarchy.

In relation to patriarchy, I use the term in a slightly different sense to denominate traditional societies with strict and fixed gender roles as well as encapsulated traditional communities within modern societies. In such societies and communities, women become part of what might be considered a 'sex-class system' based on the relation of reproduction and supported by men's ownership and control of women's reproductive power. Within this perspective, this idea may provide insight to the question of why women from Western origins might engage in suicidal activities: their motivation cannot be explained as rooted in the failed process of integration in that they were born into Western cultures. Patriarchy can be used to mean power-based relations based on the dominance of the patriarchs over members of that society, be they men or women.

Methodology

The main hypothesis of this chapter is that despite many different contexts, female suicide terrorism is likely to appear in societies that share at least five characteristics, where:

- most of the population share a perception of acute ethnic conflict or foreign invasion (which compromises or endangers ethnic or national identity);
- there are radical organizations that have chosen female suicide terrorism as a means of political action and mobilization;
- there are religious or cultural norms restricting women's access to the public sphere;
- structures of traditional society are in the process of disintegration (sometimes as a consequence of conflict);
- the process of disintegration changes the nature of authorities, values and priorities.

To explore these ideas, data from Chechnya, Dagestan and Turkey were analyzed using the relevant conflicts as an immediate social context of female suicide terrorism. The data were compared in order to identify the potential for any universal explanations. Another method used was the reconstruction of life stories, made on the basis of published material in the public domain (e.g. media reports, books, internet sources, or databases). Most open source material was written in English (although some references for Chechnya and Dagestan were in Russian).

Perception of Acute Ethnic Conflict

In any modern or traditional society a perception of, or actual, acute ethnic conflict leads to political mobilization and action for the defense of national and individual identity, values and ways of life. In patriarchal societies, such mobilization can result in more rigid gender systems and policing of such gender issues.

Female suicide terrorism in Turkey should be treated in the context of the so-called Kurdish question, which is probably the most difficult and irresolvable in the 90-year history of the Turkish republic. Radical Kurdish nationalism evolved with the Kurdish Workers' Party that was set up in 1977. The PKK started military operations in 1984, and began recruiting women for suicide bombing missions in 1996 when they lost the strategic initiative in military operations as well as diminished support from the Kurdish population (Burton and Stewart, 2007). As a result, 21 suicide operations were organized: of these, 15 were successful, and 11 of those were carried out by women. These attacks killed 19 and wounded 138 (Ergil, 2000). The PKK based these attacks on traditional Marxist ideology about gender equality, however it recognized the established gender system was based on masculine power, domination and prepotency. In other words, the transfer of the authority of the oldest man in the traditional family, whose word is a law for female members, to the political leader in a radical party could be one of the sources of the unconditional loyalty that some women demonstrated for the PKK in their readiness to sacrifice themselves for the cause.

Female suicide terrorism in Chechnya appeared during the Second Chechen War (it started on 30 September 1999, when Russian troops invaded Chechnya). During this stage, the character of the conflict evolved from a struggle for independence to a religious clash. However, Chechnya's turn to Wahhabism, fanned by Arab mercenaries, appeared to be more the product of political compromise to secure funding than of a commitment to Islamic belief (Ness, 2005). Nevertheless, this transformation was used by the new ruling elite to impose a more rigid order on the traditional gender-based system. Today many Chechen women cover their faces and heads and are victims of honour killings even though that has never been part of Chechen tradition (Berry, 2009).

After the end of the Second Chechen War, radical Islamist sects achieved stable positions in several regions of Northern Caucasus, including Dagestan. What is particularly interesting to note is that in Dagestan, suicide terrorist missions were neither so numerous nor so spectacular as those that took place in Chechnya. The percentage is relatively low and a relatively small number of people were killed or injured. It is probably due to the fact that suicide terrorism was not inspired by nationalist motives because national independence and emancipation from Russia was not strongly supported by local public opinion. As Dibirov stated, an important characteristic of Dagestani terrorism was the lack of an anti-Russian rhetoric. Instead, the economic, social, political and spiritual situation, the past and the present of Dagestani society are the causes of terrorism in the republic (Dibirov, 2006).

Radical Political Actors

Of the cases analyzed, whatever the perception of conflict, in all but one example none of the female suicide attacks were provoked directly by sentiments of indignation, revenge and anger. This only happened when radical groups and organizations appeared on local political stages and transformed personal sentiments into political action through training and indoctrination. The socialization of the norms of the 'subculture of the death' in most cases took place inside the relevant radical groups and parties. They recruited, trained and prepared female perpetrators for specific operations and often did it using a recruiter who was a close male family member.

In one example, Leila Kaplan was 17 and the youngest of the PKK female perpetrators. Leila was a daughter of a traditional patriarchal family and had seven brothers and sisters. She was arrested several times for assisting insurgents and at the time of the mission her husband was in prison, sentenced for terrorism. After his detention she joined PKK and took part in actions in rural zones. Her life story shows how the personal feelings of hate and anger were transformed by the organization she joined and were then directed towards purposeful political action. In relation to unsuccessful PKK female suicide attacks, no perpetrator acted spontaneously, following her feelings of grief and anger. All were recruited, held membership for some time and were trained for their missions. Because of lack of reliable information, it cannot be said if the family ties with members of the extremist group impacted their decision.

Of the 30 Chechen perpetrators, 24 had family ties with rebels. For example, the first Chechen female bomber, Hava Barayeva, was trained by her close relative, the field commander Anri Barayev. The recruitment of most women (at least those for whom reliable information has been found) was facilitated by a close male relative, as in the case of Alina Tumrieva who was prepared for the mission by her father. Except for one woman who was self-recruited and another one for whom no information was found, all other perpetrators received training for their missions. Only two out of six unsuccessful perpetrators in Chechnya had family relationships with the rebels. Sveta Tsagaroeva was the wife of a field commander who had been killed by Russians, and the grandmother of Zarema Muzajoeva was a sister of the field commander Shamil Basayev. Almost all (five out of six) were formally recruited by the rebels or by radical Islamist groups acting with them.

In Turkey, the evidence for successful PKK perpetrators, five women (out of seven cases) were recruited for their missions; for the other two no information was found. In six cases evidence was found of training prior to missions. At the same time, in the case of unsuccessful attacks, seven were recruited (voluntarily or formally through PKK networks) and nine were trained (inside or out of their own country). In Chechnya, amongst the unsuccessful perpetrators, five out of six were formally recruited and all were trained prior to their missions. Family bonds appeared to facilitate the recruitment process. In two cases, perpetrators had family relations with PKK (their husbands were active members). No conclusions can be drawn on this variable for the unsuccessful bombers, as no reliable information was found. In Chechnya, two of the unsuccessful bombers had family relations with rebels, while three did not. Amongst the successful perpetrators, 22 had family bonds and in Dagestan, all female bombers – successful and unsuccessful, were recruited and trained and had family bonds with the rebels.

Religious or Cultural Norms Restricting Women's Access to the Public Sphere

Restriction of women's access to the public sphere is another important feature of Kurdish, Chechen and Dagestan society. Several national and international human rights organizations have called the attention of public opinion to this issue.

'Women Living Under Muslim Laws', a human rights organization, describes the situation of Kurdish women within a traditional patriarchal society where a woman's power within the family is secured firmly below that of the man who is the income provider and leader of the household. This devaluation of women, and an orthodox Islamic lifestyle, results in problems such as domestic violence, honour killings and a lack of educational opportunities. In addition, some regions of Turkey are very poor (e.g. the south-eastern region) and in addition to general constraints, women face particular hardships due to their lack of financial independence and reliance on husbands or fathers. Furthermore, from 1984 to 1999, and with intermittent outbursts of conflict ever since, the region has been engulfed in a guerrilla war that has inflicted innumerable costs on the women of the region. These have included the destruction of their homes, the death of their husbands and sons (leaving them without income), or direct abuse by the PKK or Turkish army.

Patriarchy in Chechnya is not fundamentally different to that of Turkey. Despite changes under the communist regime, the status of Chechen women as members of an archaic society was generally lower than the status of women in Soviet society (Liborakina, 1996). A set of cultural prejudices against women's public roles has remained and was reinforced by the recent militarization of society. The decline in the status of women in Chechnya has been enforced ever since. The social infrastructure, healthcare and insurance systems were severely compromised by the Chechen authorities under the connivance of the federal government, and resources were allocated to the armed groups. Women became primarily responsible for the survival of their families and many girls in Chechnya lost the opportunities of formal education (Liborakina, 1996). After the end of the Second Chechen War, the Kadyrov regime and its patriarchal gender role system was even more restrictive such that if a woman and a man committed adultery, both of them would be sentenced to death (Cline, 2009). Kadyrov told journalists that women were the property of their husbands and their main role is to bear children. He encouraged men to take more than one wife, even though polygamy is illegal in Russia. Women and girls are now required to wear headscarves in schools, universities and government offices. Prior to this, there was no history of honour killings and faith-based murders of women in Chechnya, but Kadyrov publicly promoted and defended such actions. Muslim men were being sent a clear signal that the government approved of them treating women as disposable objects that can legitimately be destroyed when they start behaving improperly (Cline, 2009).

For some researchers, the extremely limited access of women to the public sphere provides an explanation for female suicide terrorism as a form of protest against this social injustice. It was a way of illustrating that anyone, male or female, is entitled to die for his/her beliefs and that this activity is an expression of gender equality by highlighting the rights and liberties of women. In this way, women were deliberately emancipating themselves by choosing to be suicide bombers (Israeli, 2004). In the same way, comparative cross-national and longitudinal case studies of female terrorists ascertain that female terrorists in developing societies are similar to early campaigners from the First and Second Waves of feminism. Although feminism is inherently a Western word and many female terrorists do not identify with this terminology, female terrorists are often fighting for political equality and the betterment of their gender (Sixta, 2008). By participating in political activism, which is traditionally reserved for men in their societies, these women are reaching for political equality by breaking the barriers of societal ideas of proper female political participation. Through political violence, women are participating in the political system that is so often closed to them.

Others consider that women's participation in military and political actions, as well as national liberation movements and terrorist organizations, has the potential to destabilize the gender divisions that form the basis of international relations and women's exclusion from it (West, 2004). Furthermore, the actions of female bombers as a result of victimization rather than agency reproduces a dominant femininity based on the masculine/feminine divide (West, 2004). While there are quite a lot of reasons to agree with her and there is a certain attractiveness to this feminist hypothesis, it is difficult to find any reliable proof that female suicide bombers are fighters for gender equality or that acts of female suicide terrorism enlarge the freedom of women in respective societies. Put simply, while joining radical organizations might give women a choice in how they die, it does not necessarily give them a choice in how to live.

Structures and Authorities of Traditional Society in Disintegration

Despite its restrictive character, the rigid gender roles in society offer some protection to women in a patriarchy. However, this may not be so true for societies where traditional structures are in a process of disintegration. In Turkey, in the eastern part of the country populated by Kurds, what can be seen is an agglutination and mutual penetration of traditional structures, many of which are in a process of disintegration. Economic constraints and financial dependence of women on their husbands are combined with new responsibilities, to reconstruct their homes and cope with the death of men in their families, relocating to larger cities such as Diyarbakir without the family or communal ties upon which traditional Kurdish villages are structured.

In Chechnya there is also a disintegration of traditional social organization. With the weakening of the Khel (council) that embodied the executive, a legislature, and a judiciary power lost its authority and significance. With this, it lost its capacity to impose sanctions against the lack of respect for traditional laws and moral codes. The long-lasting war caused a real gender crisis in Chechnya with serious consequences both for men and women. While not being able to fulfill their roles as defenders and winners, men became 'social elders', enjoying respect but without any real authority. This provided Chechen women with new and more dominant roles in the family and public life that further contributed to the loss of male dominance in society. As a result, women assumed new moral and physical duties (Kitaev-Smik, 2009). However, at the same time, Chechen women had no new rights. What emerged was a more rigid framework of gender roles leading to the establishment of punitive practices for women that had not been part of Chechen life before.

In Dagestan, both political and demographic conditions urged women to adopt new social roles. With a mortality rate amongst men 42 times higher than women (amongst those of an economically active age) the survival of families became the duty of women (Dronzina, 2010). But as with the other examples, the new responsibilities for women that emerged did not underpin any new recognition of their social identity and did not lead to substantial changes in the traditional gender roles system.

Individual Motivations

The personal motivation of suicide terrorists has been the focus of a lot of attention in academic debate as well as in the media. If neither religion nor ideology can explain female suicide terrorism as practiced by organizations using religion and secular ideology to legitimate violence, what then is the driving force for women who decide to conduct such activities? The hypothesis presented here is that this has to do with female identity and its specific conditions under patriarchy.

To understand this, a group of variables called 'losses, traumas and cultural norms' can provide an insight into the identity of women and its impact on the individual choice. The three variables were explored to identify how losses caused by the conflict (loss of beloved person – parents, close relatives, spouses), traumatic events not caused by a conflict (loss of relatives, parents, spouses not killed in the war, replacement, restriction of the right to movement) and sins (understood as actions or lack of actions that are not tolerated by the community or the wider society and are likely to put women in situation of social stigma) might motivate individual behavior.

In Chechnya, one of the unsuccessful bombers had lost her husband, a famous field commander, to whom she was deeply attached; another lost 38 relatives but did not detonate her explosive device. In the case of the successful perpetrators, 23 out of 30 had lost a close relative. As expected, such losses caused by the war can impact on individual choice, but they do not completely explain individual motivations. Almost all Chechens lost family members in the conflict and more than the half the population lost their homes (Bondarenko, 2003) but not all of these people became suicide bombers. It is reasonable to assume, therefore, that such losses are not the only catalysts of such extreme behaviors, but perhaps the way the losses were experienced and overcome is a factor. In relation to the women of Dagestan, all of the perpetrators had suffered losses of a close relative caused by the conflict. Maryam Sharipova's husband was killed and her brothers had been persecuted for relations with insurgents; Jennet (Jannet) Abdurahmanova's and Zaira Akaeva's husbands were eliminated; Zalina Akaeva suffered the death of her sister's husband.

Traumatic events, however, can be related not only to violent conflicts but also to other life circumstances. Could these have a negative impact on the identity and motivation of individuals? The answer is positive. Four of the unsuccessful perpetrators had suffered some individual trauma not caused by the war. Mareta Dudueva, who was a minor at the moment of her mission, had problems with her parents; the father of Zarema Inarkaeva died in 1994; Zerema Mujajoeva was abandoned by her mother when she was a small child, her father and husband were killed in a drunken brawl; Luisa Asmaeva was maltreated in a training camp. In Dagestan, Jennet Abdurahmanova also suffered a personal trauma as she was abandoned by her father.

In Chechnya, four out of six unsuccessful perpetrators committed some form of sin. Mareta Dudueva started practicing extra-matrimonial relationships at the age of 12; at 13 years old, Zarema Inarkaeva was raped by a man whom she did not marry; Zarema Mujahoeva conducted a robbery and was then prohibited from meeting her small daughter. Some of the perpetrators became victims of social stigma because they had committed sins. In Chechnya, the Hadjievi sisters did not get married because of illness; Aset Gishlurkaeva was divorced and married for a second time; Zulijan Elijadjieva maintained sexual relationships with a close relative; Marem Sharipova aborted a pregnancy.

Unfortunately, in the available data on female PKK suicide terrorists, little information on losses, traumas, and 'sins' was found. The National Police Department's report concluded that the main reasons women joined the militia arm of the terrorist organization was family pressure, having been married as a child, and to escape honor killings. I accept the conclusion of Mia Bloom that elsewhere in the world, sexual violence against women and the ensuing social stigma associated with rape in patriarchal societies, appears to be a common motivating factor for suicide attackers (Bloom, 2005). Kurdish women allegedly raped in Turkey by the military have joined the PKK, while Tamil women allegedly raped by the Sinhalese security services and military join the LTTE.

Discussion

In relation to the exploration of patriarchy as a basis for female suicide terrorism there would appear to be a number of supporting arguments. Perhaps contrary to the conventional wisdom, an armed conflict in a traditional society does not always lead to modernization in general, and of the gender roles system in particular. It can reduce the space for renegotiation of gender identities and the opportunities for the women to decide who they want to be.

What national conflicts can cause is the transformation and disintegration of structures of patriarchy rather than a modernization of female roles. One of the by-products of these processes is the weakening of traditional norms that prevent women from taking non-traditional roles, providing them with different opportunities to participate. It is almost impossible to imagine female suicide operations in a fundamentalist society (i.e. one that strictly follows patriarchal norms and prescriptions where there is no space, neither physical nor intellectual, for a female suicide act). Structures of traditional society in disintegration create opportunities for women to assume new responsibilities, but not always to achieve new rights. This specific condition obliges them to play new roles but does not give them new recognition. Traditional markers of feminine and masculine fields of action exist, but do not function. Under disintegration of traditional structures, the borders between male and female roles become permeable, but inside them the rules have not changed. More recently, this has provided a basis for Kurdish, Chechen and Dagestan women to be converted as female suicide terrorists because there is no sanction or norm or authority that prohibits such abuse or self-abuse.

Disintegration of patriarchal structures causes a deep crisis of two main authorities of traditional society: authority of age and authority of men. Typically, the elder men, embodying wisdom, moderation and experience, are substituted often by a younger military elite (e.g. field commanders) and the traditional authority of such elders is replaced by that of leaders. If men are restricted by the customary law, patriarchal tradition and customs (including attitudes and actions to women), the leader is guided by success in the political competition perhaps even when it is achieved at the expense of women who take up roles as suicide terrorists.

When these traditional patriarchal structures disintegrate, radical organizations promote female suicide terrorism as an acceptable and tolerable act. In so far as these organizations are the producers as well as the sponsors of female suicide missions, what the community learns from them about bombers are their versions and their interpretations (and not necessarily those of the women themselves). This might mean a deep alienation between personal motivation of the perpetrator and the social meaning of their acts. This is a strong argument against a feminist hypothesis for understanding female suicide terrorism as a quest for equality and access to the public sphere. The organizations (headed by men) who employ women to become suicide terrorists are not interested in gender equality. Thus, it does not seem likely that women terrorists are reconstituting conventional power relations between men and women by taking on non-traditional tasks (Knight and Narozhna, 2005). It could be argued that the opposite is true, through acts of suicide terrorism women are still confined to the gender roles that are prescribed and tolerated by patriarchy gender identities.

For the most part our identity prescribes how we are expected to live and it may also define how we should die. In particular circumstances an identity may be made more valuable by death and devalued or completely destroyed by continuing to live (Harrison, 2006). From this perspective, there emerges an explanation for female suicide terrorism as a rational act. Where suicide is chosen it minimizes the damage (e.g. the punishment and the stigma) that a woman will suffer for violating other patriarchal norms. But a rational choice, as part of minimizing the damage, also presupposes maximizing the utility. The best outcome in this case would be to change the living status of a stigmatized woman for the post-mortem status of role model or heroine who sacrifices her life for her patria or community. As far as who determines when such an act might be executed, it is usually a man who defines the time and target of the mission, and how it is then publicized and communicated back to the community.

Conclusion

What is offered here is a social commentary on the phenomenon of female suicide terrorism. As with any exploration of this kind, it has limitations. Fundamental to this kind of research is the basis and availability of empirical data. While is possible to find relatively detailed information for Chechen and Dagestan cases, this was not the case for Kurdish female terrorists. Moreover, data are often filtered by radical organizations, security services and the media. Finally, in the case of successful perpetrators, the reliability of empirical material is always questioned, because there are often limited opportunities to verify it. Nevertheless, with these limitations, it is important to analyze and evaluate female suicide terrorism as a growing and dangerous phenomenon even under conditions of incomplete information.

References

Bauman, Z. (2001). Identity in the globalizing world. *Social Anthropology*, 9(2), 121–129.

Berry, L. (2009). Chechen President Kadyrov Defends Honor Killings. *The St. Peterburg Times*, 3 March, http://www.sptimes.ru/index.php?story_id=28409&action_id=2, accessed 14 August 2014.

Bloom, M. (2005). Mother. Doubter. Sister. Bomber. *Bulletin of the Atomic Scientists*, November/December, 57.

Bondarenko, M. (2003). More than a half of Chechen women insist the troops to leave. But only 1 out of 5 thinks her motherland will be independent. *Independent Newspaper*, (Бондаренко, М. 2003. Больше половины чеченских женщин – за вывод войск. Но только каждая пятидесятая видит свою республику независимой. Независимая Газета, http://www.ng.ru/caucasus/2003-01-28/5_woman.html, accessed 14 August 2014.

Bowers, S.R., Musayev, B., and Samson, S.A. (2006). The Islamic threat to Eastern Central Asia. *Journal of Social, Political, and Economic Studies*, 31, 375–397.

Burton, F., and Stewart, S. (2007). *On the Cusp: The Next Wave of Female Suicide Bombers?* Stratfor Global Intelligence database, http://www.stratfor.com/cusp_next_wave_female_suicide_bombers, accessed 14 August 2014.

Caplan, B. (2006). Terrorism: The relevance of the rational choice model. *Public Choice*, 128, 91–107.

Cline, A. (2009). *Muslim Chechnya: 'Loose' Women Deserve to Die.* http://atheism.about.com/b/2009/03/12/muslim-chechnya-loose-women-deserve-to-die.htm.

Davis, J. (2013). *Female Suicide Bombers in Iraq*. Extremis Project. http://extremisproject.org/territory/middle-east/, accessed 14 August 2014.

Dibirov, A. (2006). Terrorism: Whose guilt and what to do? *Dagestanskaia Pravda*, N 231 (Дибиров, А. 2006. Терроризм – кто виноват и что делать? Дагестанская правда, № 231), http://dlib.eastview.com/browse/doc/10057633, accessed 10 February 2012.

Dronzina, T. (2008). *Female Suicide Terrorism*. Sofia, Bulgaria: Publishing House of Bulgarian Army (Дронзина, Т. 2008. Женският самоубийствен тероризъм. Военно издателство, София).

Dronzina, T. (2010). Female suicidal terrorism in Chechnya. In D. Antonius, A.D. Brown, T.K. Walters, J.M. Ramirez and S.J. Sinclair (Eds), *Interdisciplinary Analyses of Terrorism and Political Aggression*. Cambridge: Cambridge Scholars Publishing, 203–239.

Emirbayer, M., and Goodwin, J. (1994). Network analysis, culture, and the problem of agency. *The American Journal of Sociology*, 99(6), 1411–1454.

Ergil, D. (2000). *Countering Suicide Terrorism*. Lecture at ICT Conference. http://www.ict.org.il.

Ferrero, M. (2006). Martyrdom contracts. *Journal of Conflict Resolution*, 50, 855–877.

Ghandi, D. (2005). *Rational Choice Theory in Political Science: Mathematically Rigorous but Flawed in Implementation*. Normal, Illinois: Illinois State University.

Hafez, M. (2006). Rationality, culture and structure in the making of suicide bombers: A preliminary theoretical synthesis and illustrative case study. *Studies in Conflict and Terrorism*, 29(2), 165–185.

Harrison, M. (2006). An economist looks at suicide terrorism. *World Economics*, 7(4), 1–15.

Israeli, R. (2004). Palestinian women: The quest for a voice in the public square through 'Islamikaze martyrdom'. *Terrorism and Political Violence*, 16(1), 66–96.

Kitaev-Smyk, L. (2009). Gender crisis after many years of war in Chechnya. (Китаев-Смык Л. 2006. Гендерный кризис после многолетней войны в Чечне.), http://www.kitaev-smyk.ru/node/6, accessed 14 August 2014.

Knight, W., and Narozhna, T. (2005). Social contagion and the female face of terror: New trends in the culture of political violence. *Canadian Foreign Policy Journal*, 12(1), 141–166.

Liborakina, M. (1996). Women and the War in Chechnya. *Women plus*, http://www.owl.ru/eng/womplus/1996/gender.htm, accessed 20 August 2014.

Ness, C. (2005). In the name of the cause: Women's work in secular and religious terrorism. *Studies in Conflict and Terrorism*, 28(5), 353–373.

Sixta, C. (2008). The illusive third wave: Are female terrorists the new 'new women' in developing societies? *Journal of Women, Politics & Policy*, 29(2), 261–288.

Swidler, A. (1986). Culture in action: Symbols and strategies. *American Sociological Review*, 51(2), 273–286.

Victor, B. (2004). Equality in death. *The Observer*, 25 April. http://www.countercurrents.org/gender-victor250404.htm, accessed 14 August 2014.

West, J. (2004). Feminist ir and the case of the 'black widows': Reproducing gendered divisions. *Innovations, The Journal of Politics*, 5, 1–16.

Wintrobe, R. (2006). Extremism, suicide terror and authoritarianism, *Public Choice*, 128, 169–195.

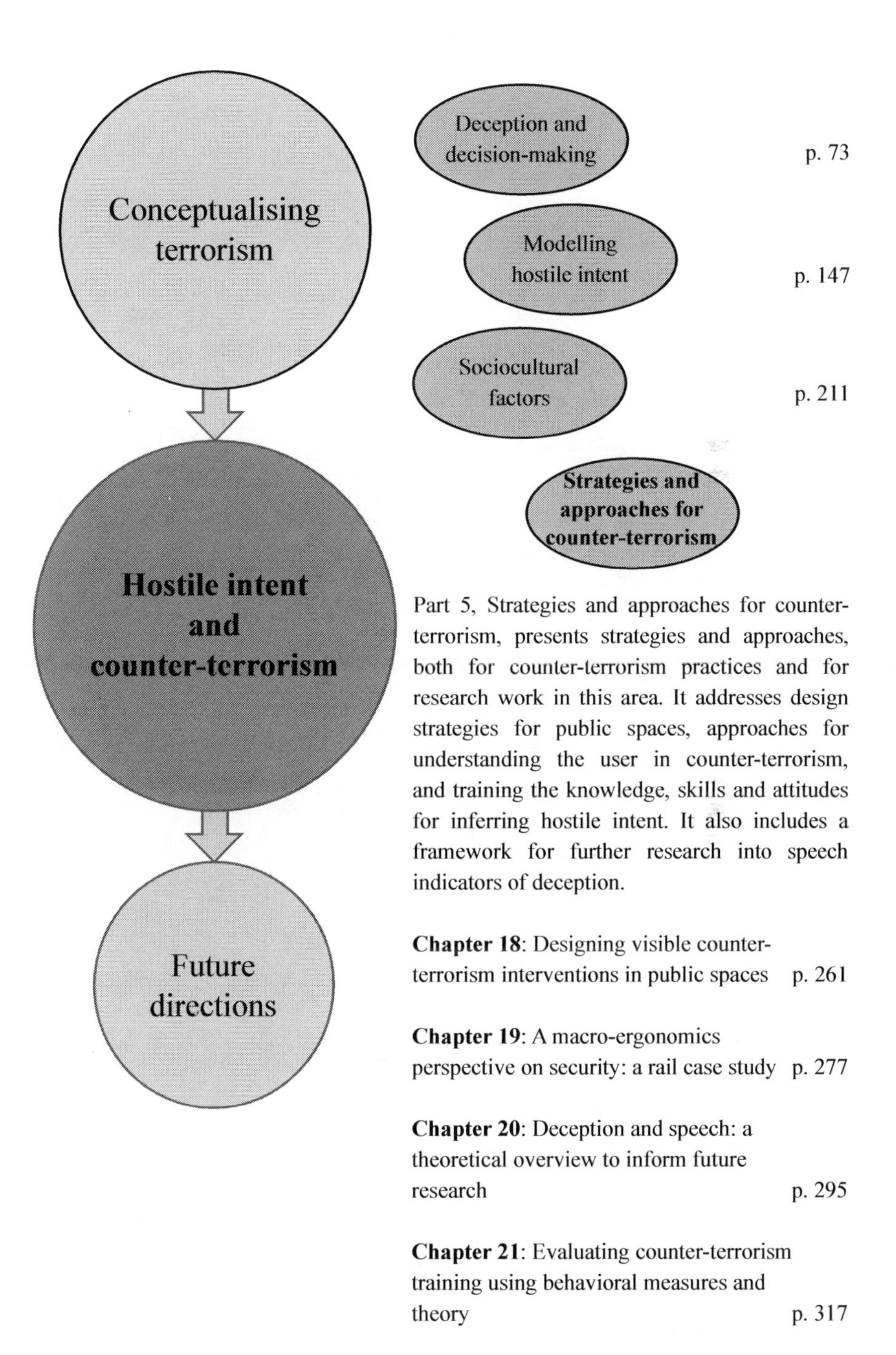

Part 5, Strategies and approaches for counter-terrorism, presents strategies and approaches, both for counter-terrorism practices and for research work in this area. It addresses design strategies for public spaces, approaches for understanding the user in counter-terrorism, and training the knowledge, skills and attitudes for inferring hostile intent. It also includes a framework for further research into speech indicators of deception.

Chapter 18
Designing Visible Counter-terrorism Interventions in Public Spaces

Ben Dalton
Faculty of Art Environment and Technology, Leeds Metropolitan University, UK

Karen Martin
CASE, Kent School of Architecture, UK

Claire McAndrew
The Bartlett School of Graduate Studies, University College London, UK

Marialena Nikolopoulou
CASE, Kent School of Architecture, UK

Teal Triggs
School of Communication, Royal College of Art, UK

Introduction

The setting is New York City where British counter-terrorism expert, Henry Moore, ponders the problem of urban security with a roomful of architects. He asks: 'So how do you think we could reduce the risk?' (Waldman, 2011, p. 53). Thus begins the discussion of safety and fear in public spaces in Amy Waldman's bestselling novel *The Submission*. The notion of 'design against terrorism' has entered into mainstream consciousness.

This chapter explores publicly visible counter-terrorism measures – uncovering the strategic role of design in creating controlled disruption in public spaces to reduce threat while at the same time reducing anxiety. Evidence of counter-terrorism security design is now essential in the planning process and projects will need to demonstrate how such issues have been addressed (Royal Institute of British Architects, 2010). This emphasis on design is highlighted in a recent Home Office report (2012) which provides advice on how to integrate such measures at different stages, from conception to development so that 'vulnerability of crowded places to terrorist attack can be tackled in an imaginative and considered way' (Home Office, 2012, p. 3). This last point is critical if we are to develop and manage public spaces in way that will not have a detrimental effect on the quality of the public realm, but will be socially responsive, enhancing a sense of vitality and well-being.

Initiatives of civic scale counter-terrorism activity are focused on critical infrastructure, crowded spaces and symbolic targets, referring to ports of entry, transport systems, shopping areas, staged events and tourist destinations which also include historical monuments. We can think of publicly salient aspects of counter-terrorism, whether of permanent or temporary nature, as communication design for two overlapping audiences. One audience is quotidian users of a space, those passing through or spending time in a location. The second audience of counter-terrorism communication design is hostile actors, those planning or attempting to carry out a terrorist attack.[1]

Pre-emptive communication strategies such as the deployment of armed police guards or controlled access points (Adey, 2004; Benton-Short, 2007) often have a multiplicity of purposes. Such interventions seek to simultaneously communicate a sense of protection and reassurance to the audience of users of a space and encourage vigilance from them, while also being designed to disrupt anyone who may be engaged in covert activities and elicit noticeable behaviour (Coaffee et al., 2008; Németh, 2010). Some counter-terrorism design is also intended to prepare people for emergencies by making them aware of communication systems and expected responses (O'Connor, Bord and Fisher, 1999).

Unfortunately, too frequently counter-terrorism design interventions have resulted in ad hoc solutions focusing on extensive disruption of covert activities, with little consideration for the resulting public realm, openness of the space and social interaction. As a result, the proliferation of security bollards and barriers has caused detrimental impact to the quality of the space in many urban environments (FEMA 430, 2007), leading to what the National Capital Urban Design and Security Plan for Washington calls the 'present intolerable environment' (NCPC, 2002).

Here we refer to both aspects of counter-terrorism communication design, that is addressing specific threats as well as the wider public, as mechanisms of disruption. By disruption we mean an intervention to abruptly interrupt routine, attention or expectation. In the case of communication to hostile actors, the aim is to disrupt reconnaissance, preparation activity or the attempts by an individual or group to hide their intentions. These interventions can have different forms, from performative demonstrations of security aimed at disrupting pre-attack activity, to provoking identifiable reactions from those conducting reconnaissance or deterring those planning an attack from choosing a particular site (Németh, 2010). Such public 'security performances' also affect the wider audience of users in the space, often trying to elicit behaviours of heightened awareness and observation, or of compliance. Activities may be aimed at engaging the public in aiding the

1 It is worth noting that there is a third potential cause for counter-terrorism activity to be communicated, and that is through leakage of visible aspects of security protocols and infrastructure not intended to be displayed.

detection of out of place behaviour that could indicate the planning or carrying out of an attack, or aimed at fostering a sense of personal responsibility for security (Coaffee et al., 2008; Fussey, 2007). While primarily addressed to the general population of a space, these activities may also be a statement to those planning an attack that such a space is under active and wide-reaching attention. It thus becomes clear that the mechanisms of disruption are framed with a number of objectives including dissuading or drawing attention to those conducting reconnaissance and dry runs, and motivating other users of these spaces to assist in identifying and reporting behaviour of concern.

A system consisting solely of disruption, constantly evolving changes and interactions runs the risk of becoming not only commonplace and ineffective, but also a tiring drain on the attention of regular users of a space. How then can visible aspects of security and counter-terrorism in public spaces be designed in an imaginative and considered way, applying effective uses of disruption while being socially responsive, fostering corresponding routine and comfort?

This chapter responds to this question, offering insights from two projects funded under the Research Councils UK 'Global Uncertainties' programme (see 'Security for All in a Changing World' one of six programmes supporting leading-edge research on significant global challenges, for more information http://www.globaluncertainties.org.uk/about/). Shades of Grey – Towards a Science of Interventions for Eliciting and Detecting Notable Behaviours (2010/13) sought to explore the relationship between environmental and interpersonal stimuli and behavioural responses in public spaces. Safer Spaces: Communication Design for Counter Terror (2008/09) examined the potential of creative approaches to reduce fear and re-engage awareness in public spaces. What these projects share are design-led insights that offer visual strategies to countering terrorism via choreographed disruptions and communication in public spaces. Conclusions from both projects recognise the importance of socially acceptable design contexts such as functionality, creativity and playfulness over authority-focused interventions for the sustainability of counter-terrorism activity.

A compendium for these critical insights, this chapter describes the desired effects of disruption in publicly visible counter-terrorism using the concepts of (i) triangulation, (ii) performance and (iii) flow. The sustainability of disruption-based intervention is reviewed by considering the issues of attention, fatigue and the disruption of civil inattention. Design for routine and usability in public space is reviewed as a vital counterpoint to disruption through the principles of foresight and communication. Implications for the broad spectrum of stakeholders implementing principles of disruption and routine are also considered. This chapter includes key examples drawn from a review of art as a means of disruption from the Shades of Grey project, and design findings from the Safer Spaces project of how one can design for disruption and routine with these principles in mind.

The Effects of Disruption in Publicly Visible Counter-terrorism

Guided by insights from Safer Spaces, this section reviews the attributes of triangulation, performance and flow described by Martin, Dalton and Nikolopoulou (2013) as currently used mechanisms for controlled disruption in public spaces. Exploring such effects frames advice on the strategic role of design later in this chapter.

Triangulation

Creating places where the unusual will be noticed and talked about is often a design goal for managers of spaces and security staff. For example, in the UK a joint partnership between police and private sector security, Project Griffin, cites a key objective to 'empower people to report suspicious activity and behaviour' (City of London Police, 2004). The challenge to designing for attention is that people employ strategies of 'civil inattention' to manage routine co-presence in public spaces that are characterised by the acknowledgement of others, followed by deliberate minimisation of contact (Goffman, 1966). Martin et al. (2013) apply the urban planning concept of 'triangulation' (Whyte, 1988) to describe the use, in counter-terrorism strategies, of disruption of civil inattention as a means of assisting in informal surveillance and reporting. Encouraging members of the public to increase their vigilance and report suspicions recurs regularly as a recommendation in counter-terrorism guidelines (Coaffee et al., 2008).

Design for what Hillier (2004) calls 'natural policing' consists of two aspects, increasing opportunities for observation and shifting a greater proportion of focus from members of the public on to observing their surroundings. Considerations of sight lines and vantage points for observation build on the field of crime prevention through environmental design by extending concepts of 'natural surveillance' from Jacobs (1961) and Newman (1972) to the context of counter-terrorism. Shifting public attention to suspicious activity has been driven by communication design techniques (Triggs and McAndrew, 2009). The US Department of Homeland Security (2010) 'If You See Something, Say Something' poster campaign offers an example of the sorts of visual communication strategy security staff have historically employed to try to heighten public attention to the unusual activity.

Triangulation can be understood as a temporary disruption of a state of civil inattention (Whyte, 1988). It is a process by which some external stimulus creates a stimulus for social interaction that prompts strangers to talk to each other. In counter-terrorism strategy, the external stimulus is the unusual activity of a potential hostile actor, or an unexpected item in the public space. Triangulation offers a framework to understand existing counter-terrorism strategies, such as signage, announcements and security as attempts to stimulate linkages between users of a space and staff who are able to respond. Taking public transport systems as its case study, Safer Spaces experimented with disruption as a means to re-engage members of the public in the protection of civic space in a non-fearful manner.

Prototyping the transformation of digital screens into mirrors (both filming the person facing it and displaying live feeds of visual activity streamed from other proximal but distinct locations) offered a form of visual disruption that connected localities, reduced anxiety through foresight (a sense of what lies ahead) and facilitated 'peer-to-peer monitoring' via the designed object (McAndrew, 2012; Triggs and McAndrew, 2009). Experimenting with the form of visual disruption through variations of scale (stable versus motion-responsive dilation) and image clarity (including mirror-representations, laplacian filters and sensor-dependent coloured filters) mediated the spectrum of responses to disruption as 'security' and/or 'play' (see the section on performance for further elaboration on the role of play).

Techniques of interaction and reflection prevalent in art are well suited to the objectives of triangulation. Technology in interactive art can in some cases allow for the disruptive effects of sculpture to be temporarily added to a space comparatively cheaply and quickly. For example, the video installation Body Movies by Rafael Lozano-Hemmer (2001) has been used in public spaces around the world. It uses floodlighting and projection to elicit participation and unexpected intimacy in night-time public spaces. In this artwork passers-by project their shadows high onto the walls with photography portraits being revealed inside these dark silhouettes. Snibbe and Raffle (2009) observe that people quickly recognise their own shadows and those of people they know. This suggests that shadows and silhouettes have the ability to provide identifiable representations while preserving an individual's sense of anonymity in the crowd.

Performance

Publicly visible counter-terrorism interventions are often intended to evoke performances from people in a space (Edwards, 2010). We can consider these performances as means to either elicit behavioural cues that help security staff identify people who have something to hide or deter unknown hostile actors who are unwilling to perform in public for fear of giving themselves away. Initiatives such as BASS (Behaviour Analysis Screening System) in the UK, and SPOT (Securing Passengers by Observation Techniques) in the US, train security staff to try to identify suspicious behaviour and body language in response to temporary, high-visibility disruption of spaces such as temporary cordons and high visibility patrols (Coaffee et al., 2008).

Martin, Dalton and Nikolopoulou (2013) give the example of the UK Protecting Crowded Places guidelines which include discussion of the role a visible security regime can play to 'deter, detect and delay' suspicious terrorist activity including hostile reconnaissance (Home Office, 2012). The guide is aimed at professionals involved in design of the built environment. It describes a case study example of an archway metal detector at a crowded venue, with staff in high-visibility jackets and signs highlighting partnership with the police as 'a potential deterrent to a large amount of criminal activity, including hostile reconnaissance' (Home Office, 2012).

While the potential impact of effective searches and metal detectors on concealed weapons is clear, the effect a 'pinch point' arch and high-visibility activity have on deterring hostile reconnaissance is less well defined. We can imagine this form of disruption could raise the stakes for an individual trying to maintain a constructed inconspicuous identity or heightening the risk of revealing concealed reconnaissance materials like a specialist map or hidden camera.

The queue and search routines of the archway metal detector require pedestrians to 'perform' in public view. For someone conducting hostile reconnaissance a performance of normality at this pinch point of scrutiny might evoke fears that their covert intentions will be found out. To understand this aspect of disruption to elicit performance we can consider the pressure to appear normal in the context of the literature on lying and deception. De Paulo et al. (2003) suggest that liars and truth-tellers all share the same goal of trying to appear honest. Granhag et al. (2004) argue that liars awaiting interview tend to compensate by planning their responses in greater detail than truth-tellers do. We should therefore expect performances of fictional honesty to demand a greater cognitive load and differ noticeably to unexpected questioning. Vrij et al. (2009) showed that when liars are presented with unanticipated questions, such as spatial or reverse chronological, they can noticeably struggle to answer them. This offers us a framework to understand unanticipated visible counter-terrorism interventions that elicit public performance as attempts to evoke physical 'unexpected questioning'.

In the context of deterring hostile reconnaissance, we are interested in design and art that can create physical examples of unexpected questioning through playful disruption or a heightened sense of being on view. The design concepts of interaction and reflection found in mirrors and shadows are not only useful for stimulating triangulation but can also be effective at encouraging performance from audiences. The interventions developed through the Safer Spaces project use the playfulness of video mirrors to entice public involvement and simultaneously enhance the sense of being on view. One participant in the study focus group commented:

> You know, it's a good thing if you are looking at it and it is looking arty … It's something you enjoy. But, it's also … it's like a dual-purpose thing. It's doing the aspect of working as a security thing, but you're also seeing different people.

O'Shea's (2009) intervention Hand From Above similarly appropriates surveillance technologies. In this artwork O'Shea interrupts routine behaviour by playing on people's tacit awareness of the ubiquity of video surveillance. Installed on the BBC's Big Screens, Hand From Above appears to show a real-time video feed of the space in which the screen is located, however, at intervals, a large hand appears on the screen and picks up, or tickles, one of the passers-by, so lifting them out of reality for a few seconds. The artist highlights how Hand From Above 'encourages us to question our normal routine when we often find ourselves rushing from one destination to another. … Passers by will be playfully transformed' (O'Shea, 2009).

Flow

With extensive public surveillance increasingly being described as forensic rather than deterring, there are concerns about the potential of CCTV systems in preventing terrorism activities, despite claims that effective CCTV systems can help prevent or even deter hostile reconnaissance. A recent review on the effectiveness of CCTV in public spaces by the Scottish Government (2009) concluded that it was only effective in deterring vehicle and property crime, with virtually no impact on reducing violent crimes, or in complex environments. In the case of counter-terrorism, the presence of widespread visible CCTV in London had no noticeable effect on the suicide bombing of the mass transit systems in 2005 (Fussey, 2007). One argument made for publicly visible surveillance infrastructure is that it has the potential to be used to manipulate the spatial patterns of hostile reconnaissance. One of the guides to 'protective intelligence' (Stratfor, 2010) suggests that overt displays of security can be used in 'heating up' key locations to attempt to repel those conducting hostile reconnaissance towards areas or routes that are less useful or 'honey pot' locations that have been prepared with covert surveillance.

The control and shaping of pedestrian movement recurs as an aspect of security planning for crowded spaces. At its simplest, permanent or temporary physical barriers are used to shape crowd flow, but flow can also be shaped without directly blocking paths through a space. For example, in the case of airport design, Adey (2008) describes manipulation of form, materials and configuration to direct the movement of passengers around departure areas. Physical and social characteristics of a space are interdependent, and changes to one element will elicit change in the other (Hillier and Sahbaz, 2009).

To understand potential mechanisms of disruption in pedestrian flow Martin, Dalton and Nikolopoulou (2013) draw on the advertising and marketing literature of shopping behaviour. Reviewing experimental evidence of techniques for shaping pedestrian movement, a number of studies have looked at how emotional states influence 'approach' – pedestrian movement towards a display – and 'avoidance' – movement away or around an area for pedestrians. Mehrabian and Russell (1974) established the importance of pleasure, arousal and dominance or personal control. More recent studies have confirmed that pleasure has a strong impact on approach (Chebat et al., 1995) and that feelings of personal control in a space are influenced by pleasure and reduced by feelings of crowding (Bateson and Hui, 1987; Hui and Bateson, 1991). To explore possible differences in flow behaviour between general users of a space and hostile actors, we can also consider discussions of personal control in criminology literature. People carrying out crime develop a 'crime template' or idealised site for their criminal act and then try to match this location with places they already know or those that they come into contact with, suggesting that a criminal's ideal crime location is one where they are comfortable and feel that they fit in (Brantingham and Brantingham, 1993). By intervening at the point where situational aspects of covert activity converge, that is the times and places where the actors, location and opportunity for criminal endeavours overlap, the intended action can be interrupted (Cornish, 1994).

The importance of pleasure and interrupted personal control in shaping movement suggest that playfulness may be a particularly effective form of disruption for counter-terrorism in public places because it can heighten crowd pleasure (approach), while reducing feelings of certainty and therefore control in those conducting hostile reconnaissance (avoidance). We therefore seek design that can heighten crowd pleasure to encourage pedestrians to approach a specified location, while simultaneously reducing feelings of personal control through disruption of routine for those conducting hostile reconnaissance. The prototype intervention in Safer Spaces intentionally played with digital screens to catch attention, elicit crowd pleasure and sense of performance, which testimonies confirmed were intriguing and held the potential to make journeys less monotonous. Design interventions of different floor patterns in Shades of Grey also had a positive effect in a range of spaces where they were used, triggering curiosity while encouraging playfulness. This is consistent with the Piano Stairs project, which demonstrated the dramatic change in patterns of flow of people by alterations to the physical space (DDB Stockholm, 2009). Transforming the stairs at the entrance to a Stockholm subway station overnight into a keyboard where each step produced a different note, with the escalator left untouched, had 66 per cent more people opting to walk up the stairs, actively changing their routine.

Designing for Routine and Usability in Public Space

The attention of the users of a space, including employees and security, is a resource that must be carefully managed. Routines and norms can be disrupted in order to heighten awareness of the surroundings and other people. However, constant disruption is both difficult to maintain and likely to become gradually more ineffective as users of the space adapt to maintain their state of civil inattention (see for example, the growing literature on 'display blindness', such as Huang, Koster and Borchers, 2008; Müller et al., 2009). There is a strong argument, therefore, for a design approach in a space that fosters daily routines and aids civil inattention, so that when attention is required, it can be evoked easily and effectively through simple disruption techniques.

This issue of habituation, that is the decrease in response to a stimulus after repeated exposure to it, can be a critical parameter in the design of interventions for counter-terrorism. From the focus group discussions of permanently positioned interventions in the Safer Spaces study, it is noted that the impression of heightened awareness that a temporary response evokes is gradually lost over time. This also seemed to be the case with the design of interventions employed in Shades of

Grey. The effect of the different floor patterns was monitored for a week and with all the designs the initial heightened awareness was reduced over the course of the week. Such findings are consistent with more traditional strategies, such as CCTV where initial deterrence fades with time (Scottish Government, 2009).

Foresight

Spaces designed for foresight encourage planning and assist predictability. Sustaining individual routines and patterns of movement in spaces is an important base condition for flow disruption techniques. Similarly, designing environments to optimise foresight heightens the effects of triangulation and performance when those disruptions are used. Designing legible spaces that pedestrians can easily plan to use and reuse acknowledges the importance of 'activity rhythms' within an environment (Lynch, 1981).

Lynch has written at length about the legibility of the city. The symbolic features of a city form narratives which can be read and understood as environmental signs. As Lynch describes: 'environmental forms may be created, or combined in new ways, to elaborate the language and thus extend our capabilities for spatial communication' (Lynch, 1981, p. 141).

More recently, the social semioticians Scollon and Scollon have looked at what they call geosemiotics – a systematic way of examining how visual language appears in the material world. They argue for a focus on the '"in place" meanings of signs and discourses and the meanings of our actions in and among those discourses in place' (Scollon and Scollon, 2003, p. 1). This places an emphasis on the 'social meanings of the material placement of signs in reference to the material world of the user of signs' (Scollon and Scollon, 2003, p. 4). This also implies a local situated-ness taking into account the characteristics and communication found within urban spaces.

Design for foresight can influence feeling of anxiety through principles of familiarity and predictability. An increasing number of practicing designers are engaging in design research focusing on the reduction of risk and anxiety (Lacy, 2008). Uniqueness and sense of place also plays a part here in route finding.

Reflecting on the design intervention to Safer Spaces, one focus group participant noted the importance of rhythm and routine in the mundane aspects of everyday activity:

> If they could work security technology into the rhythm of what you do when you enter a tube station, if they could have some sort of scanning detectors, sensors – whatever is needed to do the job, as part and parcel of the machinery forming your journey, that's fine by me.

Communication

Trust in communication is important in order to make use of unexpectedness and playful disruptions. Public art, games and interaction design bound disruption in understood and socially trusted contexts. These forms communicate that an intervention is playful and disruptive without detracting from the triangulation, performance and flow effects. When a context of play is not communicated clearly, such as when the CCTV-like screens in the Safer Spaces tests used abstracted forms instead of traditional video, people were uneasy with the sense of 'out of place', finding it disconcerting.

Artworks and interaction design that have used surveillance systems often evoke playfulness in order to foster acceptance. For example Hand From Above uses a visual language of the fairy-tale giant reaching in to tickle those under surveillance. ACCESS, by artist Marie Sester (2003) uses the design vocabulary of the theatre to evoke a sense of performance rather than persecution in the tracking, where a computer vision system is used to highlight an individual in public space by turning a spotlight on them and following them as they move around the space.

Communicating risk must strike a fine balance between providing factual, relevant information and avoiding the creation of fear (Rogers et al., 2007). Communications ought to be accurate, specific and originate from trusted sources or they may be counter-productive (Sasse, 2005; Wessely, 2005). Urging state agencies for openness and transparency, the Cabinet Office's strategy unit report *Risk: Improving government's capability to handle risk and uncertainty* (2002) has sanctioned stakeholder engagement in the development of risk communications.

Communications during terrorist incidents necessitate time-sensitive and accurate information. This assists in definition of the problem and enables informed decisions about appropriate behavioural responses (O'Connor, Bord and Fischer, 1999). Providing information that is clear, consistent and reassuring has also been implicated in reducing post-event anxiety, confusion and scapegoating (MacGregor and Fleming, 1996; Newman, Davis and Kennedy, 2006).

We propose that communication channels and associated design vocabulary should be kept separate from all disruption interventions. It is important not to undermine emergency communication strategies and channels with the techniques of disruption. This applies equally to playful and authority-focused disruption. A digital screen intended for emergency instructions should not carry generic authoritative statements such as warnings about surveillance as these are likely to train users of a space that the channel is not worth paying any further attention to. Conversely users of a space can be assisted in becoming accustomed to, and trusting emergency communication channels if they provide continuous, timely information that assists foresight in their regular routines. It is worth noting that advertising often attempts to leverage disruptive design strategies to catch attention, and so should similarly be excluded from emergency communication channels.

Advertising regularly employs design techniques intended to disrupt inattention. Eye-catching movement, colours and patterns are all used to engineer a shift in focus to the advertising. Reducing some of these channels of disruption may be required in heavily used spaces where a predictable flow and routine is needed, and similarly where it is desired that users shift their attention on to any suspicious behaviour rather than the distraction of advertising.

The Designer as Stakeholder

We have described how the disruption techniques that underpin interventions outlined in counter-terrorism guidelines can be categorised as effecting triangulation of attention, unexpected performance or shifts in crowd flow. We have argued that in order for disruption in publicly used spaces to be effective, they must be used against a background state of functional routine, comfortable inattention or low-anxiety pleasure. This design for usability and routine can be seen as reducing disruption fatigue, heightening the impact of disruptions and fostering trust in communication channels vital for emergency response.

Playfulness and performance are seen as key design contexts for disruption. Unlike authority-focused displays, they are characterised as socially positive and imply unexpectedness without significant suggestion of possible threat. Fatigue is also useful in analysing the use of authority. If displays of authority become commonplace, with repeat occurrences, they cannot be expected to retain significant impact.

Design as disruption is intended to complement existing counter-terrorism strategies. Project Griffin, for example, was introduced in London during 2004 with the objective of facilitating public trust and confidence in the capital's policing authorities (City of London Police, 2004). The impossibility of entirely preventing terrorist attacks (Fussey, 2007) lies at the heart of the tension between the normalisation of security practices and enduring public confidence in the capability of the UK authorities to protect civic spaces. In the event of terrorist attacks, public belief in the ability to deter through authority-focused mechanisms of disruption often falters, as does public confidence in the authorities to effectively police. It is in this midst of this tension that this chapter finds itself. Torn between the necessity of visible performances of authority for securing a safe future and the risk these run of routinising and trivialising the issue. There is also a need to acknowledge the complexity of broader cultural contexts when designing for disruption: that the presence of visible over-policing might encourage radicalisation as a response to the heightened presence of authorities in local communities (Fussey, 2007).

Designing performance, playfulness and unexpected pleasure in to public experience has long been a central focus of a wide range of arts practitioners and curators. We can expect these experts to play an important role in informing a sustainable programme of visible counter-terrorism. Security professionals should reposition and schedule such arts events to meet their disruption needs.

In some cases they may also wish to commission longer-term interventions specifically working with interaction designers or architects. Designers regularly respond to questions of usability, foresight and trusted communication and so are a vital resource in the design of the environments in which disruptions can be effectively used, as well as for broader multi-environment programmes of trusted communication.

We have discussed how disruption and communication are both heavily dependent on context. Unexpectedness outside a playful context can be disconcerting and confusing. Communication without a context of trust is soon ignored or misunderstood. More broadly context-specific responses also avoid what the UK Design Council has called 'bolted-on' crime prevention solutions, and instead, encourages a more integrated process to be undertaken (Wootton et al., 2003). This research seeks a more integrated approach that is responsive to the topography of public spaces, by using communication design as a tool to interweave information, space and time. The *Protecting crowded places: design and technical issues* guide from the UK Home Office (2012), calls for designers to take care to avoid creating 'bland and standardised places' in their efforts to design counter-terrorism features into civic spaces. The guide notes that 'it is important to retain or insert positive features that attract people to spaces', suggesting 'incorporating public art or locally important features' into spaces as a way to do this. Designing for Security: Using Art and Design to Improve Security illustrates how art might be integrated into New York City's security strategy:

> Artists and designers should not hesitate to use aesthetic tools as part of the arsenal of security. Light and color, changes of scale, texture – even creative use of sound or smell, temperature and climate control – can convey a sense of safety and help to engage users, staff, and the public. Site relationships, scale relationships, transparency, and opacity may be appropriated to meet expressive, functional, and security needs. (Russell et al., 2002, p. 35)

This braiding of a sense of safety and comfort with perceived security is one approach that echoed the findings of the Safer Spaces project, with participants declaring: 'Comfort consoled by security. Security's the big issue, but comfort is more important'. Design as security has also gathered momentum within UK discussions on crime prevention – see for instance UK Percent for Art which states: 'Commissioning bodies argue that good art encourages greater use of public places and increases individuals' sense of security' (Arts Council, 1991, p. 17). We would also argue that the adoption of a 'designerly way of intervening' would use the arsenal for more than just aesthetic means, that there is scope for embedding such an approach into the early stages of security' planning in urban spaces.

Acknowledgement

This research is conducted as part of the projects 'Shades of Grey – Towards a Science of Interventions for Eliciting and Detecting Notable Behaviours' project (EPSRC reference: EP/H02302X/1) and 'Safer Spaces: Communication Design for Counter Terror' (EPSRC reference: EP/F008503/1). We wish to express our gratitude to both research project teams and our partners as well as the Centre for the Protection of National Infrastructure.

References

Adey, P. (2004). Surveillance at the airport: Surveilling mobility/mobilising surveillance. *Environment and Planning A*, 36(8), 1365–1380.

Adey, P. (2008). Airports, mobility and the calculative architecture of affective control. *Geoforum*, 39(1), 438–451.

Arts Council (1991). *Percent for Art: A Review*. London: Arts Council.

Bateson, J.E.G., and Hui, M.K.M. (1987). A model for crowding in the service experience: empirical findings. In: J.A. Czepiel, C.A. Congram and J. Shanahan, (Eds) *The Services Challenge: Integrating for Competitive Advantage*. Chicago, IL: American Marketing Association, 85–9.

Benton-Short, L. (2007). Bollards, bunkers, and barriers: securing the National Mall in Washington, DC. *Environment and Planning D*, 25(3), 424.

Brantingham, P.L., and Brantingham, P.J. (1993). Nodes, paths and edges: Considerations on the complexity of crime and the physical environment. *Journal of Environmental Psychology*, 13(1), 3–28.

Cabinet Office. (2002). *Risk: Improving Government's Capability to Handle Risk and Uncertainty*. Retrieved 21 April 2012, from http://webarchive.nationalarchives.gov.uk/+/http://www.cabinetoffice.gov.uk/strategy/work_areas/ risk.aspx.

Chebat, J.C., Gelinas-Chebat, C., Vaninski, A., and Filiatrault, P. (1995). The impact of mood on time perception, memorization, and acceptance of waiting. *Genetic, Social, and General Psychology Monographs,* 121(4), 411–424.

City of London Police. (2004). *Mission*. Retrieved 18 May 2011, from http://www.projectgriffin.org.uk/index.php/mission.

Coaffee, J., Moore, C., Fletcher, D., and Bosher, L. (2008). Resilient design for community safety and terror-resistant cities. *Municipal Engineer*, 161(2), 103–110.

Cornish, D. (1994). The procedural analysis of offending and its relevance for situational prevention. *Crime Prevention Studies*, 3, 151–196.

DDB Stockholm (2009). *Piano stairs*. Retrieved from https://www.youtube.com/watch?v=2lXh2n0aPyw, accessed 14 August 2014.

Department of Homeland Security. (2010). *'If You See Something, Say Something' Campaign*. Retrieved 10 November 2012, from http://www.dhs.gov/if-you-see-something-say-something-campaign.

DePaulo, B.M., Lindsay, J.J., Malone, B.E., Muhlenbruck, L., Charlton, K., and Cooper, H. (2003). Cues to deception. *Psychological Bulletin*, 129(1), 74.

Edwards, R. (2010). Heathrow staff given body language training to spot suspected terrorists. *Telegraph, January 15*, retrieved from http://www.telegraph.co.uk/travel/travelnews/6990006/Heathrow-in-security-alert-as-two-men-arrested-on-flight.html. Accessed 14 August 2014.

FEMA 430. (2007). *Site and Urban Design for Security: Guidance Against Potential Terrorist Attacks, Providing Protection to People and Buildings*. Washington, DC: Federal Emergency Management Agency.

Fussey, P. (2007). Observing potentiality in the global city surveillance and counterterrorism in London. *International Criminal Justice Review*, 17(3), 171–192.

Goffman, E. (1966). *Behavior in Public Places*. London: Simon and Schuster.

Granhag, P.A., Andersson, L.O., Strömwall, L.A., and Hartwig, M. (2004). Imprisoned knowledge: Criminals' beliefs about deception. *Legal and Criminological Psychology*, 9(1), 103–119.

Hillier, B. (2004). Can streets be made safe? *Urban Design International*, 9(1), 31–45.

Hillier, B., and Sahbaz, O. (2009). Crime and urban design: An evidence-based approach. In R. Cooper, G. Evans, and C. Boyko (Eds), *Designing Sustainable Cities*. Chichester: Wiley-Blackwell. 163–186.

Home Office. (2012). *Protecting Crowded Places: Design and Technical Issues*. Retrieved from http://www.homeoffice.gov.uk/publications/counter-terrorism/crowded-places/design-tech-issues?view=Binary. Accessed 14 August 2014.

Huang, E., Koster, A., and Borchers, J. (2008). Overcoming assumptions and uncovering practices: When does the public really look at public displays? *Pervasive Computing*, 5013(2008), 228–243.

Hui, M. K., and Bateson, J.E.G. (1991). Perceived control and the effects of crowding and consumer choice on the service experience. *Journal of Consumer Research*, 174–184.

Jacobs, J. (1961). *The Death and Life of Great American Cities*. London: Random House.

Lacy, M. (2008). Designer security: control society and MoMA's SAFE: Design takes on risk. *Security Dialogue*, 39(2–3), 333–357.

Lozano-Hemmer, R. (2001). *Body Movies*. Retrieved from http://www.lozano-hemmer.com/body_movies.php. Accessed 14 August 2014.

Lynch, K. (1981). *Good City Form*. Cambridge, Massachusetts: MIT Press.

MacGregor, D.G., and Fleming, R. (2006). Risk perception and symptom reporting. *Risk Analysis*, 16(6), 773–783.

Martin, K., Dalton, B., and Nikolopoulou, M. (2013). Art as a means to disrupt routine use of space. *Journal of Police & Criminal Psychology*, 27(1), DOI 10.1007/s11896-013-9130-1.

McAndrew, C. (2012). Transforming public spaces. What can we learn from the ontological positioning of the 'site of the social'. *Iridescent: Icograda Journal of Design Research*, 2(1), 78–94.

Mehrabian, A., and Russell, J.A. (1974). *An Approach to Environmental Psychology*. Cambridge, Massachusetts: MIT Press.

Müller, J., Wilmsmann, D., Exeler, J., Buzeck, M., Schmidt, A., Jay, T., and Krüger, A. (2009). Display blindness: The effect of expectations on attention towards digital signage. *Pervasive Computing*, 5538(2009), 1–8.

NCPC. (2002). *The National Capital Urban Design and Security Plan*. Washington DC: National Capital Planning Commission.

Németh, J. (2010). Security in public space: an empirical assessment of three US cities. *Environment and Planning A*, 42, 2487–2507.

Newman, E., Davis, J., and Kennedy, S.M. (2006). Journalism and the public during catastrophes. In B. Raphael, Y. Neria, R. Gross, R.D. Marshall, and E.S. Susser (Eds), *9/11: Mental Health in the Wake of Terrorist Attacks*. Cambridge: Cambridge University Press, 178–196.

Newman, O. (1972). *Defensible Space: Crime Prevention Through Urban Design*. New York: Macmillan.

O'Connor, R.E., Bord, R.J., and Fisher, A. (1999). Risk perceptions, general environmental beliefs, and willingness to address climate change. *Risk Analysis*, 19(3), 461–471.

O'Shea, C. (2009). *Hand from Above*. Retrieved from http://www.chrisoshea.org/hand-from-above. Accessed 15 August 2014.

Royal Institute of British Architects (RIBA) (2010). *Guidance on Designing for Counter-Terrorism*. London: Royal Institute of British Architects.

Rogers, M.B., Amlôt, R., Rubin, G.J., Wessely, S., and Krieger, K. (2007). Mediating the social and psychological impacts of terrorist attacks: The role of risk perception and risk communication. *International Review of Psychiatry*, 19(3), 279–288.

Russell, J.S., Bershad, D., Felicella, E., Kelly, M., and Kennedy, E. (2002). *Designing for Security: Using Art and Design to Improve Security*. New York: Design Trust for Public Space Art Commission of the City of New York. Retrieved from http://www.designtrust.org/publications/publication_97security.html.

Sasse, M.A. (2005). Usability and trust in information systems. In R. Mansell and B.S. Collins (Eds), *Trust and Crime in Information Societies*. Edward Elgar: Cheltenham, 319–348.

Scollon, R., and Scollon, S.W. (2003). *Discourses in Place: Language in the Material World*. London: Routledge.

Scottish Government (2009). *The Effectiveness of Public Space CCTV: A Review of Recent Published Evidence Regarding the Impact of CCTV on Crime*.

Edinburgh: Justice Analytical Services, Police and Community Safety Directorate, Scottish Government.

Sester, M. (2003). *ACCESS*. Retrieved from http://www.accessproject.net/index.html, accessed 15 August 2014.

Snibbe, S.S., and Raffle, H.S. (2009). Social immersive media: Pursuing best practices for multi-user interactive camera/projector exhibits. In D.R. Olsen Jr, R.B. Arthur, K. Hinckley, M.R. Morris, S.E. Hudson, and S. Greenberg (Eds) *Proceedings of the 27th International Conference on Human Factors in Computing Systems*. Boston, MA, USA: ACM, 1447–1456.

Stratfor (2010). *How to Look for Trouble: A Stratfor Guide to Protective Intelligence*. CreateSpace.

Triggs, T., and McAndrew, C. (2009). Transforming policy practice in transport: Is there a space for communication design? In *Inclusive Design Into Innovation: Transforming Practice in Design, Research and Business*. Presented at Include 2009, London: Royal College of Art.

Vrij, A., Leal, S., Granhag, P.A., Mann, S., Fisher, R.P., Hillman, J., and Sperry, K. (2009). Outsmarting the liars: The benefit of asking unanticipated questions. *Law and Human Behavior*, 33(2), 159–166.

Waldman, A. (2011). *The Submission*. London: Random House.

Wessely, S. (2005). Don't panic! Short and long term psychological reactions to the new terrorism: The role of information and the authorities. *Journal of Mental Health*, 14(1), 1–6.

Whyte, W.H. (1988). *City: Rediscovering the Center*. New York: Doubleday.

Wootton, A.B., Cooper, R., Davey, C.L., and Press, M. (2003). *Think Thief: A Designer's Guide to Designing Out Crime*. London: Design Council.

Chapter 19
A Macro-ergonomics Perspective on Security: A Rail Case Study

Rose Saikayasit
Human Factors Research Group, Faculty of Engineering, The University of Nottingham, UK

Alex Stedmon
Human Systems Integration Group, Faculty of Engineering and Computing, Coventry University, UK

Glyn Lawson
Human Factors Research Group, Faculty of Engineering, The University of Nottingham, UK

Introduction

The 'human in the security system' is often the first and last line of defence in identifying, preventing and where necessary responding to threats to public safety. As with many examples of human factors integration, humans provide operational flexibility and local/tacit knowledge of their working environment that automated systems do not possess (Hancock and Hart, 2002; Saikayasit et al., 2012; Stedmon et al., 2013). However, from a systems perspective, the concept of security is fundamentally dependent upon understanding the performance issues of those humans in the system, operating in complex and challenging work environments. Through their unique understanding of their work contexts (and the ability to notice subtle patterns of behaviour and underlying social norms in those environments) security personnel represent a key asset in identifying unusual behaviours or suspicious incidents that may pose a risk to public safety (Cooke and Winner, 2008; see also Chapter 10). However, the same humans also present a potential systemic weakness if their requirements and limitations are not properly considered or fully understood in relation to other aspects of the total system in which they operate (Hancock and Hart, 2002; Wilson, Haines and Morris, 2005; Stedmon et al., 2013).

Seeking to understand the requirements of the intended end-users and involving key stakeholders in the development of new systems or protocols is an essential part of the design process and one that human factors, as a discipline, focuses on through user-centred approaches. In response to this, formal user requirements elicitation methods and participatory ergonomics have developed to support these areas of inquiry, solution generation and ultimately solution ownership (Wilson, 1995). User requirements embody critical elements that end-users and stakeholders want from a solution (Maiden, 2008). These are then mapped to system requirements that express how that solution should be designed, implemented and used (Maiden, 2008). However, these two factors are not always balanced and the resulting solutions may be developed that are not fully exploited, or used as intended. Participatory ergonomics approaches seek to incorporate end-users and wider stakeholders within work analysis, design processes and solution generation as their reactions, interactions, optimised use and acceptance of the solutions will ultimately dictate the effectiveness and success of the overall system performance.

From a human factors perspective this translates into user-centred approaches in which a variety of methods have been applied in research areas as wide as healthcare, product design, human–computer interaction and, more recently, security and counter-terrorism (Saikayasit et al., 2012). A common aim is to effectively capture information from the user's perspective so that system requirements can then be designed to support what the user needs within specified contexts of use (Wilson, 1995). Requirements elicitation is characterised by extensive communication activities between a wide range of people from different backgrounds and knowledge areas, including end-users, stakeholders, project owners or champions, mediators (often the role of the human factors experts) and developers (Coughlan and Macredie, 2002). This is an interactive and participatory process that should allow users to express their local knowledge and for designers to display their understanding, to ensure a common design base (McNeese et al., 1995; Wilson, 1995). End-users are often experts in their specific work areas and possess deep levels of knowledge gained over time that is often difficult to communicate to others (Blandford and Rugg, 2002; Friedrich and van der Poll, 2007). Users often do not realise what information is valuable to the development of solutions or the extent to which their knowledge and expertise might inform and influence the way they work (Nuseibeh and Easterbrook, 2000).

This chapter presents a systems perspective of security using a rail station case-study. It also considers the methods and approaches that are suitable for eliciting user needs in sensitive domains. The work was performed as part of the 'Shades of Grey' security consortium funded by the Engineering and Physical Sciences Research Council (EPSRC: EP/H02302X/1). This research aimed to design and develop security interventions that can be applied in crowded public spaces to amplify the signal-to-noise ratio of suspicious behaviours in order to improve the rate of real-time detection of terrorist activities. Specific attention was focused on defining requirements for security and counter-terrorism within the context of large public venues.

A User-Centred Approach to Security: Balancing Technological and Human Capabilities

Since the 9/11 and 7/7 terrorist attacks as well as subsequent high-profile attempts in aviation and public crowded spaces (e.g. the shoe-bomber Richard Reid, attacks in Boston, Oslo, Mumbai, and Nairobi) counter-terrorism initiatives have increasingly adopted technological solutions to address potential threats (e.g., CCTV, metal and explosive detectors, body scanners). As security protocols have increased (e.g. restricting the amount of fluid allowed within hand luggage), many aspects of security and threat identification in public spaces still rely upon the performance of front line security personnel who are often low-paid, poorly motivated and lack higher levels of education and training (Hancock and Hart, 2002). From these opposing perspectives, the resulting security solutions must attempt to embody complex, human-centred, sociotechnical systems in which many different users interact at different organizational levels to deliver technology-focused security capabilities.

From a macro-ergonomics perspective it is possible to explore how the factors within the overall security system contribute to the success of counter-terrorism initiatives and where gaps may exist. This approach takes a holistic view of security, by establishing the sociotechnical entities that influence systemic performance in terms of integrity, credibility and performance (Kleiner, 2006; see also Chapter 11). This approach is in contrast to micro-ergonomics, which traditionally focuses on the interaction of a single user and their immediate technology use (Reiman and Väyrynen, 2011).

By understanding macro-level issues, the complexity of sociotechnical factors can be translated into micro-level factors for more detailed analysis (Kleiner, 2006). For example, Kraemer, Carayon, and Sanquist (2009) explored issues in security screening and inspection of cargo and passengers by taking a macro-ergonomics approach. A five-factor framework was proposed that contributes to the 'stress load' of front-line security workers in order to assess and predict individual performance as part of the overall security system (Figure 19.1). This was achieved by identifying the interactions between organisational factors (e.g. training, management support and shift structure), user characteristics (i.e. the human operator's cognitive skills, training), security technologies (e.g. the performance and usability of technologies used), and security tasks (e.g. workload and the operational environment).

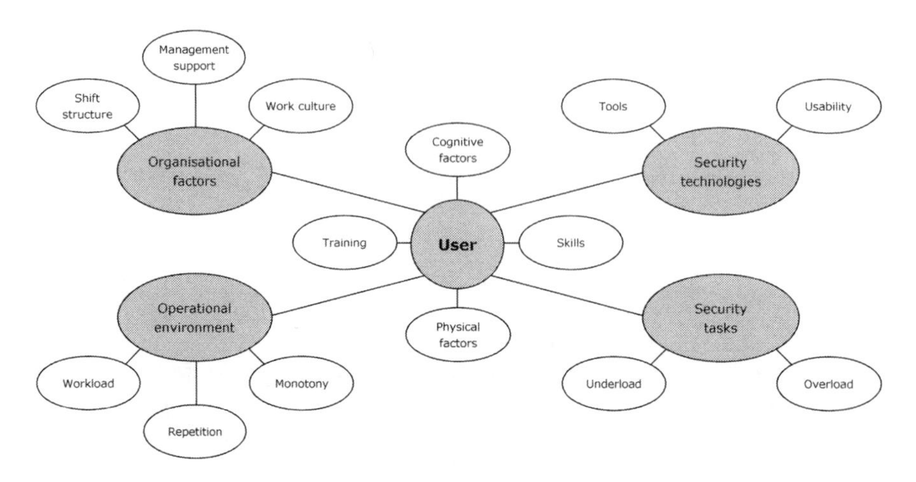

Figure 19.1 Macro-ergonomic conceptual network for security

The central factor of the framework is the user (e.g. the front-line security operator, screener, inspector) who has specific skills within the security system. The security operator is able to use technologies and tools to perform a variety of security screening tasks that support the overall security capability, however these are influenced by task and workload factors (e.g., overload/underload/task monotony/repetition). In addition, organisational factors (e.g., training, management support, culture and organisational structures) as well as the operational environment (e.g., noise, climate, temperature) also interact in delivering the overall security capability. This approach helps identify macro-ergonomic factors where the complexity of the task and resulting human performance within the security system may include errors (e.g., missing a threat signal) or violations (e.g., compromised or adapted protocols in response to the dynamic demands of the operational environment) (Kraemer, Carayon, and Sanquist, 2009). The macro-ergonomic framework has been used to form a basis for understanding user requirements within the Shades of Grey project, focusing on the interacting factors and their influence of overall performance.

Station 'X': A Security Case Study

From this macro-ergonomics perspective a user requirements elicitation exercise was conducted at a mainline railway station in the UK to understand the work and duties of front-line personnel and management as well as the more complex system issues that surround security in this arena. 'Station X' is open to the general public throughout the day and late into the evenings catering for typical commuter traffic as well as leisure users and people travelling to large-scale sports and social events within the city. To ensure anonymity and confidentiality, the exact nature of the work, staffing, location and other sensitive information are not fully disclosed.

Approach

User requirements interviews were conducted at the station in two stages. An initial series of interviews were conducted to understand the contexts of work, work experience, staff capabilities and backgrounds, as well as training experience. A total of 17 participants (8 male, 9female) took part in this first round of interviews. A second series of interviews were conducted in a similar manner approximately one month after the initial interviews. This was done to collect information across different staff shifts and to widen the timescale of the data collection. A total of 15 participants (9 male, 6 female) took part in this second round of interviews. Participants included those working as station security, fraud and revenue protection, welcome hosts, on-train hospitality, British Transport Police (BTP), shop/café assistants and station management.

A series of informal semi-structured interviews was conducted with staff members across the different security sectors with full permission from the station management. Permission was also granted to allow researchers to enter and observe work from the CCTV control room and the station manager's office. Interview questions were used as prompts to elicit information on day-to-day work activities, official and unofficial communication channels and how different sectors interact with each other. The semi-structured nature of the interviews provided the flexibility for researchers to explore several issues and past experiences of the interviewees as appropriate, while maintaining a similar overall structure between participants.

An Overview of the Station Security

It became apparent that the station relied on several security sectors working closely with one another to ensure safety and customer satisfaction. The various staff teams and their roles, including highlights of both regular and security-oriented tasks, are presented in Table 19.1.

Table 19.1 Security sectors and staff roles

Sector	Staff teams	Role
Rail operator	Customer service	• Ensure safety and welfare of passengers who require special assistance • Patrol the concourse and check ticket machines
	Station security	• Control admission between trains and station (ensure valid tickets for travel) • CCTV monitoring • Security and safety of passengers travelling through the station (monitor unattended packages) • Point of contact for customers and other tenants • Train newer staff on the job
	Fraud and revenue protection	• Board scheduled trains to check/sale tickets • Ensure all passengers travel with a valid ticket
	Lost/left luggage	• Point of contact for passengers who have left/lost belongs at the station or on trains • Document all lost property and returns • Handle all left luggage lockers and contents • Search all luggage before it is left in lockers • Report illegal substances or goods found in left luggage • Oversees public toilet entrances
	Management	• Advance planning for special events to accommodate large crowd travelling through the station • Supervise departments and the day-to-day operations at the station • Ensure all trains run on time • Cooperate with the rail operator to ensure compliance with safety and counter-terrorism rules and regulations • Design training and campaigns to promote safety for both passengers and staff
External tenants	Cafes/shops in the concourse	• All follow the rules and regulations by the train operator • Provide customer service • Report anything suspicious to the station customer service staff, including unattended baggage • Assist station customer service staff when necessary
Police (in the station)	British Transport Police	• Safety and security of the station, staff and passengers • Coordinate safety procedures with station staff • Patrol the station and act as a point of contact for passengers and station staff in emergency situations

Profiling Security Threats

The data gathered from the two rounds of interviews were used to elicit background information to represent work and responsibilities of different departments within the station. Most of the participants interviewed in both studies represented those working in close contact with the general public.

In the past, security exercises have been used to train staff and raise awareness of security issues. Two examples were identified in the interviews related to unattended and potentially dangerous packages. On one occasion a tenant noticed an unattended package that appeared to be dangerous. The staff member responded as follows:

1. The unattended package was identified – suspicious characteristics included flashing lights inside the package and wires being visible.
2. It was judged to be harmful.
3. This was quickly reported to station security staff for further assistance.
4. The suspicious package was quickly removed from the scene and the tenant who identified the package was debriefed.

In another example, a package was found by one of the lost property personnel in the car park of the station. The staff member responded as follows:

1. The unattended package was identified in the car park – suspicious characteristics included wires from the box.
2. The package was therefore judged to be harmful.
3. This was reported to the BTP office located nearby.
4. The area was closed and evacuated.
5. The package was removed by a team of specialists. Later it emerged that the box containing wires was left by accident by a team of constructors working at the station.

From these two incidents it is apparent that the lines of communication differed. However, the main goal of the station staff upon finding suspicious packages or any threats is to report them to the BTP office and wait for further assistance. The BTP officers are often able to tell whether suspicious packages are harmful and can therefore give further instructions such as telling staff to evacuate particular areas. From these initial observations, it was possible to clarify the protocol for unattended baggage.

The flow chart in Figure 19.2 illustrates the process and decisions in which the station personnel (including the external tenants) go through when unattended baggage is identified at the station.

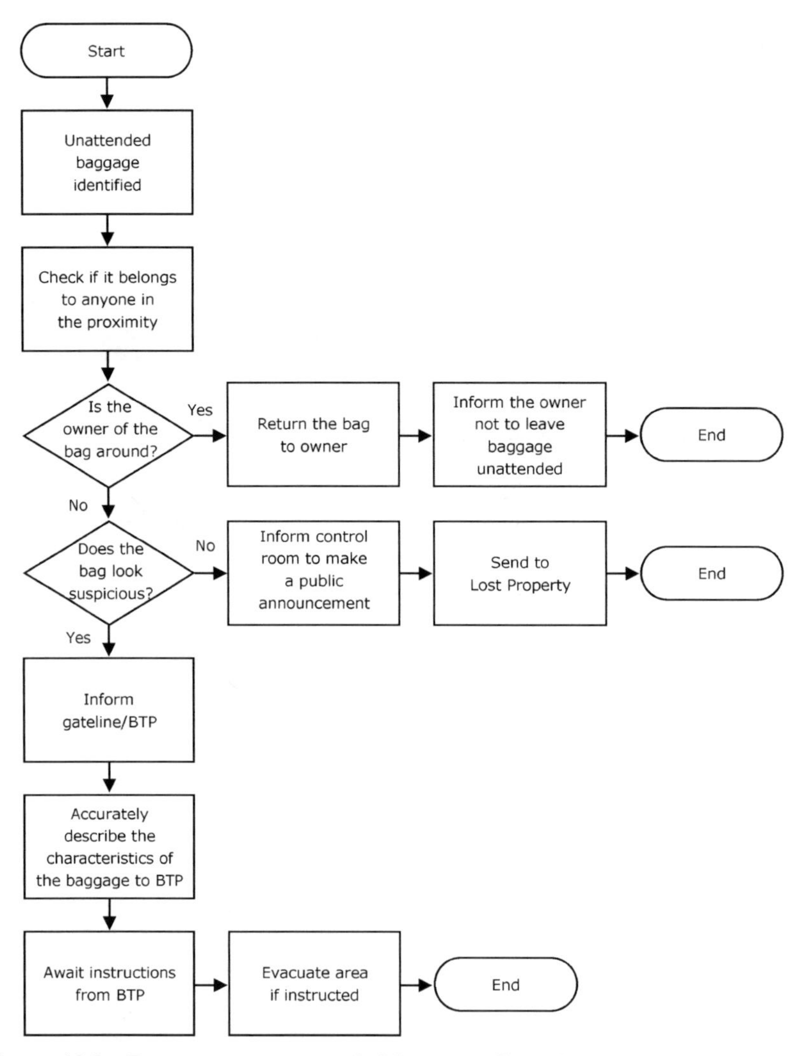

Figure 19.2 Response to unattended baggage flow chart

Security staff are also trained in conflict management as they are required to interact with customers who may have misunderstood the ticketing system or are travelling on invalid tickets. From the interviews, in most cases passengers make genuine mistakes and purchase the wrong type of travel tickets which are invalid for certain times or routes. However, ensuring that all passengers have valid tickets can lead to some form of defensive or hostile responses that can escalate into potential conflict situations. It was reported that most staff interacting with passengers have experienced difficult situations where they felt threatened in some way.

In relation to the point above and also more deliberate hostile intent, security staff have reported that they are often able to identify potential troublemakers from a crowd based on their behaviour. Typical attributes include:

- Avoiding eye contact
- Looking around to see how many security staff are within the area
- Moving away to a different parts of the station or train when staff approach during ticket checks
- Looking agitated/nervous
- Appearing drunk or intoxicated
- Loitering for no apparent reason
- Wearing clothing that looks out of place for the situation or season.

Staff reported that once they have identified a potential threat, they often confer with their colleagues or if they feel that it is safe to do so, they politely approach the individual/group in order to initiate a conversation and offer assistance. The station management team believes that approaching someone and offering help is a valuable initiative. In genuine cases it may serve an assistive role to those who may be nervous travellers and/or who might be uncomfortable in asking for help. However, in some cases, where staff are faced with anti-social behaviour (e.g. verbal abuse or physical threats) they have been trained to safely step away from the situation and report the incident directly to the BTP or inform the control room/ station manager who can further report to the BTP. Using these descriptions from the interviews it was possible to identify the protocols for dealing with suspicious behaviour (Figure 19.3).

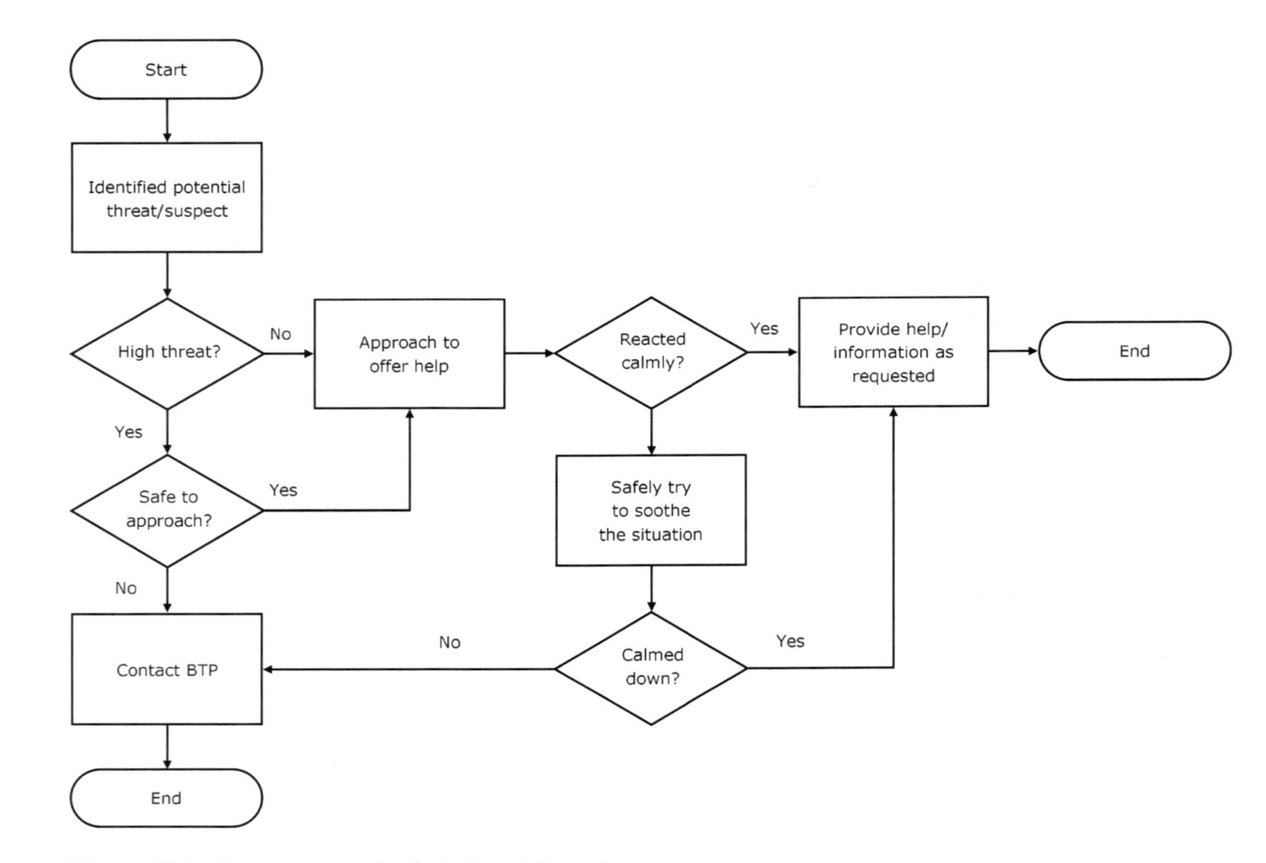

Figure 19.3 Response to physical threat flow chart

Factors Influencing Security Work

The data gathered from the interviews was used to tabulate a summary of results which illustrated the factors influencing the overall performance of 'Station X' for different groups of staff within the organisation (Table 19.2). From the data available, the focus was refined to include station security staff, external tenants and the BTP. Factors reported by Kraemer, Carayon, and Sanquist (2009) that influence human performance and the overall work system were used to categorise the observations.

Table 19.2 Systems factors applied to the rail station case study

Factors	Staff teams	Systems characteristics
Organisational factors	Station security	• One of the main departments acting as a point of contact for customers, strong presence • New staff undergo training on the job by shadowing more experienced staff • Management emphasises the importance of staying vigilant and reporting anything suspicious to the control room and the BTP
	Tenants	• Employed by external companies • Work in compliance with rules and regulations of the station • Subjected to regular compliance checks by station staff • Training provided by their employers including courses on the security of the station
	BTP	• Work closely with other stations to ensure safety of the public, staff, venues and local infrastructure • Respond to incident calls and coordinate with external emergency responses such as bomb disposal units and terrorism intelligence appropriately
User characteristics	Station security	• Some have background in security work • Many of the permanent staff have been working at the station for a couple of years • Team members know each other well • Most staff appear friendly and approachable as their job is to provide a high level of customer service • Many staff interviewed have likened the station to a small community where they know each other and often recognise regular commuters
	Tenants	• Provide customer service and represent their brands • High level of staff turnover
	BTP	• Highly trained officers • Understand the importance of communication with station staff and passengers • Open to communication with tenants

Table 19.2 *Continued*

Security technologies	Station security	• Operate and monitor CCTV cameras in the control room • Staff take turns to supervise the control room and work on the station concourse • Radio communication between the customer service staff and control room • Not all members of the station security staff will have access to radio communication but at least one team member carries a radio
	Tenants	• No dedicated communication technologies with station security, however the tenants and security staff often know each other • Tenants reported that if they need to contact the station, their main point of contact is the security team
	BTP	• Radio communication amongst officers • Telephone communication with station staff (i.e. usually get calls from the control room)
Security tasks	Station security	• High level of workload during rush hours when the station accommodates a large number of commuters • Quieter during the day, except on Thursdays through to Sunday with weekend travellers (but not as high as during rush hours) • Can be very busy when they are short of security staff as they usually need more than one staff member to assist passengers getting through the automatic barriers • High workload when there is a cancellation or rescheduling of trains • High workload during special events (i.e. when passengers travel in a large crowd to attend a sporting event)
	Tenants	• High workload during rush hours • They reported that it was often easier to deal with regular commuters who knew what they wanted to order (sometimes staff also remember what regular customers would usually order) • Workload is also high during weekends especially when dealing with travellers (domestic and foreign) who were not familiar with the menus which can cause delays
	BTP	• Similar to the station security staff, however many officers have to cover other stations during the day, which increases their level of workload • Many reported they have routines which they follow once they get on their shifts (i.e. patrol routes and schedules) to help cope with the workload • Work closely with the station staff during special events to help where possible

Table 19.2 *Concluded*

Operational environment	Station security	• The concourse is central to the station, however during the winter months, the station can be very cold and breezy • Staff reported they often have to take turns to go into the control room to warm up • Staff in the concourse area are not required to patrol the station which means a lot of standing still outside of rush hours
	Tenants	• Most of the tenants working in coffee bars are often working near a heat source (i.e. boiling water or grills) however most cafés are open and are subjected to temperature changes
	BTP	• Have their own office at the the station • Officers often patrol around the station which helps with keeping warm during the winter months

It can be seen from Table 19.2 that the main security functions are spread across the different teams of staff, with different needs, priorities and backgrounds. The station management operate a 'hands-on' philosophy with the day-to-day running of the station. Interviews with station management suggested that they are keen to encourage front-line staff to understand the thinking behind different security protocols and safety measures, so that they can assess situations for themselves. The managers empower staff to take responsibility for security and report anything they feel is 'out of the ordinary' directly to station security or BTP.

Discussion

Prior to adopting different approaches and methods for gathering user requirements, it is recommended that end-users and stakeholders are incorporated in the process and approached to assess resources (including time) and capabilities in committing to different methods of requirements elicitation. Such participatory approaches are useful in developing the required 'buy-in' and assessing the levels of commitment of end-users and front-line stakeholders (Wilson, Haines and Morris, 2005). While interviews allow an in-depth understanding of the task and context of use, they can be viewed as time-consuming by end-users if they have to incorporate them into the work breaks.

The use of interview sessions in the security domain is however particularly useful in capturing rich and complete data around particular issues. Furthermore, within this, and other sensitive domains, interviews offer an opportunity to build up trust between researchers and security organisations for future activities. By allowing participants to meet with the researchers this can reassure them about the genuine interest and basis of the research, to further increase information elicitation in an environment where information is closely guarded for the safety of the public. At the very least, informal interviews ensure that the data collected can be understood in a meaningful way (for both the researcher and user).

User requirements elicitation with users working in sensitive domains also presents issues of personal anonymity and data confidentiality (Kavakli, Kalloniatis and Gritzalis, 2005). In order to safeguard these, anonymity and pseudonymity can be used to disguise individuals, roles and relationships between roles (Pfitzmann and Hansen, 2005). In the Shades of Grey project interviews at Station X were scheduled to allow the security personnel time away from their work so as not to compromise security at the station. However, in both interview sessions the researchers had to rely on an opportunistic sample of those workers who were on shift at the time. This was part of the reasoning for conducting the interview sessions on two different occasions to expand the potential for data collection.

In sensitive domains, snowball or chain referral sampling is a preferred method of recruiting participants through introductions made by those who share knowledge or interact with others in the required domain or share specific characteristics of interest for the research (Biernacki and Waldorf, 1981). This sampling method has been used in the areas of drug use and addiction research where information is limited and the snowball approach is initiated with a personal contact or through an informant (Biernacki and Waldorf, 1981). However, one of the problems with this approach is that the eligibility of participants can be difficult to verify as researchers rely on the referral process and the interpretation of research needs by other stakeholders. Another problem is that the sample includes only one sub-set of the relevant user population that to an extent self-selects itself through active involvement in the research process. In some cases it may be useful to gain information from those agencies that do not want to participate in the process.

In relation to privacy and confidentiality issues there are added ethical challenges to the process of user requirements elicitation, where researchers must ensure that:

- end-users and stakeholders are comfortable with the type of information they are sharing and how the information might be used
- end-users are not required to breach their professional agreements and obligations with their employers or their associated organisations (Kavakli, Kalloniatis and Gritzalis, 2005).

In addition, observation data gathered in the field allows richer insights into real work practices. Observations are particularly useful when supplemented by field interviews as they allow for the exploration of issues that have arisen during the observation. However, it is not always possible to gain access to such environments as organisations have different outlooks on allowing external bodies to observe their actual work practices. This is often to protect the privacy of their staff or avoid disruption to highly demanding work. Therefore researchers should work closely with the organisation when designing an observation session to establish a mutual understanding of the process, the duration of the observation and access issues (i.e. access to control rooms that might usually be restricted).

Using a variety of human factors methods in this case study provided the researchers with a clearer understanding of how security, as a process, operates within the current protocols that have been designed for different situations. Without the use of interviews and field observation, indirect methods such as questionnaires would have been insufficient to gather the data presented. In addition, without using the tools for communicating the findings the process would have been incomplete and end-users and other stakeholders would have missed an opportunity to learn about their security system as well as contribute further insights into their roles.

While a range of human factors methods exist to elicit requirements in different research domains (Preece, Rogers and Sharp, 2007), methods for identifying and gathering user needs in the security domain are underdeveloped and little research exists on understanding the work of security personnel and systems (Hancock and Hart, 2002; Harris, 2002; Kraemer, Carayon, and Sanquist, 2009). This leads to the lack of case studies or guidance on how methods have been used or can be adopted in different security settings. There is a lack of guidance on how methods can be adopted or have been used in different security settings (Hancock and Hart, 2002; Kraemer, Carayon, and Sanquist, 2009). As a result it is necessary to revisit the fundamental issues of conducting user requirements elicitation that can then be applied to security research.

Conclusion

This chapter has presented a case study of user requirements elicitation conducted for the Shades of Grey research project to illustrate the use of different human factors methods in the security domain. User-centred processes are prominent in the field of human factors and ergonomics where the input of end-users and stakeholders is a key part of the design process. The role of human factors is to ensure that user needs, limitations, expectations and contexts of use are translated into meaningful design criteria that are then incorporated into a final system specification. As illustrated in this chapter, the involvement of users in the design process can be extended to research in security settings.

The case study has shown the importance in understanding the context of work and other factors contributing to the overall performance of a system of security through user requirements-gathering exercises. Without interviews and observation work *in situ*, the lack of understanding and appreciation of the work context can lead to misinformed understandings of the nature of security work in this context and ultimately the inappropriate design of security interventions. This case study also highlighted the presence of different end-user groups within the same organisation whose needs and requirements all need to be addressed in the design and implementation of future security solutions.

Acknowledgement

The research in this chapter was funded by the Engineering and Physical Sciences Research Council (EPSRC) as part of the 'Shades of Grey' project (EP/H02302X/1). The authors would also like to thank all the end-users and stakeholders from 'Station X' who took part in the user requirements elicitation exercise.

References

Biernacki, P., and Waldorf, D. (1981). Snowball sampling. Problems and techniques of chain referral sampling. *Sociological Methods & Research*, 10(2), 141–163.
Blandford, A., and Rugg, G. (2002). A case study on integrating contextual information with analytical usability evaluation. *International Journal of Human–Computer Studies*, 57, 75–99.
Cooke, N.J., and Winner, J.L. (2008). Human factors of homeland security. In D.A. Boehm-Davis (Ed.), *Reviews of Human Factors and Ergonomics (3)*. Human Factors and Ergonomics Society, Santa Monica, CA, 79–110.
Coughlan, J., and Macredie, R.D. (2002). Effective communication in requirements elicitation: A comparison of methodologies. *Requirements Engineering*, 7, 47–60.
Friedrich, W.R., and van der Poll, J.A. (2007). Towards a methodology to elicit tacit domain knowledge from users. *Interdisciplinary Journal of Information, Knowledge and Management*, 2, 179–193.
Hancock, P.A., and Hart, S.G. (2002). Defeating terrorism: What can human factors/ergonomics offer? *Ergonomics in Design*, 10, 6–16.
Harris, D.H. (2002). How to really improve airport security. *Ergonomics in Design*, 10, 17–22.
Kavakli, E., Kalloniatis, C., and Gritzalis, S. (2005). Addressing privacy: Matching user requirements to implementation techniques. In I. Lypitakis (Ed.) 7th Hellenic European Research on Computer Mathematics & its Applications Conference (HERCMA 2005), 22–24 September 2005. Athens, Greece: LEA Publishers.
Kleiner, B.M. (2006). Macroergonomics: Analysis and design of work systems. *Applied Ergonomics*, 37, 81–89.
Kraemer, S., Carayon, P., and Sanquist, T.F. (2009). Human and organisational factors in security screening and inspection systems: Conceptual framework and key research needs. *Cognition, Technology and Work*, 11(1), 29–41.
Maiden, N. (2008). User requirements and system requirements. *IEEE Software*, 25(2), 90–91.
McNeese, M.C., Zaff, B.S., Citera, M., Brown, C.E., and Whitaker, R. (1995). AKADAM: Eliciting user knowledge to support participatory ergonomics. *International Journal of Industrial Ergonomics*, 15, 345–363.

Nuseibeh, B., and Easterbrook, S. (2000). Requirements engineering: A roadmap. In C. Ghezzi, M. Jazayeri, and A. Wolf (Eds) *Proceedings of International Conference of Software Engineering (ICSE-2000)*, 4–11 July 2000, ACM Press, Limerick, Ireland, 37–46.

Pfitzmann, A., and Hansen, M. (2005). *Anonymity, Unlinkability, Unobservability, Pseudonymity and Identity Management: A Consolidated Proposal for Terminology*. from http://dud.inf.tu-dresden.de/Anon_Terminology.shtml, version v0.25, December 6, 2005, accessed 11 November 2013.

Preece, J., Rogers, Y., and Sharp, H. (2007). *Interaction Design: Beyond Human-computer Interaction*, 2nd edn. John Wiley & Sons Ltd: Hoboken, NJ.

Reiman, A., and Väyrynen, S. (2011). Review of regional workplace development cases: A holistic approach and proposals for evaluation and management. *International Journal of Sociotechnology and Knowledge Development,* 3, 55–70.

Saikayasit, R., Stedmon, A.W., Lawson, G., and Fussey, P. (2012) User requirements for security and counter-terrorism initiatives. In P. Vink (Ed.), *Advances in Social and Organisational Factors*. CRC Press: Boca Raton, FL. 256–265.

Stedmon, A.W., Saikayasit, R., Lawson, G., and Fussey, P. (2013). User requirements and training needs within security applications: Methods for capture and communication. In B. Akhgar and S. Yates (Eds), *Strategic Intelligence Management*. London: Elsevier.

Wilson, J.R. (1995). Ergonomics and participation. In J.R. Wilson and E.N. Corlett (Eds), *Evaluation of Human Work: A Practical Ergonomics Methodology*, 2nd and revised edition. Taylor & Francis: London.

Wilson, J.R., Haines, H., and Morris, W. (2005). Participatory ergonomics. In J.R. Wilson and E.N. Corlett (Eds), *Evaluation of Human Work: A Practical Ergonomics Methodology*, 3rd edn. Boca Raton, FL: CRC Press, 933–963.

Chapter 20

Deception and Speech: A Theoretical Overview to Inform Future Research

Christin Kirchhübel
Audio Laboratory, Department of Electronics, University of York, UK

David M. Howard
Audio Laboratory, Department of Electronics, University of York, UK

Alex Stedmon
Human Systems Integration Group, Faculty of Engineering & Computing, Coventry University, UK

Introduction

This chapter presents an overview of the theoretical foundations pertinent to research on deception. In particular, it focuses on the relevance of different theoretical frameworks for the investigation of deception and speech. Using a parameter-centred description the reader is presented with a summary of the acoustic and temporal correlates of various affective and cognitive states. This overview will provide an informative point of departure for those wishing to embark on the study of deception-related speech characteristics.

It is acknowledged that general information can be gained about a human speaker from their speech signal alone, including but not limited to age, gender, regional and social background, speech- or voice-based pathology, voice/language disguise, speaking style (reading vs. spontaneous speech) and influence of alcohol intoxication (French and Harrison, 2006). The voice can also provide specific information about a speaker's mental or affective state. Listening to a third party conversation, lay listeners are usually able to tell whether the speaker is stressed, happy, sad, angry or experiencing cognitive workload, for example. While at an interpersonal level it is possible to accurately perceive these psychological and emotional states, empirical research has only been moderately successful in establishing the associated acoustic and temporal characteristics (Beckford Wassink et al., 2006; Hammerschmidt and Jürgens, 2007; Jessen, 2006; Köster, 2001; Lively et al., 1993; Meinerz, 2010; Scherer, 2003; Yap et al., 2011). This is largely attributable to methodological and conceptual obstacles. Hence, rather than referring to the correlations that have been discovered so far as 'reliable acoustic indicators', it is more appropriate to regard them as 'acoustic tendencies'.

Initial research on deception was motivated by the wish to locate a cue or behaviour that would reliably indicate that a deception was taking place. The thinking behind this was analogous to 'Pinocchio's nose' where there would be an observable trait or sign indicating the deception in reality (Collodi, 2005). However, no such unambiguous behaviour has been found to date and instead of searching for behavioural correlates of deception per se, some researchers have reasoned that it would be more fruitful to investigate the emotional, cognitive and communicative processes that tend to accompany deception (see Chapters 6 and 7 for reviews of non-verbal cues and detecting deception). A number of theoretical frameworks have been developed to predict and account for the way that deception might be identified and detailed descriptions can be found in DePaulo et al. (2003); Ekman (1985); Miller and Stiff (1993, pp. 52–55) and Vrij (2008). The most relevant to research on speech is the Multi-Factor Model developed by Zuckerman, DePaulo and Rosenthal (1981) and which provides the focus of this chapter. The model makes use of three theoretical frameworks:

- *Emotional approach (psychological stress)* – the emotional approach assumes that liars are more stressed and aroused than truth-tellers. The likely cause for this arousal is the presence of heightened emotions, the most common being fear, guilt and 'duping delight' (Ekman 1985). Liars may feel guilty either for a transgression that they are trying to hide or for the act of lying itself. They may experience fear of being caught or of the consequences if their lies are discovered. They may also experience excitement in the prospect of fooling someone. The strength of the emotional involvement may be reflected in liars' demeanour and the stronger the emotions experienced the more likely they might become apparent through cue leakage (Ekman, 1985). There are weaknesses with this theory in that there is no direct link between being deceptive and being emotionally aroused or stressed (Lykken, 1998). Certainly, there will be liars who manifest stereotypical characteristics of nervousness and stress. At the same time, however, truth-tellers may also exhibit anxiety and tension, especially if they are in fear of not being believed. Furthermore, liars might not conform to the stereotypical image described above but rather display a composed and calm countenance if they are well trained or have a well-practised cover story.
- *Content complexity approach (cognitive load)* – the foundation of the content complexity approach rests on the fact that lying can often be cognitively more demanding compared to truth-telling. Deceivers need to suppress the truth, which is an automatically activated process. In order to be successful in their lies, deceivers need to make their stories plausible and consistent and they constantly need to monitor their own and their target's behaviour. Yet, within the context of social relations, lying might

actually be cognitively less challenging compared to telling the truth as, for example, illustrated by insincere compliments. As with emotional arousal, there may be situations that prompt hesitation and uncertainty in truth-tellers and therefore, caution needs to be taken in how behavioural signs are interpreted. In addition to this, personality traits may determine a person's non-verbal appearance. Schlenker and Leary (1982), for example, found that socially anxious people are naturally slower in their verbal responses.

- *Attempted control theory (hyper-control)* – people often have stereotypical views about how liars behave (Hocking and Leathers, 1980). In order to present a truthful demeanour people may try to suppress or control behaviours that they associate with lying and consequently would expect their target to be associating with lying. This control may be conscious but can also be the result of subconscious mechanisms. Paradoxically, this type of control can result in a less natural, more rigid appearance. In order to make an honest impression, deceivers need to be able to manage their behaviour. Ekman and Friesen (1969) predicted that people might control different communication channels to differing degrees. While they tend to be skilful at controlling the face (as the face is such an important means of non-verbal communication), they are likely to be less adept at controlling the body or voice. Ekman et al. (1991) modified the notion of a 'leakage hierarchy', reporting that in situations where people are highly motivated to deceive they will attempt to control all possible channels: however they might only be successful at controlling some of them. Once more, it must be remembered that signs of control are not necessarily unique to an attempted deception. Truth-tellers may well adopt a controlled and careful demeanour in order to ensure that their message is judged to be sincere or if they are paying attention to communicating a complex argument.

Research on Speech under Psychological Stress

The concept of stress is multifaceted and a single definition would not be sufficient in recognising its multidimensionality. Accordingly, 'speech under stress' research has comprised a range of aspects, from looking at physical stressors such as physical exercise and temperature extremes, chemical stressors such as drugs or alcohol and psychological stressors including emotion and workload. The reader is directed to Kirchhübel, Howard and Stedmon (2011) for a more in-depth discussion of the theoretical underpinnings of speech under stress. For the purpose of this review, it is more appropriate to take a relatively narrow perspective, focusing on those studies that have investigated the effects of psychological stress on speech.

In order to discover potential correlations between speech and stress the majority of empirical research has employed laboratory experimental designs to induce stress which is usually referred to as emotional stress, situational stress, laboratory stress, task-induced stress, cognitive stress and physical stress. Despite these different labels, all can be considered as factors on the psychological stress spectrum. Some authors differentiate between situational stress on the one hand and laboratory stress or task-induced stress on the other (Hicks, 1979; Meinerz, 2010). Laboratory stress refers to stress induced in a highly controlled laboratory setting, such as solving cognitive tasks under various conditions (Fernandez and Picard, 2003; Hecker et al., 1968; Jessen, 2006; Ruiz et al., 1996; Tolkmitt and Scherer, 1986), being subjected to environmental factors such as noise and temperature extremes (Vilkman and Manninen, 1986) or receiving electric shocks (Hicks 1979). Situational stress, in comparison, is not caused by the experimenter inducing a stressor but rather by the naturally stressful nature of the setting itself. Examples of the latter include making a public speech (Hicks, 1979), completing an oral examination (Fuller, Horri and Conner, 1992; Sigmund, 2006), completing a simulated job interview (Meinerz, 2010), deceiving an interviewer (Eachus, Stedmon and Baillie, 2013; Streeter et al., 1977), watching gruesome slides (Ekman, Friesen and Scherer, 1976) and completing a dangerous mission (Johannes et al., 2000).

A number of studies have been conducted using recordings of stressful situations from real life such as catastrophic aeroplane crashes or emergencies (Benson, 1995; Brenner et al., 1983; Hausner, 1987; Kuroda et al., 1976; Ruiz et al., 1996; Williams and Stevens, 1969). Williams and Stevens (1972) analysed the speech of a radio reporter during the Hindenburg crash and Streeter et al. (1983) examined the recording of a system administrator and his superior during the New York power cut in 1977. However, few studies have concentrated on the investigation of the acoustic changes that are perceived to be indicative of stress (Protopapas and Lieberman, 1997).

The following presents a selection of key research with the focus on the acoustic and temporal characteristics of stress; a parameter-centred description is adopted. A more comprehensive overview can be found in Jessen (2006).

Fundamental Frequency (F_0), Jitter and Intensity

The rate of vocal fold vibration is the parameter most widely studied in research on stress in speech. Inducing psychological stress in a laboratory setting has often resulted in an increase in mean F_0 measures (Brenner et al., 1983; Hicks, 1979; Jessen, 2006; Johannes et al., 2000; Ruiz et al., 1996; Scherer et al., 2002; Sigmund, 2006; Streeter et al., 1977). There are studies, however, which did not show the expected rise in mean F_0. Tolkmitt and Scherer (1986) did not observe an elevation in mean F_0 and the results of Hecker et al. (1968) also failed to show an increase in F_0 with stress. When looking at the extent of the rise in F_0, interesting differences can be observed between studies that have investigated real-life stress and those that have induced stress in the laboratory where typically, the F_0 increase is much greater in the former compared to the latter. However, there is generally a high degree of interpersonal variance with respect to the nature of the reactions to the same stressors.

The findings with regard to jitter are also inconsistent. Some research reported a decrease in jitter in various stress conditions (Brenner et al., 1983; Mendoza and Carballo, 1998; Vilkman et al., 1987), whereas other research indicated that jitter increased with stress (Fuller, Horii and Conner, 1992). Jitter also increased in a cognitive stress condition as well as in a real stress situation (Ruiz et al., 1996), however, based on their experimental findings, Jessen (2006) and Protopapas and Lieberman (1997) could not arrive at any correlation between jitter and stress.

The intensity of the overall acoustic output is also a common measurement despite the difficulties involved in ensuring that reliable measurements are obtained. Some studies report measurements of overall amplitude of the speech data but the results from these are inconsistent. Streeter et al. (1983) observed a rise in overall amplitude for some speakers but no such tendency for others. The findings of Hecker et al. (1968) revealed a significant decrease in overall amplitude in a stress condition for three participants, a significant increase for one and no correlation for another six participants. Hicks (1979) observed a slight increase in a physical stressor condition but a decrease in a condition where participants were asked to make a public speech on stage.

More recently, rather than measuring the overall amplitude, other studies have tended to determine the amplitude of different frequency bands. Although different techniques were employed to achieve these calculations, there is general agreement that amplitudes are greater in higher frequencies (i.e., above 1000Hz in stress conditions compared to control conditions) (Benson, 1995; Fuller, Horii and Conner, 1992; Mendoza and Carballo, 1998; Scherer et al., 2002). Ruiz et al. (1996) analysed spectral balance frequency in vowels and tentatively suggested that this might be a promising parameter that is sensitive to stress.

Temporal Characteristics (Speaking Tempo and Hesitations)

Speaking tempo was found to increase as stress increased in Hausner (1987) and while one employee in Streeter et al. (1983) reduced his speaking tempo, another showed no remarkable change. The majority of laboratory studies converged on the finding that speaking tempo (as measured in articulation rate – AR) increases as a result of stress (Hecker et al., 1968; Hollien, Saletto and Miller, 1993; Karlsson et al., 2000; Siegman, 1993). Nevertheless, studies exist that did not identify a correlation between stress and AR (Jessen, 2006; Scherer et al., 2002). Meinerz (2010) reported a decrease in speaking rate (SR) when participants had to complete a mock interview. Similarly, a slight but not statistically significant reduction in SR occurred when participants had to make a speech in the study of Hicks (1979). In that same study, however, no change in SR occurred when participants were given electric shocks.

Relatively little investigation has been directed at how stress might influence pauses and hesitation patterns. One reason for this could be that often there is insufficient data in order to measure these phenomena quantitatively. Studies that have attempted to examine the former conclude that the number of pauses decreases with stress (Benson, 1995; Hollien, Saletto and Miller, 1993; Siegman, 1993). Hicks (1979) observed an increase in the number of pauses in a laboratory stress condition (i.e. receiving electric shocks) but a decrease in a situational stress condition (i.e., making a public speech). In terms of hesitations Hicks (1979) and Siegman (1993) observed an increase whereas Benson (1995) reported the absence of these in a stress condition.

Formants and Voice Quality

Formants have also received relatively little research attention. The few studies that have reported on formant measurements show contradictory findings. Hecker et al. (1968) observed that under stress, vowel targets were often not reached by some participants and Karlsson et al. (2000) reported a centralisation of the vowel space in stressful conditions. Hecker et al. (1968) suggested increased speaking tempo as the cause for target undershoot. Benson (1995) observed an increase in F_1 in high vowels which again could be caused by imprecision of articulation. Having analysed formant frequencies as well as formant bandwidths, Sigmund (2006) concluded that only the former showed significant changes. The author noted an increase in F_1 and F_2 but, unfortunately, it was not specified which vowel phonemes were analysed. Similar results were reported by Meinerz (2010), whose analysis showed an increase in F_1 and F_2 in three different stress conditions compared to modal speech (F_3 was not significantly affected by stress). Ruiz et al. (1996) also found significant variation in formant values but these were vowel-dependent and did not apply to F_1, F_2 and F_3 to the same extent. A more complex picture is presented in Tolkmitt and Scherer (1986) in that women tended to show target overshoot or hyper-articulation in a cognitive stress condition whereas articulation precision was reduced in a condition involving emotional stress. For men, the variation in formant values as a function of stress was insignificant.

Evidence of an auditory assessment of voice quality can be found in Hecker et al. (1968), who observed the presence of low frequency vibration during a stress condition which could be indicative of 'creaky voice'. Van Lierde et al. (2009) observed female voices to be more breathy and strained under stress. Using acoustic analysis, Waters et al. (1995) observed that there are effects on different measures of the glottal pulse when speech is produced under stress. Without naming the direction of change they reported significant modifications in pitch, rising slope and closing slope of the glottal pulse. In addition, they discovered a greater amount of inter-speaker variability between the relationship of various glottal parameters and stress. A number of studies have all remarked that voicing irregularities or 'voicing breaks' occur when people speak under stress (Hecker et al., 1968; Hollien, 1990; Jessen, 2006; Williams and Stevens, 1972).

Research on Speech and Cognitive Load

The same challenge that was faced with determining a general concept of stress or psychological stress applies equally to cognitive load or cognitive workload. As Baber et al. (1996, p. 38) state, 'given the range of demands which can have a bearing upon workload, it would be difficult to provide a unified definition of the term; indeed there are many definitions of workload'. Cognitive load/workload has often been confounded with psychological stress, especially in those studies that have employed a task-induced stressor such as solving cognitive tasks under time pressure. In these instances it is difficult to differentiate between whether changes, if present, were generated by the effects of stress or by cognitive loading. Cognitive load/workload may well lead to stress in some contexts and some individuals, but nevertheless, cognitive load and stress are individual entities and should not be confused. The reader is directed to Berthold (1998) and Khawaja (2010) for a comprehensive discussion encompassing the many different aspects of the concept. In the same way as a narrow reading was taken for the concept of 'speech under stress', the discussion of cognitive load will primarily focus on those studies that have clearly differentiated between task-induced cognitive stress and cognitive load.

Different methods have been employed to induce cognitive load in the laboratory. One of the most popular techniques is the multiple task experiment. These commonly involve participants completing a primary task such as real or simulated flying or driving, playing computer games or talking to an interviewer, as well as simultaneously completing a secondary task such as memorising words or digits, solving mathematical equations or holding constant eye contact with an interlocutor. At times, an additional time-pressure factor is included to achieve even higher levels of cognitive load. Studies involving time-constrained tasks are likely to produce cognitive/psychological stress as well as cognitive load and therefore have largely been excluded from the present review. Span tests have also been used to raise cognitive load levels. In reading span experiments, for example, participants were required to read fluently while answering comprehension questions (Daneman and Carpenter, 1980). An example of a speaking span experiment entailed testing the maximum number of words for which participants could successfully compose a grammatically correct sentence (Daneman, 1991). Word or digit span tests are similar to the multi-task experiments already mentioned in that they intersperse a primary word/digit memory task with a demanding secondary task (Conway et al., 2005). The 'Colour Word test' developed by Stroop (1935) has proven to be an effective means of inducing cognitive load without simultaneously stimulating stress. Participants are presented with words naming colours; however, there is a mismatch between the ink of the word and the colour it refers to. Participants are asked to direct their attention to naming rather than reading (i.e., name the colour of the font used). Moving outside of the laboratory a number of studies have analysed recordings of real-life events containing high and even extreme levels of cognitive load. Examples include communication of fighter controllers and fighter pilots during combat flights and communicative exchanges during Australian bushfire management (Khawaja et al., 2009). As pointed out above, findings from real-life studies need to be treated with caution as they are likely to demonstrate a confounding of stress and cognitive load.

Fundamental Frequency (F_0) and Intensity

The majority of studies that have investigated cognitive load on speech have reported a tendency for an increase in mean F_0 (Griffin and Williams 1987; Huttunen et al., 2011; Scherer et al. 2002). In the case of Lively et al. (1993) only one of the five speakers showed a significant increase in mean F_0 and Huttunen et al. (2011) detail a mean F_0 increase ranging from an average of 7 Hz to 12 Hz in the most extreme cases. All studies agree on a reduction in F_0 variability (Johnstone and Scherer 1999; Lively et al. 1993) and Huttunen et al. (2011) specify this decrease to be 5 Hz on average in their investigation.

Not as many studies could be located that investigated the intensity of the speech output. This is likely to be an effect of the difficulties surrounding the obtainment of accurate amplitude measurements. When producing speech under cognitive workload as compared to a control condition, Lively et al. (1993) observed an increase in amplitude and amplitude variability. Similarly, Griffin and Williams (1987) and Huttunen et al. (2011) stated an increase in mean intensity, the latter stipulating it to be 1 dB in their results. Scherer et al. (2002) describe an increase in higher frequency energy in their findings.

Temporal Characteristics (Speaking Tempo and Filled/unfilled Pauses)

A major focus of analysis of speech under cognitive load has been the investigation of temporal parameters. There is agreement across published investigations into cognitive load that SR and AR decrease in cognitively demanding situations (Bromme and Wehner, 1987; Huttunen et al., 2011; Kowal and O'Connell, 1987). As one might expect, those studies which have looked at the concept of cognitive load in terms of cognitive or time-induced stress noted a rise in speech tempo (Baber et al., 1996; Lively et al., 1993; Scherer et al., 2002; Silberstein and Dietrich, 2003). Rather than accrediting the increase in SR to cognitive stress, Baber et al. (1996) provide an alternative explanation based on the observed increase in speech errors. All studies agree that frequency and duration of unfilled/silent pauses increase under cognitively demanding conditions (Baber et al., 1996; Bromme and Wehner, 1987; Jou and Harris, 1992; Goldman-Eisler, 1968; Greene, Lindsey and Hawn, 1990; Kowal and O'Connell, 1987; Müller et al., 2001; Roßnagel, 1995). In addition to a rise in silent pauses, Rummer (1996) and Oomen and Postma (2001) registered a higher amount of filled pauses/hesitations in the cognitive load group. Yap et al. (2011) is one of the surprisingly few studies that investigated phoneme duration, concluding that vowels were longer with cognitive load. Lively et al. (1993) on the other hand noted a reduction in phrase and segment durations.

Analysis of response onset time (ROT) is also relatively sparse but the general tendency appears to point to an overall increase when speaking under cognitive load (Greene, Lindsey and Hawn, 1990; Roßnagel, 1995).

Formants and Voice Quality

Spectral vowel characteristics appear to have been rather less at the centre of investigation for research of speech under cognitive load. The findings of Lively et al. (1993) and Yap et al. (2011) correspond in that both studies did not uncover changes in mean formant frequencies F_1, F_2 and F_3. Huttunen et al. (2011) detected a slight tendency for F_1 and F_2 to centralise in their analysis. It is clear that a considerable amount of work needs to be invested into this aspect of cognitive load research.

Even less investigated are parameters related to voice quality. Yap et al. (2010) reported that cognitive load might be related to more creaky voice quality. However, they also draw attention to the presence of individual variation, revealing that a few speakers adopted a more breathy glottal setting when speaking with cognitive load.

Research on Hyper-controlled Speech

No clear definition exists in the research literature of what exactly constitutes controlled or planned speech. In fact, references to the notion of speech under 'behavioural control' are rare. As a consequence less strict classification criteria need to be applied when investigating this type of speech context. 'Clear' speech (Picheny, Durlach and Braida, 1986) or 'hyper-speech' (Johnson, Flemming and Wright, 1993; Lindblom, 1990) could present possible points of departure. Lombard speech, which refers to speech produced under high background noise, may also be viewed as potentially relevant. Indubitably, speech under background noise is connected to clear speech but for the purposes of the present review, it was considered too far removed from the speech situation in question.

People may choose to speak 'clearer' for a variety of reasons and in a number of contexts. For example, clear speech may be employed when speaking in noisy environments, when communicating with non-native speakers or hearing-impaired persons or when addressing small children. People may also pay increased attention to their speech when communicating with computerised spoken language systems, such as car navigation systems or automated transport reservation systems. In light of this, clear speech is usually regarded as an attempt to increase communicative intelligibility. The question arises whether additional reasons exist that motivate people to pay more attention to their speech, reasons that are connected to peoples' self-presentation and image management rather than purely for the smooth flow of the communicative exchange. In accordance with the previous discussion on speech and psychological stress and speech and cognitive load, the question of whether there are different types of clear speech is a reasonable one to ask. Krause and Braida (2004) briefly discuss this aspect but additional dialogues regarding this issue were difficult to locate.

The literature on the acoustic and phonetic characteristics of clear speech is relatively sparse compared to other areas of affective speech research. The following overview attempted to cover most findings but more comprehensive discussions can be found in Smiljanić and Bradlow (2009) and Uchanski (2005). Experimentally, clear speech tended to be elicited by instructing participants to speak clearly or asking them to speak as though they were conversing with a person who had a hearing loss or who was not a native speaker of the language (Bradlow, Kraus and Hayes, 2003; Krause and Braida, 2002; Picheny, Durlach and Braida, 1986; Smiljanić and Bradlow, 2005). Referring back to Lindblom's Hyper-Hypo theory of speech production, Hazan and Baker (2011) highlight the importance of communicative intent. Future studies, they stress, should distinguish between clear speech that is elicited via formal instruction and more spontaneously produced clear speech used to ensure successful and effective communication with a conversation partner. Other methods involved degrading the communication channel between interlocutors by introducing various types of noise over headphones (Cooke and Lu, 2010; Granlund, Baker and Hazan, 2011; Grynpas, Baker and Hazan, 2011). As explained above, studies that used binaurally applied masking noise in order to elicit clear speech production were largely ignored from the present review as they were considered too broad.

Fundamental Frequency (F_0) and Intensity

Of all the studies reviewed, every one that measured F_0 in connection with clear speech arrived at the same result, which was an increase in the average frequency of vocal fold vibration (Bradlow and Bent, 2002; Chen, 1980; Clark, Lubker and Hunnicun, 1988; Krause and Braida, 2004; Köster, 2001; Picheny, Durlach and Braida, 1986). In conjunction with an increase in the mean value of F_0, greater F_0 variability, usually expressed in terms of F_0 range, was also reported (Bradlow and Bent, 2002; Krause and Braida, 2004; Picheny, Durlach and Braida, 1986).

In the same way as there was relatively little investigation into the intensity parameters for speech under stress and cognitive load research, reports on intensity modifications associated with clear speech are sporadic. Despite this infrequency however, the results are compatible in terms of the observed trend of an increase in amplitude when people speak 'clearly' (Chen, 1980; Clark, Lubker and Hunnicun, 1988; Picheny, Durlach and Braida, 1986). Picheny, Durlach and Braida (1986) go even further and specify the overall increase to be 5–8 dB.

Temporal Characteristics (Speaking Tempo, Segment Durations and Pauses)

In addition to fundamental frequency and intensity, the temporal aspects of clear speech seem to have also received foremost attention in the research. In this respect, it is the durational aspect that comprises the core interest. A number of studies inform the reader of increased speech segment durations, be it vowels or consonants (Bond and Moore, 1994; Chen, 1980; Clark, Lubker and Hunnicun, 1988; Ferguson and Kewley-Port, 2002, 2007; Köster, 2001; Lindblom, 1990; Picheny, Durlach and Braida, 1986). In addition to phoneme duration, Chen (1980) also commented on an increase in voice onset time (VOT) and aspiration for 'clear' speech. Liu et al. (2004) illustrated the effect of speech element lengthening in terms of longer sentence duration for clear speech. With regards to SR, the general tendency of a reduction in rate becomes visible in a range of studies (Bradlow, Torretta and Pisoni, 1996; Bradlow and Bent, 2002; Moon and Lindblom, 1994; Picheny, Durlach and Braida, 1986; Smiljanić and Bradlow, 2005; Uchanski et al., 1996). Caution needs to be employed when comparing across studies as it transpires that different methodologies were used when measuring speaking rate, some calculating words per minute (Picheny, Durlach and Braida, 1986), others determining average sentence duration (Liu et al., 2004) and others counting phonemes per second (Krause and Braida, 2004). There appears to be a lack of findings on the effect of clear speech on AR.

In terms of pausing behaviour, a similar trend as was observed with cognitive load emerges. When speaking clearly, people tend to increase the amount and length of silent periods in their speech (Bradlow and Bent, 2002; Cutler and Butterfield, 1990, 1991; Krause and Braida, 2004; Picheny, Durlach and Braida, 1986). Both Picheny, Durlach and Braida (1986) and Krause and Braida (2004) addressed the connection between an increase in speech rate and the occurrence and nature of pauses and indeed, when rate was controlled, the difference in pausing behaviour between baseline speech and clear speech disappeared (Krause and Braida, 2004).

Formants and Voice Quality

Compared to the research on speech under stress or cognitive load described above, investigations into clear speech have devoted a substantial amount of attention to vowel formant frequencies. Using, at times, different methods and verbal descriptions, almost all of the studies agree that vowels tend to be more distinct in clear speech. In the majority of cases this conclusion is based on vowel space area measurements deduced from centre formant frequencies for F_1 and F_2. Clear speech was reported to show vowel space expansion in various studies (Bradlow, Krause and Hayes, 2003; Bradlow and Bent, 2002; Chen, 1980; Ferguson and Kewley-Port, 2002, 2007; Johnson, Flemming and Wright, 1993). Picheny, Durlach and Braida (1986) argued that the observed increase in vowel space dimension was a result of F_1 and F_2 moving closer to their respective target values and Lindblom (1990) expresses the phenomenon in terms of vowels becoming more distinct. Bond and Moore (1994) approached the issue from a slightly different angle, comparing the speech characteristics of the least intelligible talkers, who were assumed to be producing 'unclear' speech, with the most intelligible talkers producing 'clear' speech. They concluded that unclear speech showed the least differentiated and most constricted vowel space while clear speech was characterised by markedly distinguished vowels. In addition to the peripheralisation of vowels, Chen (1980) also remarked on a tighter clustering of formants. He argued that this was a possible result of the consonant-vowel-consonant (CVC) stimulus data used. Monosyllabic wordlists are likely to have provided less room for contextual variability (Chen, 1980). Ferguson and Kewley-Port (2002, 2007) revealed a little more detail, stating that the overall expansion was not uniform across vowels. More specifically, it was the F_2 values of the front vowels that were most heavily involved while the back vowels failed to show a significant difference. Furthermore, it was the F_1 of the low vowels that tended to have a more substantial increase, the high vowels being more resistant to change. In contrast to this, Köster (2001) only observed a periperalisation in the back vowels, the front vowels being less affected. Krause and Braida (2004) did not concur with the finding of an overall enlargement of vowel space area, reporting that there was no significant change in formant frequencies. The authors did perceive a narrower formant bandwidth however, which could be viewed as supporting the notion that vowels are more distinct in clear speech. Given that the amount of research on voice quality in relation to speech under stress/cognitive load was sparse, it comes to no surprise that no literature could be located which addressed changes or lack of changes in view of this parameter from the perspective of clear speech production.

Summary

As will have become apparent from the above compilation of literature, psychological stress, cognitive load, and hyper-control have been shown to exert a range of effects on speech and although the exact nature of these effects has not been completely established, tendencies can be identified. Table 20.1 lists the known tendencies for the three affective states under consideration and also relates them back to the three theoretical processes potentially underlying deception.

As Vrij (2008) points out, the vocal cues connected to the three behavioural states may suggest that deception is occurring. However, the emphasis is on 'may suggest' as people could also experience heightened arousal, cognitive load or the impulse to consciously control their behaviour in connection to contexts or situations devoid of deception. Conversely, people may be lying without displaying any of the signs associated with the three theories. While all of the theoretical frameworks mentioned have been substantiated by empirical evidence (Vrij and Heaven, 1999; Vrij, Semin and Bull, 1996; Walczyk et al., 2003, 2005) some have received more support than others. Furthermore, it is apparent that no single approach is sufficient in explaining deceptive behaviour. Although the different deceptive processes were summarised in separate fashions above it ought to be noted that they should not be viewed as mutually exclusive. Rather, it would be more appropriate to expect that, if they occur, they occur simultaneously and to varying degrees. As a consequence we might find contradictory behavioural effects during deception (as illustrated with formants, speaking tempo and pauses).

If it is possible to deduce a speaker's emotional condition from listening to their voice, could it also be viable to make judgements about their sincerity from speech as well? To date no method/machine has been established that is capable of achieving this. However, in recent times, claims have been brought forward concerning deception/stress detection based on voice analysis. Psychological Stress Evaluators (PSE) and Voice Stress Analysers (VSA) are said to measure so-called 'micro-tremors' in the voice, which are believed to be physiological indicators of stress and by extension of deception. Layered Voice Analysis- (LVA) based products are promoted as relying on a link between certain types of brain activity and deception related tremors in the voice (Eriksson and Lacerda, 2007). Scientific reliability testing of these products has resulted in exclusively negative evaluations (Bhatt and Brandon, 2008; Damphousse et al., 2007; Eriksson and Lacerda, 2007; Harnsberger et al., 2009). Despite the lack of scientific support, parts of the non-scientific community (e.g. some local governments, insurance companies, law enforcement agencies, the military and the popular media) still believe in the merit of these tools and large budgets have been spent buying into these systems (Lykken, 1998).

Table 20.1 Vocal tendencies related to affective states

	Arousal theory (psychological stress)	**Cognitive theory (cognitive load)**	**Attempted control theory (hyper-articulation)**
Mean F_0	↗	↗	↗
F_0 variability	↗	↘	↗
Mean intensity	↗	↗	↗
Formants F_1 and F_2	↘	?	↗
Formant F_3	?	?	?
Vowel space area	↘	?	↗
Speaking rate	↘	↘	↘
Articulation rate	↗	↘	?
Silent pauses	↘	↗	↗
Filled pauses	↗	↗	?
Response latency	?	↗	?
Voice quality	Creaky, tense, breathy	Creaky, breathy	?

While testing of these products is a necessary part of their evaluation, a more fundamental step has been overlooked. It should be ascertained whether the assumptions on which such products are based are valid. In order to assess validity, it needs to be established whether a relationship exists between deception, truthfulness and speech, and if so, what the nature of any such relationship is. Surprisingly, very little research has been conducted on the acoustic and phonetic characteristics of deceptive speech. There are a number of studies that have analysed temporal features such as speaking rate, pauses, hesitations and speech errors (Benus et al., 2006; Feeley and deTurck, 1998; Stroemwall et al., 2006), but only a few studies have investigated frequency-based parameters such as mean F_0 and amplitude (Anolli and Ciceri, 1997; Ekman et al. 1991; Rockwell et al., 1997). Evidence for the analysis of vowel and consonant articulation and voice quality in connection with deceptive speech is rare. Recently completed work by Enos (2009) is one of the first attempts to analyse deceptive speech using spoken-language processing techniques. The review in this chapter provides a basis in this complex area but more research is needed within the subject matter in order to improve our understanding of deceptive speech and consequently, to assess whether differentiating truthfulness and deception from the speech signal is a realistic and reasonable aspiration.

Conclusion

This chapter has presented an overview of the knowledge base surrounding deception and speech. While there are clear theoretical principles upon which deceptive speech is based, deception itself is a complex notion, dependent on contextual factors and the specific motivations of the deceiver. In many ways the literature is inconclusive, suggesting that reliable and consistent correlations are hard to find. However, the findings are still of interest since they point to some potential limitations when attempting speech analysis for deception detection purposes. Future researchers should exploit the possible behavioural overload that deceivers are likely to experience in tailoring their methodological designs around these aspects.

Acknowledgement

This work was made possible through EPSRC Grant number: EP/ H02302X/1.

References

Anolli, L., and Ciceri, R. (1997). The voice of deception: vocal strategies of naïve and able liars. *Journal of Nonverbal Behavior*, 21(4), 259–284.

Baber, C., Mellor, B., Graham, R., Noyes, J.M., and Tunley, C. (1996). Workload and the use of automatic speech recognition: The effects of time and resource demands. *Speech Communication*, 20, 37–53.

Beckford Wassink, A., Wright, R.A., and Franklin, A.D. (2006). Intra-speaker variability in vowel production: An investigation of motherese, hyperspeech, and Lombard speech in Jamaican speakers. *Journal of Phonetics*, 35, 363–379.

Benson, P. (1995). Analysis of the acoustic correlates of stress from an operational aviation emergency. In I. Trancoso and R. Moore (Eds), *Proceedings ESCA-NATO Tutorial and Research Workshop on Speech under Stress*, INESC: Lisbon, 61–64.

Benus, S., Enos, F., Hirschberg, J., and Shriberg, E. (2006). Pauses in deceptive speech. *Speech Prosody*, 18(2006), 2–5.

Berthold, A. (1998). *Repräsentation und Verarbeitung sprachlicher Indikatoren für kognitive Ressourcenbeschränkungen* (Representation and processing of linguistic indicators of cognitive resource limitations). PhD thesis, submitted to Universität des Saarlandes.

Bhatt, S., and Brandon, S.E. (2008). *Review of Voice Stress-based Technologies for the Detection of Deception.* Retrieved 12 April 2010 from http://www.polygraph.org/section/review-voice-stress-based-technologies-detection-deception.

Bond, Z.S., and Moore, T.J. (1994). A note on the acoustic–phonetic characteristics of inadvertently clear speech. *Speech Communication*, 14, 325–337.

Bradlow, A.R., and Bent, T. (2002). The clear speech effect for non-native listeners. *Journal of the Acoustical Society of America*, 112(1), 272–284.

Bradlow, A.R., Kraus, N., and Hayes, E. (2003). Speaking clearly for learning-impaired children: sentence perception in noise. *Journal of Speech, Language, and Hearing Research*, 46, 80–97.

Bradlow, A.R., Torretta, G.M., and Pisoni, D.B. (1996). Intelligibility of normal speech I: Global and fine-grained acoustic-phonetic talker characteristics. *Speech Communication*, 20, 255–272.

Brenner, M., Shipp, T., Doherty, E.T., and Morrissey, P. (1983). Voice measures of psychological stress – laboratory and field data. In I.R. Titze and R.C. Scherer (Eds), *Vocal Fold Physiology: Biomechanics, Acoustics and Phonatory Control*. Denver, CO: Denver Center for Performing Arts, 239–248.

Bromme, R., and Wehner, T. (1987). Zum Zusammenhang von Sprechgeschwindigkeit und Sprechfehlem mit der Aufgabenschwierigkeit beim lauten Denken (On the connection between speech rate and speech impairment with the task difficulty in thinking aloud). *Zeitschrift für experimentelle und angewandte Psychologie*, 34, 1–16.

Chen, F.R. (1980). *Acoustic Characteristics and Intelligibility of Clear and Conversational Speech at the Segmental Level.* Master's Dissertation, submitted to Massachusetts Institute of Technology.

Clark, J., Lubker, J., and Hunnicun, S. (1988). Some preliminary evidence for phonetic adjustment strategies in communication difficulty. In R. Steele and T.

Threadgold (Eds), *Language Topics: Essays in Honour of Michael Halliday*. Amsterdam: John Benjamins, 161–180.

Collodi, C. (2005). Pinocchio: The tale of a puppet. *The Project Gutenberg EBook of Pinocchio*, http://www.gutenberg.org/files/16865/16865-h/16865-htm, accessed 15 August 2014.

Conway, A.R.A., Kane, M.J., Bunting, M.F., Hambrick, D.Z., Wilhelm, O., and Engle, R.W. (2005). Working memory span tasks: A methodological review and user's guide. *Psychonomic Bulletin and Review*, 12(5), 769–786.

Cooke, M. and Lu, Y. (2010). Spectral and temporal changes to speech produced in the presence of energetic and informational markers. *Journal of the Acoustic Society of America*, 128, 2059–2069.

Cutler, A., and Butterfield, S. (1990). Durational cues to word boundaries in clear speech. *Speech Communication*, 9, 485–95.

Cutler, A., and Butterfield, S. (1991). Word boundary cues in clear speech: A supplementary report. *Speech Communication*, 10, 335–353.

Damphousse, K.R., Pointon, L., Upchurch, D., and Moore, R.K. (2007). Asessing the validity of voice stress analysis tools in a jail setting. Report submitted to the U.S. Department of Justice. Retrieved 28 January 2011 from http://www.ncjrs.gov/pdffiles1/nij/grants/219031.pdf.

Daneman, M. (1991). Working memory as a predictor of verbal fluency. *Journal of Psycholinguistic Research*, 20, 445–464.

Daneman, M., and Carpenter, P.A. (1980). Individual differences in working memory and reading. *Journal of Verbal Learning and Verbal Behavior*, 19(4), 450–466.

DePaulo, B.M., Lindsay, J.J., Malone, B.E., Muhlenbruck, L., Charlton, K., and Cooper, H. (2003). Cues to deception. *Psychological Bulletin*, 129(1), 74–118.

Eachus, P., Stedmon, A.W., and Baillie, L. (2013). Hostile intent in crowded places: A field study. *Applied Ergonomics*, 44(5), 703–709.

Ekman, P. (1985). *Telling Lies: Clues to Deceit in the Marketplace, Politics and Marriage*. New York: W.W. Norton.

Ekman, P., and Friesen, W.V. (1969). The repertoire of nonverbal behavior: Categories, origins, usage, and coding. *Semiotica*, 1, 49–98.

Ekman, P., Friesen, W.V., and Scherer, K.R. (1976). Body movement and voice pitch in deceptive interaction. *Semiotica*, 16, 23–27.

Ekman, P., O'Sullivan, M., Friesen, W.V., and Scherer, K.R. (1991). Face, voice, and body in detecting deceit. *Journal of Nonverbal Behavior*, 15(2), 125–135.

Enos, F. (2009). *Detecting Deception in Speech*. PhD thesis, submitted to Columbia University.

Eriksson, A., and Lacerda, F. (2007). Charlatanry in forensic speech science: A problem to be taken seriously. *International Journal of Speech, Language and the Law*, 14(2), 169–193.

Feeley, T.H., and deTurck, M.A. (1998). The behavioural correlates of sanctioned and unsanctioned deceptive communication. *Journal of Nonverbal Behaviour*, 22(3), 189–204.

Ferguson, S.H., and Kewley-Port, D. (2002). Vowel intelligibility in clear and conversational speech for normal-hearing and hearing-impaired listeners. *Journal of the Acoustical Society of America*, 112, 259–271.

Ferguson, S.H. and Kewley-Port, D. (2007). Talker differences in clear and conversational speech: acoustic characteristics of vowels. *Journal of Speech, Language, and Hearing Research*, 50, 1241–1255.

Fernandez, R., and Picard, R.W. (2003). Modelling drivers' speech under stress. *Speech Communication*, 40, 145–159.

French, J.P., and Harrison, P. (2006). Investigative and evidential application of forensic speech science. In H.A. Armstrong, E. Shepherd, E.G. Gudjonsson and D. Wolchover, (Eds), *Witness Testimony: Psychological, Investigative and Evidential Perspectives*. Oxford: Oxford University Press, 247–262.

Fuller, B.F., Horii, Y., and Conner, D.A. (1992). Validity and reliability of nonverbal voice measures as indicators of stressor-provoked anxiety. *Research in Nursing and Health*, 15, 379–389.

Goldman-Eisler, F. (1968). *Psycholinguistics-Experiments in Spontaneous Speech*. London: Academic Press.

Granlund, S., Hazan, V., Baker, R. (2011). Acoustic-phonetic characteristics of clear speech in bilinguals. *Proceedings of the 17th International Congress of Phonetic Sciences (ICPhS)*, 17-21 August 2011, Hong Kong, 763–766.

Greene, J.O., Lindsey, A.E., and Hawn, J.J. (1990). Social goals and speech production: Effects of multiple goals on pausal phenomena. *Journal of Language and Social Psychology*, 9, 119–134.

Griffin, G.R., and Williams, C.E. (1987). The effects of different levels of task complexity on three vocal measures. *Aviation, Space, and Environmental Medicine*, 58(12), 1165–1170.

Grynpas, J., Baker, R., Hazan, V. (2011). Clear speech strategies and speech perception in adverse listening conditions. *Proceedings of the 17th International Congress of Phonetic Sciences (ICPhS)*, 17-21 August 2011, Hong Kong, 779–782.

Hammerschmidt, K., and Jürgens, U. (2007). Acoustical correlates of affective prosody. *Journal of Voice*, 21(5), 531–540.

Harnsberger, J.D., Hollien, H., Martin, C.A., and Hollien, K.A. (2009). Stress and deception in speech: Evaluating layered voice analysis. *Journal of Forensic Science*, 54(3), 642–650.

Hausner, M. (1987). Sprechgeschwindigkeit als eine Funktion von Stress: Eine Fallstudie (Speech rate as a function of stress: A case study). *Language and Style*, 20, 285–311.

Hazan, V., and Baker, R. (2011). Acoustic-phonetic characteristics of speech produced with communicative intent to counter adverse listening conditions. *Journal of the Acoustical Society of America*, 130(4), 2139–2152.

Hecker, M.H.L., Stevens, K.N., von Bismarck, G., and Williams, C.E. (1968). Manifestations of task-induced stress in the acoustic speech signal. *Journal of the Acoustical Society of America*, 44, 993–1001.

Hicks, J.W., Jr. (1979). *An Acoustical/temporal Analysis of Emotional Stress in Speech*. PhD Dissertation, University of Florida.

Hocking, J.E., and Leathers, D.G. (1980). Nonverbal indicators of deception: A new theoretical perspective. *Communication Monographs*, 47, 119–131.

Hollien, H. (1990). *The Acoustics of Crime. The New Science of Forensic Phonetics*. New York: Plenum Press.

Hollien, H., Saletto J.A., and Miller, S.K. (1993). Psychological stress in voice: A new approach. *Studia Phonetica Posnaniensia*, 4, 5–17.

Huttunen, K., Keränen, H., Väyrynen, E., Pääkkönen, R., and Leino, T. (2011). Effect of cognitive load on speech prosody in aviation: Evidence from military simulator flights. *Applied Ergonomics*, 42, 348–357.

Jessen, M. (2006). *Einfluss von Stress auf Sprache und Stimme. Unter besonderer Beruecksichtigung polizeidienstlicher Anforderungen* (Influence of stress on language and voice. With special regard to Police requirements). Idstein: Schulz-Kirchner Verlag GmbH.

Johannes, B., Salnitski, V.P., Gunga, H-C., and Kirsch, K. (2000). Voice stress monitoring in space: Possibilities and limits. *Aviation, Space, and Environmental Medicine*, 71, 58–64.

Johnson, K., Flemming, E., and Wright, R. (1993). The hyperspace effect: Phonetic targets are hyperarticulated. *Language*, 69(3), 505–528.

Johnstone, T., and Scherer, K.R. (1999). The effects of emotions on voice quality. In *Proceedings of the 14th International Congress of Phonetic Sciences*, San Francisco, 2029–2032. Available at: http://brainimaging.waisman. wisc.edu/ wtjohnstone/0602.pdf, accessed 15 August 2014.

Jou, J., and Harris, R.J. (1992). The effect of divided attention on speech production. *Bulletin of the Psychonomic Society*, 30, 301–304.

Karlsson, I., Banziger, T., Dankovicová, J., Johnstone, T., Lindberg, J., Melin, H., Nolan, F., and Scherer, K. (2000). Speaker verification with elicited speaking styles in the VeriVox project. *Speech Communication*, 31, 121–129.

Khawaja, A. (2010). *Cognitive Load Measurement using Speech and Linguistic Features*. PhD thesis, submitted to University of New South Wales.

Khawaja, A., Chen, F., Owen, C., and Hickey, G. (2009). *Cognitive load measurement from user's linguistic speech features for adaptive interaction design*. In T. Gross, J. Gulliksen, P. Kotze, L. Oestreicher, P. Palanque, R.O. Prates, and M. Winckler (Eds) *Human–Computer Interaction – INTERACT 2009. Lecture Notes in Computer Science: Information Systems and Applications*. Berlin, Heidelberg: Springer, 485–489.

Kirchhübel, C., Howard, D.M., and Stedmon, A.W. (2011). Acoustic correlates of speech when under stress: Research, methods and future directions. *The International journal of Speech, Language and the Law*, 18(1), 75–98.

Köster, S. (2001). Acoustic-phonetic characteristics of hyper-articulated speech for different speaking styles. *Proceedings International Conference on Acoustics, Speech and Signal Processing*, Salt Lake City, Piscataway, NJ: IEEE 873–876.

Kowal, S., and O'Connell, D. (1987). Some temporal aspects of stories told while or after watching a film. *Bulletin of the Psychonomic Society*, 25, 364–366.

Krause, J.C., and Braida, L.D. (2002). Investigating alternative forms of clear speech: The effects of speaking rate and speaking mode on intelligibility. *Journal of the Acoustical Society of America*, 112, 2165–2172.

Krause, J., and Braida, L.D. (2004). Acoustic properties of naturally produced clear speech at normal speaking rates. *Journal of the Acoustical Society of America*, 115(1), 362–378.

Kraut, R.E. (1980). Humans as lie detectors: Some second thoughts. *Journal of Communication*, 30, 209–216.

Kuroda, I., Fujiwara, O., Okamura, N., and Utsuki, N. (1976). Method for determining pilot stress through analysis of voice communication. *Aviation, Space and Environmental Medicine*, 47, 528–533.

Lindblom, B. (1990). Explaining phonetic variation: a sketch of the H&H theory. In W. J. Hardcastle and A. Marchal (Eds), *Speech Production and Speech Modelling*. The Netherlands: Norwell, MA: Kluwer Academic, 403–439.

Liu, S., Del Rio, E., Bradlow, A.R., and Zeng, F-G. (2004). Clear speech perception in acoustic and electric hearing. *Journal of the Acoustical Society of America*, 116(4), 2374–2383.

Lively, S.E., Pisoni, D.B., Van Summers, W., and Bernacki, R.H. (1993). Effects of cognitive workload on speech production: Acoustic analysis and perceptual consequences. *Journal of the Acoustical Society of America*, 93, 2962–2973.

Lykken, D. (1998). *A Tremor in the Blood: Uses and Abuses of the Lie Detector*. Reading, MA: Perseus Publishing.

Meinerz, C. (2010). *Effekte von Stress auf Stimme und Sprechen: Eine phonetische Untersuchung auf der Grundlage ausgewählter akustischer und sprecherdynamischer Parameter unter Berücksichtigung verschiedener Stressklassen* (Effects of stress on voice and speech: A phonetic investigation on the basis of selected acoustic and dynamic speaker parameters considering different stress classes). Norderstedt: Books on Demand GmbH.

Mendoza, E., and Carballo, G. (1998). Acoustic analysis of induced vocal stress by means of cognitive workload tasks. *Journal of Voice*, 12, 263–273.

Miller, G.R., and Stiff, J.B. (1993). *Deceptive Communication*. Newbury Park, CA: Sage.

Moon, S.J., and Lindblom, B. (1994). Interaction between duration, context, and speaking style in English stressed vowels. *Journal of the Acoustical Society of America*, 96, 40–55.

Müller, C., Großmann-Hutter, B., Jameson, A., Rummer, R., and Wittig, F. (2001). Recognizing time pressure and cognitive load on the basis of speech: An experimental study. In J. Vassileva, P. Gmytrasiewicz, and M. Bauer (Eds) *UM2001, User Modelling: Proceedings of the Eighth International Conference*. Berlin: Springer, 24–33.

Oomen, C.C.E., and Postma, A. (2001). Effects of divided attention on the production of filled pauses and repetitions. *Journal of Speech, Language, and Hearing Research*, 44, 997–1004.

Picheny, M.A., Durlach, N.I., and Braida, L.D. (1986). Speaking clearly for the hard of hearing II: intelligibility differences between clear and conversational speech. *Journal of Speech and Hearing Research*, 29, 434–446.

Protopapas, A., and Lieberman, P. (1997). Fundamental frequency of phonation and perceived emotional stress. *Journal of the Acoustical Society of America*, 101, 2267–2277.

Rockwell, P., Buller, D.B., and Burgoon, J.K. (1997). The voice of deceit: refining and expanding vocal cues to deception. *Communication Research Reports*, 14(4), 451–459.

Roßnagel, C. (1995). Übung und Hörerorientierung beim monologischen Instruieren: Zur Differenzierung einer Grundannahme (Exercise and listener orientation in monological instructing: Differentiating a basic assumption). *Sprache und Kognition*, 14, 16–26.

Ruiz, R., Absil, E., Harmegnies, B., Legros, C., and Poch, D. (1996). Time- and spectrum-related variabilities in stressed speech under laboratory and real conditions. *Speech Communication*, 20, 111–129.

Rummer, R. (1996). *Kognitive Beanspruchung beim Sprechen* (Cognitive load in speech). Weinheim: Beltz.

Scherer, K.R. (2003). Vocal communication of emotion: A review of research paradigms. *Speech Communication*, 40(2003), 227–256.

Scherer, K.R., Grandjean, D., Johnstone, T., Klasmeyer, G., and Bänziger, T. (2002). Acoustic correlates of task load and stress. *Proceedings of the International Conference on Spoken Language Processing (ICSLP)*, Denver, USA: Interspeech, 2017–2020.

Schlenker, B.R., and Leary, M.R. (1982). Social anxiety and self-presentation: A conceptualisation and model. *Psychological Bulletin*, 92, 641–669.

Siegman, A.W. (1993). Paraverbal correlates of stress: implications for stress identification and management. In L. Goldberger and S. Breznitz (Eds) *Handbook of Stress. Theoretical and Clinical Aspects*, 2nd edn. New York: The Free Press, 274–299.

Sigmund, M. (2006). Introducing the database ExamStress for speech under stress. *Proceedings of the 7th Nordic Signal Processing Symposium*, Rejkjavik, Piscataway, NJ: IEEE, 290–293.

Silberstein, D., and Dietrich, R. (2003). Cockpit communication under high cognitive workload. In D. Dietrich (Ed.), *Communication in High Risk Environments*.Hamburg: Helmut Buske Verlag, 9–56.

Smiljanić, R., and Bradlow, A.R. (2005). Production and perception of clear speech in Croatian and English. *Journal of the Acoustical Society of America*, 118(3), 1677–1688.

Smiljanić, R., and Bradlow, A.R. (2009). Speaking and hearing clearly: Talker and listener factors in speaking style changes. *Language and Linguistics Compass*, 3, 236–264.

Streeter, L.A., Krauss, R.M., Geller, V., Olson, C., and Apple, W. (1977). Pitch changes during attempted deception. *Journal of Personality and Social Psychology*, 35, 345–350.

Streeter, L.A., Macdonald, N.H., Apple, W., Krauss, R.M., and Galati, K.M. (1983). Acoustic and perceptual indicators of emotional stress. *Journal of the Acoustical Society of America*, 73, 1354–1360.

Stroemwall, L.A., Hartwig, M., and Granhag, P.A. (2006). To act truthfully: Nonverbal behaviour strategies during a police interrogation. *Psychology, Crime and Law*, 12(2), 207–219.

Stroop, J.R. (1935). Studies of interference in serial verbal reactions. *Journal of Experimental Psychology*, 18(6), 643–662.

Tolkmitt, F.J., and Scherer, K.R. (1986). Effect of experimentally induced stress on vocal parameters. *Journal of Experimental Psychology*, 12(3), 302–313.

Uchanski R.M. (2005). Clear speech. In D.B. Pisoni and R.E. Remez (Eds), *Handbook of Speech Perception*.Malden, MA: Blackwell Publishers, 207–235.

Uchanski, R.M., Choi, S.S., Braida, L.D., Reed, C.M., and Durlach, N.I. (1996). Speaking clearly for the hard of hearing IV: Further studies of the role of speaking rate. *Journal of Speech and Hearing Research*, 39, 494–509.

Van Lierde, K., van Heule, S., De Ley, S., Mertens, E., and Claeys, S. (2009). Effect of psychological stress on female vocal quality: A multi-parameter approach. *Folia Phoniatrica et Logopaedica*, 61, 105–111.

Vilkman, E., and Manninen, O. (1986). Changes in prosodic features of speech due to environmental factors. *Speech Communication*, 5, 331–345.

Vilkman, E., Manninen, O., Lauri, E.-R., and Pukkila, T. (1987). Vocal jitter as an indicator of changes in psychophysiological arousal. In M. Jessen, *Einfluss von Stress auf Sprache und Stimme. Unter besonderer Beruecksichtigung polizeidienstlicher Anforderungen.* (Influence of stress on language and voice. With special regard to Police requirements) Idstein: Schulz- Kirchner Verlag GmbH, Proceedings 11th International Congress of Phonetic Sciences, Tallin, 188–191.

Vrij, A. (2008). *Detecting Lies and Deceit: Pitfalls and Opportunities*, 2nd edn. Chichester: Wiley.

Vrij, A., and Heaven, S. (1999). Vocal and verbal indicators of deception as a function of lie complexity. *Psychology, Crime and Law*, 5(3), 203–215.

Vrij, A., Semin, G.R., and Bull, R. (1996). Insight into behavior displayed during deception. *Human Communication Research*, 22, 544–562.

Walczyk, J.J., Roper, K.S., Seemann, E., and Humphrey, A.M. (2003). Cognitive mechanisms underlying lying to questions: Response time as a cue to deception. *Applied Cognitive Psychology*, 17, 744–755.

Walcyyk, J.J., Schwartz, J.P., Clifton, R., Adams, B., Wei, M., and Zha, P. (2005). Lying person-to-person about live events: A cognitive framework for lie detection. *Personnel Psychology*, 58, 141–170.

Waters, J., Nunn, S., Gillcrist, B., and Von Colln, E. (1995). The effect of stress on the glottal pulse. In I. Trancoso and R. Moore (Eds), *Proceedings of the ESCA-NATO Tutorial and Research Workshop on Speech under Stress*. 14–15 September 1995, Lisbon: European Speech Communication Association, 9–11.

Williams, C.E., and Stevens, K.N. (1969). On determining the emotional state of pilots during flight: An exploratory study. *Aerospace Medicine*, 40, 1369–1372.

Williams, C.E., and Stevens K.N. (1972). Emotions and speech: Some acoustical correlates. *Journal of the Acoustical Society of America*, 52, 1238–1250.

Yap, T.F., Epps, J., Choi, E.H.C., and Ambikairajah, E. (2010). Glottal features for speech-based cognitive load classification. *Proceedings of the IEEE International Conference on Acoustics, Speech and Signal Processing*, 5234–5237.

Yap, T.F., Epps, J., Ambikairajah, E., and Choi, E.H.C. (2011). Formant frequencies under cognitive load: Effects and classification. *EURASIP Journal on Advances in Signal Processing*, Article ID 219253, doi: 10.1155/2011/219253.

Zuckerman, M., DePaulo, B.M., and Rosenthal, R. (1981). Verbal and nonverbal communication of deception. In L. Berkowitz (Ed.), *Advances in Experimental Social Psychology*.New York: Academic Press, 14(1989), 1–59.

Chapter 21
Evaluating Counter-terrorism Training Using Behavioural Measures Theory

Joan H. Johnston
Naval Air Warfare Center Training Systems Division, Orlando, USA

V. Alan Spiker
Anacapa Sciences, Inc., Santa Barbara, USA

Introduction

Since 2008, the US Department of Defense has placed Irregular Warfare (IW) on an equal footing with conventional warfare in future military planning and operations (Department of Defense, 2008). Among IW mission objectives are developing Counter-Terrorism (CT) competencies to identify people with hostile intent before events become lethal. As part of IW mission readiness, small units are called upon to execute a full range of kinetic and non-kinetic operations, often within a single day. Whether practiced by military ground units or law enforcement personnel (e.g., Customs and Border Patrol), the knowledge, skills, and attitudes (KSAs) to read the human and physical terrain is an essential element of CT training. Constructing the behavior profiles necessary to read terrain and infer hostile intent is now considered every bit as important as body armor and weaponry (Kobus and Williams, 2010), so that profiling KSAs have become a valuable addition to small unit tactics, techniques, and procedures (TTPs).

The CT behavior profiling KSAs receive special emphasis in the Combat Hunter (CH) training conducted by the US Marine Corps School of Infantry. CH is a 10-day course that is taught in two 5-day blocks, Combat Tracking and Combat Profiling, by subject matter experts in the respective disciplines. Each block has a classroom academic and a scenario-based field exercise segment. The academic segment introduces core concepts to the students, while the field scenarios reinforce the material through repeated practice and feedback. The typical class size is around 40 students, who are drawn from the same regiment though different platoons and squads.

In Combat Tracking, students receive academic instruction on the fundamentals of tracking (e.g., footprint interpretation, maintaining a track line) while field scenarios have students track 'quarry' (role-playing instructors) as five-person tracking teams. During the field exercises, students learn to read quarry footprints, interpret environmental cues, build social/biometric profiles of their quarry, anticipate their quarry's actions by acquiring their mindset, and incorporate their own unit's TTPs into their newly-acquired tracking skillset. Over days, the field scenarios increase in complexity as the terrain becomes more difficult, the quarry more 'skilled' (i.e., the role-players use more deceptive tactics), and more intricate team maneuvers are employed.

Combat Profiling is concerned with perceiving, analyzing, and articulating critical events within the human terrain. Its main goal is to identify pre-event indicators of human behavior before a destructive event occurs, by training individuals to look for behavioral anomalies beyond the baseline of a culture or a particular location. Through profiling, students learn to be more situationally aware and accurately interpret subtle cues that forewarn a critical event. In academics, students are exposed to the basic concepts of optics fundamentals, pattern recognition, reasoning by analogies, forming prototypes, ethical-moral decision-making, and the six domains of combat profiling: heuristic, geographics, proxemics, atmospherics, biometrics, and kinesics (Kobus and Williams, 2010). The field segment splits students into teams, where they occupy observation posts to observe role-players engaged in varying types of behavior within a village mockup. Observations are made at a distance (via optics), where they must distinguish instances of neutral and potentially hostile behavior in the context of increasingly challenging scenarios. Training culminates in a four-hour final exercise where all teams deploy as a maneuver unit into the village using insights gained from the previous scenarios.

While the 10-day course is the typical CH offering, training in tracking and profiling has been re-packaged for other venues, including a shorter version, a train-the-trainer course, and a special 'gold standard' course (for Army and law enforcement officers) called Border Hunter (Institute for Training and Simulation, 2010). Since its initial offering in 2007, there has been intense interest within US DOD to demonstrate the effectiveness of CH training in order to justify the time/ resource commitments in sending units to the training and expanding throughput so more Marines can receive the training prior to deployment (Kobus et al., 2009). However, serious training challenges must be overcome to acquire the diverse and complex KSAs that comprise Combat Tracking and Profiling. Equally great are the methodological demands for measuring those KSAs in both academic and field settings. For a successful evaluation, a multi-measure, multi-method approach is needed where one 'triangulates' on a conclusion if multiple indices point in the same direction (Cook and Campbell, 1979).

To that end, Table 21.1 lists ten classes of behavioral measures that have been used in various evaluation studies of CH training effectiveness. In the remainder of the chapter, we describe four studies that have used these measures in varying combinations and venues. Before presenting the study results, we briefly discuss the key dimensions on which these measures vary.

Table 21.1 Ten classes of behavioral measures used to evaluate CH training effectiveness

M#	Measure Type	Description	Employed by:	Study
1	Competency-requirements matching	Functional performance requirements are elicited from job subject matter experts and matched to the primary underlying cognitive or behavioral skill area.	Irregular Warfare Training Symposium (2009)	1
2	Objective indices	Observers score performance using objective indices of counting, speed, and accuracy.	Goulding (2008); Hilburn (2007)	2
3	Protocol accounts	Trainees provide a written account of a visual scene where protocols are scored for accuracy, completeness, detail, and interpretation.	Kobus et al. (2009)	2
4	Event-based checklist	Observer scores whether a particular action occurred during the scenario at the appropriate time.	Johnston, Poirier and Smith-Jentsch (1998); Spiker et al. (2010)	2
5	Exemplar-primed KSA matrix	Observer scores the presence of a class of profiling behaviors based on the occurrence of specific marker behaviors.	Spiker and Johnston (2010a)	3
6	Adherence to established criteria based on scientific principles	Observers rate the extent to which training content and delivery adheres to specific criteria that were derived from principles of learning.	Spiker and Johnston (2010a)	3
7	Anecdotal field reports	After-action reports from recently returned troops on the utility of profiling training they had received prior to their deployment.	Williams (2009); Spiker and Williams (2010)	3
8	Perceptual Functional Field Of View (FFOV) measures	Perception tests of trainee FFOV via accuracy of reports of information on periphery of a scene.	Kobus et al. (2010)	4
9	Situational Judgment Tests (SJTs)	Scenario-based written tests of trainee judgment and decision-making administered before and after field scenario exercises.	Spiker and Johnston (2010b)	4
10	Behavioral Observation Checklists (BOCs)	Rated field performance on basic and advanced profiling behaviors.	Spiker and Johnston (2010b)	4

The measures vary on a number of dimensions, including degree of objectivity (M#2 and #8 require little interpretation); extent to which the resulting data can be quantified, as with counting or rating scales (M#2, 3, 8, 9, 10); and whether they are used to assess trainer (M#6, 7) or trainee (the others) performance. Other distinguishing dimensions include whether they: can be used to assess the impact of academic learning (M#3, 6) vs field performance; require a high (M#4) vs low (#9, 10) degree of scripting to be useful; are more analytic (M#1) than empirical; and require portability (M#8 is a lab setup). This list by no means exhausts the possible ways to evaluate training effectiveness using behavioral measures, but reflects the ones used to evaluate CH.

As we discuss these measures, we will show how theory-based research can result in more well-developed concepts and more practical solutions to some of the challenging problems in training effectiveness evaluation. To this end, the theoretical approaches espoused in the Navy's Tactical Decision Making Under Stress (TADMUS) program (Smith-Jentsch et al., 1998) and Klein's Naturalistic Decision-Making model (Klein, 2008), among others, have guided the development of psychometrically-sound behavioral measures. Importantly, these models not only tell us what to measure, they also tell us why certain strategies of training are effective and others are not.

Study 1: Identification of CT Competencies

The first step in establishing critical job competencies is typically accomplished through structured interviews with job subject matter experts (SMEs). Functional performance requirements are elicited from SMEs and then matched to the primary underlying cognitive or behavioral skill areas of interest. Study 1 describes how the competency-requirements matching (M#1) method was used with participants during the 2009 Irregular Warfare Training Symposium to identify small unit (squad) skill areas and learning requirements. The first author used research findings from the literature (e.g., cultural cognition: Johnston et al., 2011; TADMUS: Cannon-Bowers and Salas, 1998) to categorize skills into the four competencies of decision-making, teamwork, stress resilience, and cultural cognition. Overall, decision skills had the most number of job requirements. Decision-making skills include: using effective observation, tracking, and pattern recognition (e.g., analysis of atmospheric and body language); using all the human senses; demonstrating tolerance of information ambiguity (e.g., partial information); understanding when to act and when not to act; and considering the positive and negative consequences of moral, legal, and ethical decisions in terms of second- and third-order effects. Teamwork involves participating in a team decision-making process, wherein team members are sensors that coordinate and pass information to create a common operational picture; and being proactive and adaptive in order to make rapid shifts in missions. Stress skills involve using effective attention and energy management in order to be resilient to stress, and recognizing when unit members need to rest so they can 'recharge their batteries'. Cultural cognition includes understanding the impact of cultural differences and how the enemy exploits that to their own advantage; employing survival language skills; understanding how to work with interpreters; employing effective negotiation strategies; and recognizing cultural and language competencies in subordinates and assigning duties accordingly. Our focus in this chapter is primarily on the decision-making competency, with the other three competencies addressed to a lesser extent in some of the studies.

Study 2: Development of CT Skills

Study 2 is a cumulative analysis of a series of individual effectiveness studies that were conducted during the early stages of CH implementation. They include three Limited Objective Experiments (LOEs) conducted by the US Marine Corps Warfighting Laboratory (MCWL) in 2007 (Goulding, 2008), a CH trainer course evaluation conducted by Kobus et al. (2009) in 2009, and interviews with CH instructors also conducted in 2009 by Spiker et al. (2010). With regard to the LOEs, initial administrations of the course were evaluated by comparing trainees' ability to accurately and quickly detect hostile intent within the context of semi-structured field scenarios. Experimental assessments were made by comparing the performance of Marines who had not received any CH training (the control) with those who had received either the Profiling or Tracking blocks of CH.

The LOE results from Goulding (2008) are presented in terms of how they highlight the use of objective indices (M#2). In this regard, the LOE evaluators collected data on whether, how quickly, and at what distance each team detected a number of potentially hostile entities within a series of increasingly more complex scenarios. Examples of hostile intent included trucks (potentially with explosives), a possible IED factory, snipers, prowlers, and suspicious characters with cell phones, to name a few. The scenarios increased in complexity, starting with stationary observations and graduating to day-time patrol, night patrol, urban patrol, and various team maneuvers. Importantly, notable differences in favor of the experimental groups were found in the accuracy of detection (as a percentage), time to detect hostile threats, and distances at which threats were detected. For each objective index, the differences were quite pronounced, where depending on the scenario, Marines who received CH training exhibited threat detection superiority on the order of 20 to 50 percent compared to the control group. Needless to say, the results were taken as a positive indication of CH effectiveness and the MCWL commissioned administration of CH training on a broader scale starting in 2008.

In a subsequent assessment of CH training effectiveness, Kobus et al. (2009) used a pre-/post-test design to compare trainees' ability to describe the important aspects of a scene before and after CH Profiling training. For this assessment, trainees were shown a picture slide of a Third World village (e.g., children standing at a corner surrounded by broken pavement and litter, or a meeting between military and local leaders) for 45 seconds. The projector was then turned off and the trainees were given 5 minutes to write down information they would report to higher command based on their observations. These written protocols (M#3) were collected before training began and at the end of training, where different slides (equated for difficulty) were used. Content analysis performed by subject matter experts was used to score each protocol with regard to descriptiveness (e.g., number of children in the picture), meaningfulness (e.g., dirt barrier that might impede traffic), and use of CH terminology. The results indicated that trainees' scores on descriptiveness, meaningfulness, and terminology all increased significantly ($p < 0.01$) from pre-test to post-test, thus indicating that CH training aided trainees' ability to interpret and extract useful information from a visual scene. In a separate analysis, the protocol accounts were reviewed by an expert intelligence officer with combat experience to gauge their value from an intelligence perspective. Again, there was a significant increase in the intelligence value of trainees' pictorial scene descriptions from pre-test to post-test.

While objective indices and written protocols are useful for gauging the magnitude of CH training effectiveness, they provide only a limited capability to specify what skills are being acquired or how. For this more in-depth qualitative assessment, an Event-Based Checklist or EBC (M#4) can be used. An EBC is very useful for scoring performance by a trained observer when the scenario being observed has been sufficiently scripted in advance so events can be prerecorded in sequence on a checklist along with notes to assist in scoring and interpreting the scenario action. Table 21.2 describes an event to be scored during one of the CH LOEs described by Goulding (2008). With scripting, it is known approximately when the suspicious villager will be talking on the cell phone. As can be seen, the checklist provides prompts for the observer to note whether and when the event is detected. If it is, then further elements of the checklist allow for an evaluation of communication and decision adequacy, where these are to be judged by the observer on a simple three-point scale. Less subjective are the indications of who in the team makes the communication and/or decision. Using these indices as building blocks, a summary index of how well an event was handled can be scored on a five-point scale (where 0 = event not detected). Comparison of CH-trained Marine performance with Marines that did not receive the training demonstrated a consistent advantage for CH training, with average event-rating differences on the order of 1.3 in one of the LOEs.

Table 21.2 Example of an event-based checklist

Elements of event	Element measure
Event to be observed	Suspicious villager talking on a cell phone
Time anomaly detected	From scenario start
Distance at which detected	In meters from location
Who detected anomaly first	Point/flank/team leader/other
How communicated	Radio, verbal, signal, other
Time communicated	From scenario start
Evaluation of communication	Good/adequate/poor
Who made decision	Point/flank/team leader/other
Adequacy of decision	Good/adequate/poor

Study 3: Strategies for CT Training

While the measures in Study 2 paint a fairly comprehensive picture of trainee performance, they do not directly address the quality and nature of the training strategies that are being used to instill CT KSAs during CH. Study 3 assessed CH training practices by observing two successive course offerings in their entirety, including all academic lessons and field scenarios (Spiker and Johnston, 2010a). The authors collected data via extensive field notes and structured observation forms, portions of which are discussed below as Measures 5 and 6. We supplemented these with interviews with returning Marines from Afghanistan concerning the utility of their CH training during their deployment (Spiker and Williams, 2010).

To discern the qualitative aspects of performance, we constructed a 33-item exemplar-primed KSA matrix (M#5) that captures how trainees' tracking and profiling KSAs 'emerged' during the field scenarios. With this instrument, each higher-order KSA is instantiated with specific marker behaviors. The KSAs were developed to reflect the program of instruction utilized in each curriculum block, so data collected using this measure speaks directly to whether the course objectives are being accomplished. The matrix was converted into a scoring instrument by having the researcher review their field notes and pinpoint which behaviors occurred, and when, during discrete points in training. To facilitate usability, the KSAs were organized into five categories: use of enhanced observation techniques, identification of critical event indicators, proactive analysis, dynamic decision-making, and employment of cognitive discipline. Table 21.3 illustrates two of these KSAs, one from the Cognitive Discipline area (KSA #18) and the second from Synthesis of Ambiguous Information (KSA #28). Below each KSA are illustrative data extracted from field notes. Each KSA applies to both course blocks, only the particular marker behaviors change. This framework was an efficient way to capture the richness of the researchers' field notes and document the qualitative aspects of KSAs emerging over training. Specifying concrete behaviors for each KSA allowed rapid mapping of notes to KSAs. Analysis of the resultant qualitative data showed that all KSAs were evident during the field scenarios in both course segments, with the frequency and complexity of the behaviors increasing over the course of training.

Table 21.3 Examples of a behavior-primed KSA matrix

	Profiling		Tracking	
KSA	**Behavior marker**	**Field note**	**Behavior marker**	**Field note**
#18. Keep an open mind to the unexpected (recognize there are unknown variables in the situation).	Do they consider the possibility that insurgents might use new tactics (e.g., different IED emplacing) or attempt something completely different than anything that has been tried before?	*Individual team members work through and talk through the courses of action that the Prince might take following the extensive violence in his village, including negotiation, fleeing, fighting, or calling in the US.*	Do they consider that the hostiles might consider something completely different, like splitting up to rejoin at a rally point further down the track line?	*Team Leader decides to have team follow main group if quarry splits off; after quarry moved from initial sighting point, team attempted to predict which ravine they might have headed for.*
#28 Imagining alternative courses of action or alternative event outcomes by what-if mental simulations.	Does the profiler try to 'think through' what might be happening in an unfolding event (e.g., a possible complex ambush) by running through different alternative outcomes?	*The team begins discussion of what kind of signs might be present to indicate a possible attack on the neutral nomadic village, including sniper-looking individuals lurking by the tents and the possibility that some of them might become more active in village affairs, such as attending a wedding with the Prince.*	Does the tracker or team leader try to 'think through' what the quarry might be doing (ahead of them) based on the track pattern they are looking at?	*Team stops and looks at action indicator, examining the quarry's formation to determine if they are splitting up and planning an ambush.*

Spiker and Johnston (2010a) also developed a more direct index of CT training in order to assess adherence to established training principles (M#6). An Instructional-Behavior Observation Criteria (I-BOC) checklist of 25 items was drawn from research on decision-making (TADMUS), teamwork (Smith-Jentsch et al., 1998), critical thinking (Cohen, Freeman and Thompson, 1998), and stress resilience (Driskell and Johnston, 1998). Example items from the I-BOC assessed whether:

- Critical incident interviews were used to identify critical thinking training requirements
- Critical incidents were used for generating demonstration, practice, and test materials for training
- Trainees were taught how to challenge one's own personal biases and consistently reevaluate the situation as new information is received
- Trainees were taught how to frame problems in order to identify the essential elements in a situation
- Team members were encouraged to provide leadership and show initiative in behaviors that provide direction for the team
- Simulation exercises provided an opportunity to demonstrate critical thinking skills and provide feedback in real-time.

The I-BOC was then converted into numeric form by rating each criterion using a three-point scale, where the training: did not cover (0), partially covered (1), or completely covered (2) the criterion. Ratings were assigned during observations of CH training described under Measure 5. Follow-up interviews were conducted with the lead instructors to ensure observations were recorded accurately. Results showed that a majority of the criteria were completed satisfactorily, with an overall I-BOC score of 88 percent (Spiker and Johnston, 2010a), supporting adherence of the CH training curriculum with established training principles.

Perhaps the most direct, though less rigorous, index of CH training utility is anecdotal field reports (M#7) of Marines returning from deployment, principally Afghanistan. Classified as After Action Reports or AARs, these field reports have value, despite their anecdotal nature, in pinpointing those aspects of CT training that Marines found operationally useful. Of particular value are reports from Marines who have had multiple deployments, as they can contrast their earlier deployments (without CH training) with their most recent post-CH training deployment.

A number of CH AARs have now been conducted, all with very positive results. In a post-deployment survey of one of the first battalions to receive CH training (Marine Corps Center for Lessons Learned, 2008), the median rating (on a 10-point scale) of the utility of CH training given by 30 respondents was 8, indicating that most respondents found the training useful for serving in theater. Furthermore, Williams (2009) reported that of 62 Marines who received CH training in-country, all respondents gave glowing reports on the quality and value of the instruction.

Spiker and Williams (2010) interviewed 12 Marines from a battalion that had just had a very successful deployment in Afghanistan. Respondents included the Battalion Commander, Executive Officer, Operations Officer, and a Company Commander, among others. Once again, every one of the 12 Marines extolled the virtues of CH training and cited numerous instances where training in profiling and tracking made the difference between success and failure under high threat conditions.

Most recently, Bancroft (2012) reported summary statistics from AARs of four Marine battalions that had deployed during the past year. Surveys were used to collect the data. Sixty percent of the respondents indicated they had used CH techniques at least every quarter while deployed. The same percentage reported that CH training increased their knowledge of the technique from a baseline of 'none or very little' to 'above average' or 'excellent.' Moreover, over 80 percent of those surveyed reported that the quality of the CH training was above average.

While the high percentages of reported utility and training value support CH effectiveness, the comments themselves give the best indication of what aspects of CH training are useful for CT operations. Content analysis of the extensive comments is beyond the scope of this paper, but an overview will shed light on select aspects of CH training that appear to transfer well to the operational IW environment. For example, the modal responses from the original (Marine Corps Center for Lessons Learned, 2008) AAR survey concerning CH training benefits included counter IED work, tracking and positively identifying enemy personnel, discerning situations where danger was imminent, and understanding the use of optics. In Williams (2009), areas that benefited from CH training were taking someone else's perspective, anticipating what will happen next, using prototypes for more rapid identification, and finding signature locations (anchor points, habitual areas) to streamline visual search. Importantly, these skills can be acquired and used by anyone in-country – drivers, cooks, civilian contractors – not just Marine ground units. Spiker and Williams (2010) noted that CH benefits commonly cited included consistent terminology to improve combat reporting, increased confidence in countering a terrorist act ('staying left of bang'), being more aware of behavioral indicators of high value individuals (e.g., leaders), appreciating the value of tactical patience, and being more aware of the need to look for cues earlier in the entire cycle of unfolding behavior (e.g., steps in planting an IED).

Study 4: CT Train the Trainer Assessment

In this fourth study, 41 students took a 'graduate level' version of CH called Border Hunter (BH). BH was twice as long as CH and intended as a train the trainer course for experienced (average 9 years in service) Army, Border Patrol, and other law enforcement personnel (Fautua et al., 2010). A specialized team of 14 Tracking and Profiling instructors was provided, who had a combined experience of nearly 400 years. The training was held at Fort Bliss Texas in April 2012, under the auspices of the Joint Task Force North. A 13-person research team participated in collecting observational and laboratory data from trainees and instructors throughout the 20-day training period.

A core objective of BH is to improve trainees' observational skills so they are able to 'see more' and extract more information from a complex scene, whether studying tracks on the ground or profiling the behaviors of possible threats. To that end, Kobus et al. (2010) administered a selective attention test to students at the beginning of BH and again at the end to see whether their functional field of view (FFOV) (M#8) had increased, an expected outcome if BH training is having its intended effect. Trainees were seated individually in front of a computer monitor where they were shown simple stimuli (smiling or frowning schematic faces) and asked to indicate the location (quadrant) of a matching stimulus displayed a short time after the target stimulus. The visual angle between the two stimuli was systematically varied, where targets were surrounded by an array of distracters. Statistical testing showed a significant ($p < 0.01$) increase in FFOV over the course of training. Subtests revealed this effect was present for all subgroups, regardless of prior experience. Thus, the FFOV results provide evidence that BH training improves participants' underlying observational skills, allowing them to extract more information from a complex scene than before training.

While the FFOV test is an excellent way to assess improvements in basic perceptual processes, it has historically been difficult to pinpoint corresponding gains in higher-order processes such as decision-making and judgment resulting from field training exercises. Although one can administer tests of declarative knowledge to assess improvements due to academic training, it is difficult to document field training effectiveness since the observation conditions are demanding, the curriculum is not always standardized, its objectives are not always well-specified, and instructional delivery is highly variable.

To fill this gap, situational judgment tests (SJTs) (M#9) were developed and administered to participants at the beginning and end of field training for both the Tracking and Profiling blocks. SJTs are low- to moderate-fidelity work sample simulations that ask respondents to assess the effectiveness of various response options (Gessner and Klimoski, 2006). The scenarios are intentionally written so that not all situational cues are known, which requires a balance between analysis and intuition, where good judgment is the ability to go beyond the information given and rely on broader knowledge and experience. If trainees have been acquiring this experience in the field, they should exhibit improved performance on the SJTs between pre-test and post-test. The SJT is a particularly useful tool because its realistic scenario items reasonably approximate the types of cognitive process improvements expected from repeated field scenarios. By using instructor responses to the same test as the 'answer key', we assessed how trainees' mental representations of real-world problems began to resemble those of the instructors.

Each test consisted of six items; an example from the Tracking SJT is presented below.

Your team has been tracking an experienced, well-armed band of insurgents for several days. The time/distance gap has been slowly closing to where it is now about 8 hours. You come upon where their tracks should be, but they have been obliterated by tracks of local cattle that cut through the ground spoor from several directions.

Please rate the effectiveness of the following six decision options using this 5-point scale. Don't hesitate to use the entire scale in judging these choices.

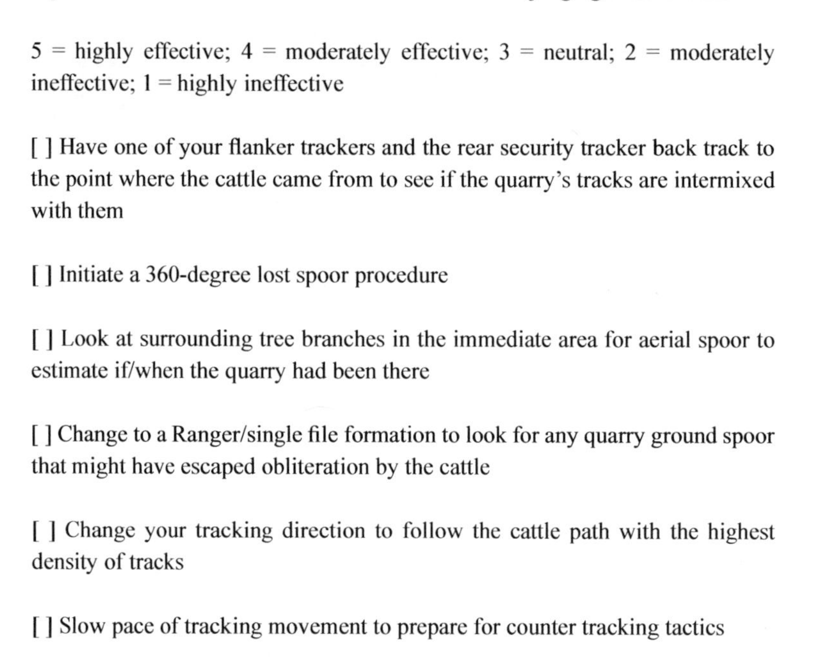

5 = highly effective; 4 = moderately effective; 3 = neutral; 2 = moderately ineffective; 1 = highly ineffective

[ ] Have one of your flanker trackers and the rear security tracker back track to the point where the cattle came from to see if the quarry's tracks are intermixed with them

[ ] Initiate a 360-degree lost spoor procedure

[ ] Look at surrounding tree branches in the immediate area for aerial spoor to estimate if/when the quarry had been there

[ ] Change to a Ranger/single file formation to look for any quarry ground spoor that might have escaped obliteration by the cattle

[ ] Change your tracking direction to follow the cattle path with the highest density of tracks

[ ] Slow pace of tracking movement to prepare for counter tracking tactics

The test items were constructed to cover the core skills for each course segment. The scenario setup is brief, reducing the reading requirement and leaving key information omitted. Trainees provided a 1 to 5 rating for each of six possible response options. The options were designed so that two were fairly good, two mediocre, and two poor. We calculated the Euclidean distance of the student's responses from the instructor's to generate a total score, where lower scores reveal a closer match and hence are better.

The SJT results are presented in Figure 21.1 for both blocks. The figure depicts the number of students whose post-test–pre-test difference score fell into one of the bins of size 10. Negative scores indicate a learning effect, as students' deviation (from the instructors) score was smaller on post-test, a desirable outcome. There was statistical evidence for improvement in judgment during Tracking, as students' post-test scores were lower than pre-test ($t = 2.229$, $p < 0.011$, $df = 41$). Though failing to reach significance, the trend for the Profiling SJTs was in the right direction. Looking at the distribution, the field training experience appears to 'calibrate' those students whose initial (pre-test) judgments were discrepant from the instructors' judgments; where the number of such low-scoring students decreased dramatically from pre test to post test.

The Behavioral Observation Checklist (BOC) (M#10) was also used to assess the effectiveness of the field training exercises. BOCs offer a structured method to collect quantitative and qualitative data on individual and team performance during field exercises. Separate BOCs were created for Tracking and Profiling, where each instrument was created in a two-column layout for portability. The Tracking BOC consisted of a series of three-point and five-point rating scales. The three-point scales covered basic procedural skills for Tracking (e.g., not walking on spoor, recording starting point, keeping in visual contact with other team members) or Profiling (e.g., covering an observation sector, distributing duties, constructing accurate profiling baseline). The five-point rating scales addressed higher-order behaviors, some of which are common (communication, mindset of threat, decision-making) while others are unique to Tracking (dynamics of footprint, team control) or Profiling (anticipating events, tactical patience). A further column was used for noting problems, emerging skills, and areas of team strength and weakness.

Throughout field training, trainees were divided into five eight-person teams. Teams were observed by at least one field researcher who was responsible for completing the BOC for that block. For Tracking, a researcher followed the team on a not-to-interfere basis, using the BOC to score behaviors on each of 10 days. Similarly, a field researcher accompanied each team in Profiling as they observed a village mockup from afar at one of the pre-established observation platforms. For select scenarios, two researchers observed the same team to gauge inter-rater reliability. Separate Kappa statistics (Cohen, 1960) were computed for

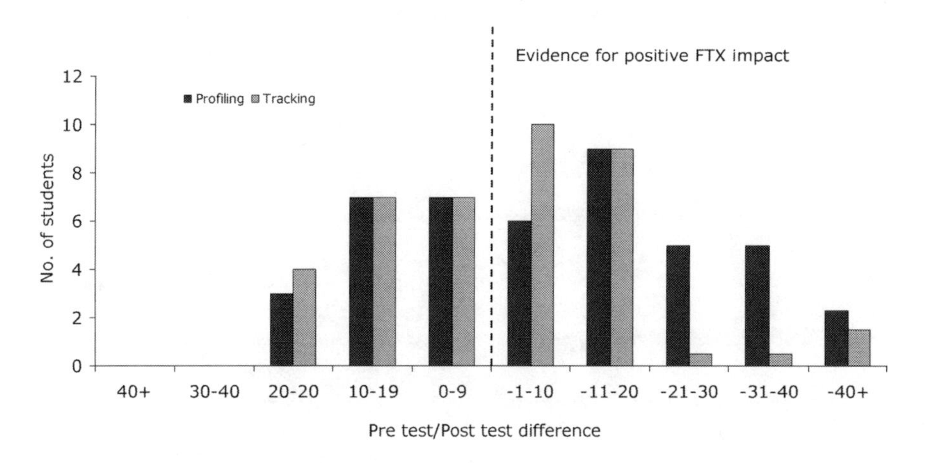

Figure 21.1 Frequency distribution of SJT pre-test and post-test difference scores

the procedural level and higher order behavior categories; these were 0.590 and 0.553, respectively. Both correspond to 'moderate agreement' on the Landis and Koch (1977) scale, missing 'substantial agreement' (0.61–0.80) by only a few percentage points.

BOC ratings were pooled across teams and days to generate stable quantitative trends showing how student CT behaviors improved. A typical result is depicted in Figure 21.2, where performance ratings for three higher-order Tracking behaviors – reading footprint dynamics, adopting a 'quarry mindset' and tactical decision-making – are plotted across days. Statistical t-tests on the average ratings revealed that, despite starting out at fairly high levels, all three measures showed significant increases ($p < 0.025$) from Days 2–3 to Day 10.

Further analysis of the quantitative data revealed that most procedural and higher-level behaviors exhibited similar increases across days, indicating a training effect. For Tracking, most procedural skills increased significantly across days, such as not walking on the spoor line, maintaining visual contact, and marking the starting point of a track. For high-level behaviors, students' performance on situation awareness, communication, and team control increased significantly ($p < 0.05$) over days. For Profiling, most procedural behaviors (e.g., spreading observations across team members, establishing a stable baseline, using criteria to make a positive ID) improved over scenarios. All Profiling high-level behaviors, such as detecting basic events, adopting insurgent mindset, tactical patience, also improved.

The qualitative BOC data, obtained from researcher comments, yielded insights concerning content of student behaviors, emerging problems, and effective instructional techniques; detailed findings are in Spiker and Johnston (2010b). For example, the Tracking BOCs showed that early in training students kept their heads mostly down so they could pick up the details of individual tracks (micro-tracking). While effective for seeing detail, it was slow. Later, students adopted

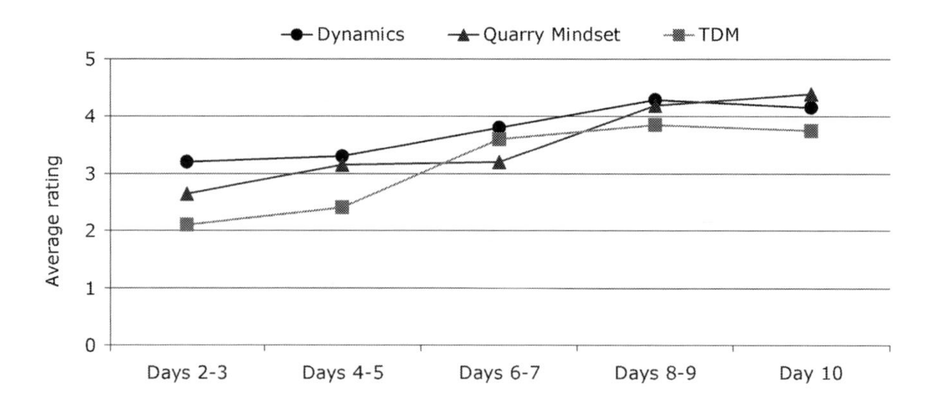

Figure 21.2 Average performance ratings for three higher-level tracking behaviors

the more efficient macro-technique of looking up frequently, using the track line to discern where the quarry is likely headed. For Profiling, students exhibited various skill gains as they accrued scenario experience, such as improved ability to identify high-value individuals, synthesize events ('connect the dots'), and predict events from early signs. In later scenarios, more complex skills emerged, such as scenario recreation ability, trust-building, and adopting the mindset of other cultures.

Conclusions

CH – and its BH offshoot – represent a continuing, major effort by DOD to provide training in CT operations to US Marines, the other military services, and law enforcement. A wide range of perceptual, behavioral, and cognitive skills are trained, practiced, and reinforced during entertaining academic lectures and demanding field scenarios. Key skill areas include observation, distributed attention, decision-making, situation awareness, and judgment, among many others. Over the course of four studies, and additional experiments and demonstrations, the ten behavioral measures discussed in this chapter provide substantial, and converging, evidence supporting the effectiveness of CH training.

Accordingly, Study 1 documented the performance requirements and skills necessary for CT success, which mesh well with the current CH curriculum. Study 2 demonstrated significant skill acquisition over the course of training, as gauged by objective (time, accuracy) indices of trainee performance, trainee verbal descriptions, and reports by observers. Study 3 provided additional evidence of training effectiveness in the form of qualitative shifts in behavior, adherence of the curriculum to established training principles, and anecdotal reports of CH training value from post-deployment Marines. Finally, Study 4 documented the effectiveness of field training through demonstrations of improved perceptual skills, increased situational judgment, and steady acquisition of basic procedural and higher-order behaviors in both combat tracking and profiling. The empirical evidence compiled from these ten classes of measures supports the quantitative impact of CH training on participants' ability to recognize hostile intent earlier in the threat cycle and the qualitative shifts in the behaviors that comprise this capability.

Followup interviews with CH instructors at the Marine Corps School of Infantry – West resulted from two themes that had emerged from CH training observations: the need for skill maintenance and the benefits of simulation-based training exercises. The interview results are described in more detail in Spiker and Johnston (2010a), and summarized here:

1. cultivate profiling naturals
2. encourage student-developed job aids
3. employ training on DVD and online
4. employ game-based training, and

5. employ unit level simulation-based training.

Clearly, a major outcome from this multiple measurement strategy is the ability to tailor the learning requirements, training content, and training design in order to optimize effective decision-making skills and reach a wider trainee audience. We recommend the ten behavioral described in this chapter can be applied to any type of IW training that is targeted to enhancing the warfighter's advanced cognitive skills.

References

Bancroft, J.J., MAJ. (2012). *Effectiveness of Combat Hunter Saturation. Final Report*. Quantico, VA: Operations Analysis Division USMC.

Cannon-Bowers, J.A., and Salas, E. (1998). *Making Decisions Under Stress: Implications for Individual and Team Training*. Washington, DC: APA.

Cohen, J. (1960). A coefficient of agreement for nominal scales. *Educational and Psychological Measurement*, 76(5), 378–382.

Cohen, M.S., Freeman, J.T., and Thompson, B.B. (1998). Critical thinking skills in tactical decision making: A model and a training strategy. In J.A. Cannon-Bowers and E. Salas (Eds), *Making Decisions Under Stress: Implications for Individual and Team Training*. Washington, DC: APA, 155–189.

Cook, T.D., and Campbell, D.T. (1979). *Quasi-Experimentation: Design & Analysis for Field Settings*. Chicago, IL: Rand-McNally.

Department of Defense (2008) *Directive 3000.07: Irregular Warfare, December, 2008*. Available from http://www.dtic.mil/whs/directives/corres/pdf/300007p.pdf, accessed 28 June 2010.

Driskell, J.E., and Johnston, J.H. (1998). Stress exposure training. In J.A. Cannon-Bowers and E. Salas (Eds), *Making Decisions Under Stress: Implications for Individual and Team Training*. Washington, DC: APA, 191–217.

Fautua, D., Schatz, S., Kobus, D.A., Spiker, V.A., Ross, W., Johnston, J.H., Nicholson, D., and Reitz, E. (2010). *Border Hunter Research Technical Report*. Norfolk, VA: US Joint Forces Command.

Gessner, T.L., and Klimoski, R.J. (2006). Making sense of situations. In J.A. Weekley and R.E. Ployhart (Eds), *Situational Judgment Tests*. Mahwah, NJ: LEA, 13–38.

Goulding, V.J. Jr., COL. (2008). DO. *Marine Corps Gazette*, April, 77.

Hilburn, M. (2007). Combat hunter. *Seapower*, 60–62.

Institute for Training and Simulation (2010). *Border Hunter Research Technical Report*, S. Schatz and D. Fautua (Eds). Suffolk, VA: US Joint Forces Command.

Irregular Warfare Training Symposium (2009) *The Future of Small Unit Excellence in Immersive Cognitive Training*, September. Retrieved April 1, 2012 from www.teamorlando.org/conference/images/iwts-Conference-Report-FINAL.PDF.

Johnston, J.H., Paris, C.R., Wisecarver, M.M., Ferro, G., and Hope, T. (2011). *A Framework for Cross-cultural Competence and Learning Recommendations*. Presentation at the 2011 HSCB FOCUS Conference, Chantilly, VA.

Johnston, J.H., Poirier, J., and Smith-Jentsch, K.A. (1998). Decision making under stress: Creating a research methodology. In J.A. Cannon-Bowers and E. Salas (Eds), *Making Decisions Under Stress: Implications for Individual and Team Training*. Washington, DC: APA, 39–59.

Klein, G. (2008). Naturalistic decision making. *Human Factors*, 50(3), 456–460.

Kobus, D.A., Palmer, E.D., Kobus, J.M., and Ostertag, J.R. (2009). *Assessment of the Combat Hunter Trainer Course (CHTC): Lessons Learned. PSE Report 09-08*. San Diego, CA: Pacific Sciences and Engineering Group.

Kobus, D.A., Palmer, E.D., Kobus, J.M., Ostertag, J., and Kelly, M.R. (2010). *Border Hunter Training: Assessments and Observations. PSE Report 10-11*. Norfolk, VA: US Joint Forces Command.

Kobus, D., and Williams, G. (2010). Training tactical decision making under stress in cross-cultural environments. *Proceedings of the Conference on Cross-Cultural Decision Making*, Miami, FL [CD-ROM].

Landis, J.R., and Koch, G.G. (1977). The measurement of observer agreement for categorical data. *Biometrics*, 33, 159–174.

Marine Corps Center for Lessons Learned. (2008). *2nd Battalion, 7th Marines Combat Hunter, Lessons and Observations from Operation Enduring Freedom (OEF)*. Quick Look Report, US Marine Corps: Washington DC.

Smith-Jentsch, K.A., Zeisig, R.L., Acton, B., and McPherson, J.A. (1998). Team dimensional training. In J.A. Cannon-Bowers and E. Salas (Eds), *Making Decisions Under Stress: Implications for Individual and Team Training*. Washington, DC: APA, 271–297.

Spiker, V.A., and Johnston, J.H. (2010a). *Limited Objective Evaluation of Combat Profiling Training for Small Units. Technical Report*. Suffolk, VA: US Joint Forces Command.

Spiker, V.A., and Johnston, J.H. (2010b). *Border Hunter: Evaluation of Field Training, Technical Report*. Norfolk, VA: US Joint Forces Command.

Spiker, V.A., and Williams, G. (February 2010). *Summary of interviews with Combat Hunter graduates from the 1/5, Technical Report*. Suffolk, VA: US Joint Forces Command.

Spiker, V.A., Johnston, J.H., Williams, G., and Lethin, C. (2010). Training tactical behavior profiling skills for irregular warfare. *Proceedings of the 30th Interservice/Industry Training Systems and Education Conference. Orlando, FL* Washington DC: National Training and simulation association, 3234–3244.

Williams, G. (2009). *MTC Afghanistan Deployment 2009 First Look. Draft Report*. Suffolk, VA: US Joint Forces Command.

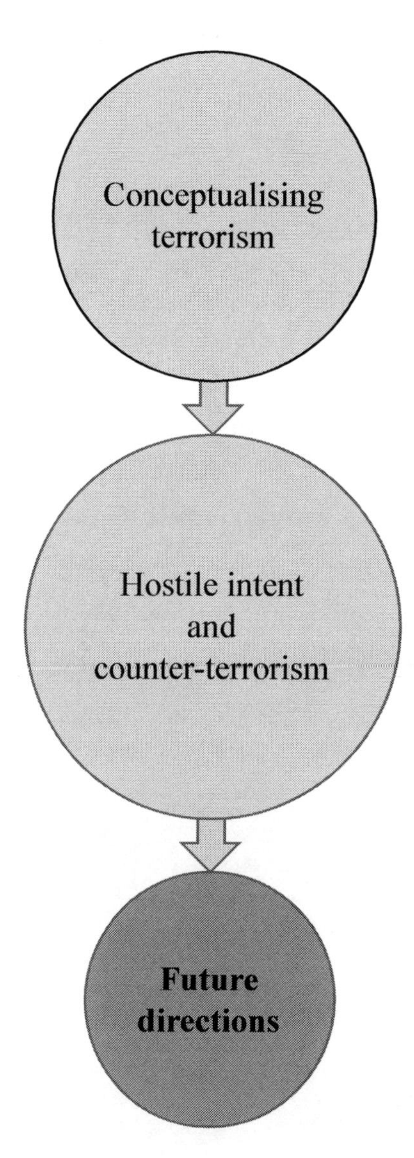

Part 6 Future directions

From this collection of research we see emergent themes and areas worthy of future research. These are reported in Chapter 22. They include areas such as: the importance of multidisciplinary research collaborations and benefiting from advances in associated fields; counter-terrorism and its impact on the daily life of the general public; ethics, with particular regard to privacy; and users of counter-terrorism interventions and the role of design.

Chapter 22
Hostile Intent and Counter-terrorism: Future Research Themes and Questions

Alex Stedmon
Human Systems Integration Group, Faculty of Engineering & Computing, Coventry University, UK

Glyn Lawson
Human Factors Research Group, Faculty of Engineering, The University of Nottingham, UK

Introduction

In setting out to write this book, one of the aims was to provide readers not only with a unique insight of how human factors and associated disciplines can inform the research, practice and policy of counter-terrorism, but also to provide a starting point for new research agendas.

Quite often with edited books the reader is given an exemplary collection of work on a particular topic area and yet the individual chapters are presented as stand-alone contributions. While developing the initial idea for this book, we felt it was important to provide a final chapter that draws together aspects from the different contributions. To achieve this, we obtained open questions for the specific topic areas covered within each chapter. From these questions it was then possible to identify themes that cut across the book and serve to provide the basis for the summary presented here.

The scope of the book has been purposefully wide-reaching so that it provides a resource that is accessible to a wide readership. Contributions were invited and chosen that expand typical definitions of human factors, representing diverse disciplines that contribute knowledge to this area, such as psychology, criminology, sociology, social and scientific philosophy, political science, ethics, art and design, neuroscience, behavioural science, engineering and computer science. The chapters have addressed the issues surrounding hostile intent in different ways ranging from social and political theory, applied and social psychology, interaction design, and user-centred approaches. In providing a multidisciplinary resource for researchers and practitioners alike, the underlying theme across all chapters is the inclusion of end-users (e.g. security personnel, police, general public) within the various debates on hostile intent and counter-terrorism. From this foundation the book has explored a vast array of factors associated with hostile intent and counter-terrorism factors in five key areas:

- *Contextualising terrorism* – it is imperative to understand the issues surrounding terrorism, and the contexts in which it occurs, in order for counter-terrorism initiatives to succeed. The chapters in this section of the book developed our understanding of the contexts in which terrorism might occur and where science and the fundamentals of criminology can provide new ways of approaching counter-terrorism.
- *Deception and decision-making* – central to counter-terrorism initiatives is a better understanding of hostile intent and the underlying deceptions that are employed by terrorists in order to exploit weaknesses in security. This section focused on research that can help identify cues, both verbal and non-verbal, to deception. Other challenges include understanding deception and decision-making when the research laboratory differs to the contexts in which terrorism occurs in the real world.
- *Modelling hostile intent* – in developing new counter-terrorism initiatives it is possible to draw expertise and learn from associated disciplines or application areas. Similarly, taking a multidisciplinary approach and considering the systems in which terrorism operates can bring benefits not seen when issues are considered in isolation. Further, modelling can afford us with tools to predict, and thus proactively address, terrorism.
- *Sociocultural factors* – many factors have an influence on the occurrence and nature of terrorism. This section addressed aspects such as home-grown terrorists and conflict-engaged citizens; terrorism targeting schools; and female suicide terrorism. It also contained an innovative chapter that challenged the views of existing models of terrorism by drawing analogies to parasitic infection.
- *Strategies and approaches for counter-terrorism* – this section presented a range of initiatives both for counter-terrorism practices and for research in this area. It addressed design strategies for public spaces, approaches for understanding user requirements in counter-terrorism, and training the knowledge, skills and attitudes required for identifying hostile intent.

Thematic Analysis

A thematic analysis was conducted on the research questions and issues that were identified for each chapter. The themes were identified in a cumulative manner (so that new themes were added when no current descriptors were suitable) and revised in an iterative fashion so that a comprehensive assessment was achieved. From this activity a range of cross-cutting themes emerged that illustrate the higher-level issues within hostile intent and counter-terrorism as well as the potential for the different areas represented in the book to inform each other and offer novel multidisciplinary research collaborations in the future. At the outset of this book the ideas, for example, that neuroscience might inform art and design; that art and design might inform ethics; that engineering might inform behavioural science; or that political science might inform computer science, would have been difficult to perceive. However, it is now possible to illustrate the collective power of the chapters and the combined strength of taking on a broad range of perspectives to seek solutions to the singular focus of counter-terrorism.

The themes are presented in Table 22.1 with a range of descriptors and the relevant chapters that are linked to them. The themes were drawn from a wide range of areas but they also illustrate where human factors can provide support to counter-terrorism initiatives and where multidisciplinary research can provide more enriched solutions in the future.

Table 22.1 Thematic analysis of future research

Theme	Descriptors	Chapters
Psychology of terrorism	Fear, psychological violence, intimidation, habituation effects, public vulnerabilities, situational bias	2, 3, 4, 5, 6, 9, 10, 12, 15, 16, 18, 19
Theory and approaches	Applicability of theories, individuals and team planning, research in sensitive domains, contextualising suspicious behaviours	3, 6, 7, 9, 12, 13, 15, 16, 18, 19, 21
Counter-terrorism policy	Planning choices, decision points about who is suspicious, authority and playfulness, common agreed taxonomies	2, 3, 4, 5, 7, 11, 15, 16, 18, 19, 21
Perpetrator perspectives	Terrorist motivations, extremist recruitment, anonymity, stereotypes, deviance, deception, lying	3, 7, 9, 10, 12, 15, 19
Real-world contexts	Usefulness of theories, application of theory to real contexts, situated research	6, 10, 12, 18, 19, 21
User-centred design	User requirements elicitation and communication, public crowded spaces, enhancing the user experience while frustrating hostile intent	6, 7, 9, 10, 18, 19

Table 22.1 *Concluded*

Public confidence	Technologies and counter-terrorism initiatives might inadvertently undermine public confidence	2, 4, 5, 8, 18, 19
Intelligence analysis	Open source analysis, social media, collecting and distributing intelligence across pan-European networks, real-time data	8, 10, 11, 12, 13, 15
System vulnerabilities	Parasitic analogies and frameworks, systems approaches to security, terrorism as a rogue agent in the system	8, 10, 13, 18, 19
Research and mixed methods	Triangulation of methods, manipulating status in interviews, levels of stakes in research, transfer of findings, securing research outputs	7, 12, 18, 19, 21
Privacy and ethics	Balancing privacy and public safety, neuro-ethics, surveillance, tracking suspects, civil liberties, responsible science	4, 5, 19
Low-cost solutions	Low-cost methods for standardised training, reconfigurable solutions	18, 19, 21
Contingency and resilience	Translating biological analogies to inform system resilience, graceful degradation of systems, modelling for contingency, course of action decision-making	13, 15, 19
Stakeholder engagement	Designer as a stakeholder, identifying relevant stakeholders, stakeholders vs end-users	13, 15, 19
Infrastructure	Vulnerabilities of open security networks, monitoring public spaces	8, 10, 19
ICT skills and training	Core skills for training, co-located and distributed strategy planning, refresher training and skill fade	11, 21

Future Research Questions

From the themes identified above, each chapter is presented with a brief summary and, where appropriate, specific questions or open topics are listed that identify prominent issues and provide a frame of reference for future research.

Chapter 2: The Role of Fear in Terrorism

This chapter provided an overview of the ways in which the academic literature has tended to characterise the role of fear and psychological violence in the process of political terrorism.

- How can counter-terrorism policies be adapted to best engage the public?
- What counter-terrorism practices have the perverse effect of undermining public confidence?
- How enduring and widespread are public fears of terrorism after attacks occur?

Chapter 3: Understanding terrorism through criminology? Merging crime control and counter-terrorism in the UK

In this chapter enduring practices and underpinning principles of counter-terrorism were identified and questions raised over their applicability to new, mutable and dynamic forms of terrorism. The key issues include:

- Understanding similarities and differences in crime and terrorism through perpetrator perspectives.
- Examining the extent, applicability and appropriateness of criminological theories of transgression across counter-terrorism practice.

Chapter 4: Analysing the Terrorist Brain: Neurobiological Advances, Ethical Concerns and Social Implications

This chapter focused on neurobiological research into terrorism and the underlying ethical and societal issues that unravel as research is conducted and the findings are communicated back to a wider audience. Further work should aim to:

- Further analyse the trade-off of liberty vs. security, especially for what concerns counter-terrorism
- Open up a debate on the social implications and ethical concerns deriving from the perceived infallibility of science
- Investigate the effects from the sense of false security that derives from advances in neuroscience.

Chapter 5: Ethical Issues in Surveillance and Privacy

Chapter 5 explored issues involved in balancing public security against individual privacy. It made specific reference to supporting mass public transport security, with surveillance exercises in complex travel interchanges across major European cities.

- What can legitimately be regarded as 'suspicious behaviour'?
- How can individuals engaging in such behaviour be categorised and identified?
- Who makes this judgement call? (What 'qualifies' them to do so?)
- How far is it possible to anticipate levels of threat or risk?
- If privacy is not to be preserved at all costs, then under what conditions can it be compromised?

Chapter 6: Non-verbal Cues to Deception and their Relationship to Terrorism

Chapter 6 outlined the progress of research related to the use of non-verbal cues in deception detection within the specific context of terrorist activities.

- Are the theories of deception to date useful in explaining how terrorists may plan and carry out attacks?
- Are the 'real-world' contextual factors involved in terrorism activity sufficiently taken into account in academic research?
- Are the interventions developed in scientific research user-friendly and can they be easily adopted by end-users?

Chapter 7: Deception Detection in Counter-terrorism

This chapter drew attention to the recent emphasis on interviewing techniques that elicit and enhance cues to deception. In particular it focused on the settings that are relevant for terrorism, such as lying about intentions; examining people when they are secretly observed; and interviewing suspects together.

- Future work lies in the development of a memory-based lie-detection technique, as a new cognitive load lie-detection technique. People do not know what they remember and typically overestimate their memory.
- In the area of collective interviewing, what will happen if the pairs of interviewees have different status (a superior with a subordinate)?
- Which aspects of sketching makes the use of drawings a successful tool to detect deceit?

Chapter 8: A Field Trial to Investigate Human Pheromones Associated with Hostile Intent

Chapter 8 presented evidence that stressed individuals secrete a volatile steroid-based marker that could form the basis for remote detection of deception.

- Why have terrorists never attacked the mainline railway infrastructure in the UK?
- How safe are we from a Chemical, Biological, Radiological, Nuclear and high-yield Explosives (CBRNE) attack by terrorists?
- How can intelligence analysis be improved?
- How can novel techniques such as pheromone detection be incorporated into mainstream counter-terrorism applications?

Chapter 9: On the Trail of the Terrorist: A Research Environment to Simulate Criminal Investigations

This chapter reported on the collective movements and communications of persons working together and co-operating in the planning and execution of a major terrorist event.

- How does deceptive strategy planning differ between individuals and teams?
- How does deceptive strategy planning differ between known players and anonymous players?
- How does co-location versus disparate locations affect deceptive strategy planning?

Chapter 10: Safety and Security in Rail Systems: Drawing Knowledge from the Prevention of Railway Suicide and Trespass to Inform Security Interventions

This chapter provided an overview of the problems associated with rail trespass and suicide in the context of national prevention programmes and the transfer of knowledge to the wider arena of railway security.

- What motivates people to access and threaten safety and operations on the railway?
- Can better data be collected and used in the prevention of incidents?
- How can relationships and collaboration between relevant stakeholders be strengthened?
- What are suspicious behaviours in the rail context?

Chapter 11: Tackling Financial and Economic Crime through Strategic Intelligence Management

Chapter 11 described an approach for strategic intelligence management that would provide a tool capable of capturing a spectrum of criminal activities.

- How can the EU Local Enforcement Agencies (LEAs) go about creating a common agreed taxonomy of Serious Organised Economic Crime (SOEC) and fraud, and what will be incorporated into the taxonomy?
- How can information be sourced, acquired, homogenised and federated in a pan-EU monitoring system?
- What tools and techniques are required for LEAs to make use of this federated information?

Chapter 12: Competitive Adaptation in Militant Networks: Preliminary Findings from an Islamist Case Study

Chapter 12 provided an interdisciplinary framework from which to study the behaviour of militant groups that either carry out acts of political violence themselves or support the use of violence by others.

- How might the social network analysis and mixed methods used in this chapter be applied to the study of other violent non-state actors, including terrorist organisations?
- How do the text-mining and network analytic techniques used in this chapter need to be adapted to apply to other types of open source information such as social media?
- How can the bias in the terror networks extracted from various news sources be estimated?

Chapter 13: Evaluating Emergency Preparedness: Using Responsibility Models to Identify Vulnerabilities

This chapter investigated the notion of responsibility in sociotechnical systems that encompass both organisations and infrastructure.

- How can responsibility modelling be incorporated into the development and maintenance of contingency plans?
- What tools are needed to support responsibility modelling that can help automate (at least some of) the process of analysing a system to identify vulnerabilities?

Chapter 14: Unintended Consequences of the 'War on Terror': Home-grown Terrorism and Conflict-engaged Citizens Returning to Civil Society

Chapter 14 presented a series of case studies in terrorism investigations, focusing on the emerging threat of home-grown terrorism as well as individuals returning from conflict abroad. It concluded by highlighting that there are many challenges associated with the return of conflict-engaged citizens from theatres of war.

Chapter 15: Parasites, Energy and Complex Systems: Generating Novel Intervention Options to Counter Recruitment to Suicide Terrorism

This chapter challenged conventional explanations of terrorism and elaborated around the analogy of parasitic infection shaping or driving undesirable behaviour.

- What other parasitic mechanisms can be implemented in a framework to aid our understanding of forms of recruitment to extremist or other deviant behaviour?
- What can be learned about the defence of compromised systems to parasitic infection that could make systems (cyber sociotechnical; groups and individuals) detect, reject and become resilient to infection?
- At the molecular and particle level, how can the analogy be translated into practical activities for the UK's Prevent Strategy or support intelligence profiling of 'clean skins'?
- How could bridges be made to other research directions to add value and develop an integrated suite of innovative approaches?

Chapter 16: Terrorist Targeting of Schools and Educational Establishments

Chapter 16 focused on armed assaults on educational institutions, the frequency of which has increased sharply since 2003. A comprehensive chronology of all armed assaults on educational institutions since 1980 was created and analyses revealed that attacks can be classified on an expressive-instrumental continuum.

- Very little is currently known about the decisions made by terrorists in relation to targeting. Further research should be directed towards understanding planning choices would benefit both theory development and policy-making.
- Although described through case studies in the literature, very little is known about the psychological mechanisms and practical effects of public protest (i.e. 'backlash') on the cessation or change in terrorist activities.

Chapter 17: Female Suicide Terrorism as a Function of Patriarchal Societies

Chapter 17 argued that there is a relationship between patriarchy and female suicide terrorism. It also highlighted that female suicide is a growing and dangerous phenomenon. Future research may seek to address the limitations in the availability of empirical data.

Chapter 18: Designing Visible Counter-terrorism Interventions in Public Spaces

This chapter explored publicly visible counter-terrorism measures, uncovering the strategic role of design in creating controlled disruption in public spaces to reduce threat while at the same time reducing anxiety.

- There needs to be a better understanding of the disruptive qualities of design interventions from a temporal perspective: on both quotidian users of public spaces and those with mal-intent.
- How best can we integrate design into security dialogue/practice?
- How can we translate knowledge into guidelines for communication design? What frameworks can be put in place?
- Interventions that use the language of authority are only a small subset of the total design space of interventions that can evoke the desired responses in public spaces. The integration of 'playfulness' is being recognised as a significant resource in intervention design.

Chapter 19: A Macro-ergonomics Perspective on Security: A Rail Case Study

This chapter outlined user requirements methods/approaches applied to a rail case study in order to identify organisational responses to security threats.

- How can we develop a systems understanding of security?
- How can we conduct research in sensitive environments?
- How can we best communicate user requirements to different stakeholders?

Chapter 20: Deception and Speech: A Theoretical Overview to Inform Future Research

This chapter presented an overview of the theoretical foundations pertinent to the investigation of deception and speech. Can future research in deception in speech exploit the possible behavioural overload that deceivers are likely to experience?

Chapter 21: Evaluating Counter-terrorism Training using Behavioral Measures and Theory

Chapter 21 considered the knowledge, skills, and attitudes (KSAs) to read the human and physical terrain and infer hostile intent as an essential element of counter-terrorism training.

- Can a core set of KSAs be identified that are central to all types of counter-terrorism training, satisfying both civilian and military requirements? If so, can we identify relatively low-cost yet effective methods for conducting standardised training across both populations?
- How much repetitive practice is required to demonstrate a significant and sustained acquisition of the core counter-terrorism KSAs for novices vs trainees having some expertise? What kind of refresher training will be needed for these KSAs and what would the skill decay rates over time look like?
- Is it possible to develop broadly applicable converging methods to 'triangulate' on precise estimates for how much KSA increase has occurred in a given counter-terrorism training setting? How much variation in the methods will be needed for successful application to diverse training settings?

Final Thoughts from the Editors

This chapter has identified gaps in knowledge and key questions about hostile intent and counter-terrorism that provide a basis for future research. However, the issues that have been raised are not exhaustive and the threats of terrorism continue to evolve. As demonstrated with the attack on the Westgate shopping mall in Nairobi by Al-Shabab terrorists and hostage killings by Islamic State (IS) militants those with hostile intent continue to operate and threats to public safety continue to exist.

This book has illustrated the unique insights that the discipline of human factors can provide in developing our understanding of counter-terrorism measures from both user-centred and socio-technical systems perspectives. In this way, and to echo the words of Professor John Wilson, human factors can really make a difference.

Index